/ X✓✓

THE DIARY OF THOMAS TURNER

1754–1765

THE DIARY OF
THOMAS TURNER
1754–1765

Edited by
DAVID VAISEY

Oxford   New York

OXFORD UNIVERSITY PRESS

1984

Oxford University Press, Walton Street, Oxford OX2 6DP

London New York Toronto
Delhi Bombay Calcutta Madras Karachi
Kuala Lumpur Singapore Hong Kong Tokyo
Nairobi Dar es Salaam Cape Town
Melbourne Auckland

and associated companies in
Beirut Berlin Ibadan Mexico City Nicosia

Oxford is a trade mark of Oxford University Press

British Library Cataloguing in Publication Data

Turner, Thomas
The diary of Thomas Turner, 1754–1765.
1. East Hoathly (East Sussex)—Social life and customs
I. Title II. Vaisey, David
942.2'51 DA690.E136
ISBN 0–19–211782–3

Set by Hope Services, Abingdon
Printed in Great Britain by
St. Edmundsbury Press Ltd
Bury St. Edmunds, Suffolk

Preface

HISTORIANS and antiquarians have known about the diary of Thomas Turner for a century and a quarter. It was in 1859 that R. W. Blencowe and M. A. Lower published extracts from it in the *Sussex Archaeological Collections*. It has been unfortunate that these extracts, which the editors in many cases mistranscribed and often realigned chronologically in order to fit the drift of their article, became fossilized as 'the diary of Thomas Turner'. Subsequent editors, including Florence Maris Turner and G. H. Jennings, did not have access to the original 111-volume manuscript, and historians have relied on their versions, despite Dean K. Worcester's excellent monograph, *The life and times of Thomas Turner of East Hoathly*, which in 1948 drew attention to the partial and not always accurate snippets which Blencowe and Lower had reproduced. In this present edition the original manuscript has been revisited, Blencowe and Lower's text has been checked and put in its correct sequence, and far fuller extracts have been made in an attempt to provide both the general reader and the social historian in the last quarter of the twentieth century with a more representative selection from this unique source and with far more information than was given by the miscellaneous bits and pieces in which Blencowe and Lower considered the value of the diary lay. In the original manuscript, they wrote, 'the entries are very multifarious, and, for the most part, trivial and uninteresting'; and they therefore cut it down and provided a selection which could lead J. B. Priestley to write in his introduction to Florence Maris Turner's edition in 1925 that 'what his Diary lacks in length, fullness and historical importance, it makes up for in richness, quaintness and a comical-pathetic naïveté'.

The historical viewpoint changes over the years and we can now recognize that *naïveté* is not necessarily either comical or pathetic and that it is precisely from the reporting of the trivial details by 'a little figure from the past' and 'an eighteenth-century nonentity' (to use two more of Priestley's phrases) that we can piece together a portrait of day-to-day life in a mid-eighteenth-century English village.

Much, unfortunately, has still had to be omitted from this edition. More than a third of a million words have been reduced to around

130,000. Much of the omitted material consists of the diarist's moralizing reflections after having heard a striking sermon, read a stirring passage from some improving work, or suffered pangs of conscience after either a quarrel with his wife or a heavy evening's drinking. Such passages tend to be tediously repetitive both in word and in sentiment and it was felt that honour was satisfied by reproducing one or two representative samples and omitting the rest. Most of the remainder of the omitted passages concern two other aspects of Turner's life about which he wrote assiduously in his journal. First, his financial dealings: he used his diary to record an enormous number of small payments and receipts, loans and borrowings. It was judged that in view of the need to reduce the repetitive bulk, many of these entries could be discarded without seriously damaging the impact of the diary. Secondly, his diet: almost every day Turner recorded the food eaten by himself and his household, including what was eaten at home if he himself was away. A representative selection of these references is reproduced in this edition, but many more are recorded in the original diaries. The most frequent entry under this heading (and at one time it appears for eleven days in succession) is 'we dined on the remains of yesterday's dinner', and the next most frequent is 'we dined on the remains of yesterday's dinner with the addition of . . .'. When a pig was killed, the diary records how day after day for perhaps a month each part of it was consumed. For the most part the common fare consisted of puddings and casseroles, but scattered through the diaries are references to eating most meats together with pigeon, woodcock, blackbirds, teal, turkey, and all kinds of fish including sturgeon, which Turner himself served at a Christmas entertainment. The multifarious entries in the diary on both these aspects of village life in the 1750s and 1760s are omitted with great regret since it is by their repetitious nature that they make their impact, but those wishing to pursue these topics are referred to the original manuscript, a microfilm copy of which is available in the Bodleian Library (MSS. Film 1673–6).

The original manuscript of the diary is now in the Stirling Library at Yale University and I am very grateful to the authorities at Yale University Library who granted permission to Oxford University Press to publish a new edition. I am further most grateful to Dr Henry Hardy of the Press, who tracked down the original manuscript and prepared a draft selection for my consideration, and to his wife Dr Anne Hardy, who suggested to him that I should make the new selection and prepare this edition. I am afraid that I have kept Dr Henry Hardy and his colleague Mr Peter Sutcliffe waiting far too long for the finished text; that this new series of extracts can now appear in 1984—precisely 125 years after the extracts which it supersedes—is due to the kindness of the Curators of the Bodleian Library who granted me a period of study leave in 1983 during which I could work on the edition uninterrupted by the calls of my Library duties.

I owe a great debt, too, to those people who have helped me over particular points in the diary: my colleagues in the Department of Western Manuscripts at the Bodleian, in particular, Mrs Mary Clapinson who came to my aid with much advice and practical assistance with the index; Dr Molly Barratt and Mr Steven Tomlinson for helping me in discussion on points to do with eighteenth-century parochial administration; my secretary, Mrs Joan Leggett, and her successor, Miss Caroline Garner, who with great cheerfulness produced a splendid clean typescript from some really dreadful drafts; Dr Joan Thirsk for help on agricultural matters; Dr Michael L. Ryder for sharing his knowledge of sheep breeds; Mr John R. Millburn for information about astronomical instruments; Mr D. C. Gibson of the Kent Archives Office; Mr Harold P. Melcher Jr. of Wilton, New Hampshire, who generously put his undergraduate work on Turner's reading at my disposal; Mr John Farrant of the University of Sussex for allowing me to draw on his great knowledge of Sussex history; and to the staff of the East Sussex Record Office and in particular Mr Christopher Whittick for calmly and expertly answering one question after another about the collections in their charge. The following have most kindly allowed me to include illustrations of items from their collections: The East Sussex Record Office (Plates 1, 4–7), The Bodleian Library, Oxford (Frontispiece, Plates 2a, 8), and Yale University Library (Plate 3), Mr. J. Meakin of Hailsham took the photograph in Plate 2b.

One person above all, however, deserves more than my gratitude: he deserves the gratitude of all who read this edition. Dean K. Worcester Jr., while an undergraduate at Yale in the 1940s and subsequently as a postgraduate at Cambridge, did an enormous amount of work on the diary and on Thomas Turner. He made a transcript of the diary and compiled notes on it in the hope that it would find a publisher: when it failed to do so he deposited all his work in Yale University Library for the use of any scholar who in the future might wish to work on a new edition. I have drawn heavily on his work and this present edition rests four-square on his endeavours. In correspondence with me he has been consistently generous and encouraging and he has contributed

Appendix E on the history of the manuscript. I cannot thank him enough for helping me in this way and he, in his turn, wishes to acknowledge the help of Judith A. Schiff, the chief archivist at Yale University, and of Nicholas Leonard of East Hoathly. If Thomas Turner's diary now assumes a new interest and importance both for the general reader and for the historian it is almost entirely due to Dean K. Worcester's work.

Finally to my wife Maureen and to my daughters, Katie and Libby, I would like to offer my thanks for allowing me for so long quietly to get on with editing Turner's diary and never pointing out that he was preventing me from pulling my proper weight in family matters. This book is dedicated to them.

Bodleian Library, Oxford
November 1983

Contents

List of Illustrations

1. Turner's part of Sussex in the mid-eighteenth century: from the New Map of Sussex made by John Rocques in 1761. *Bodleian Library, Oxford, Gough Maps Sussex 10. facing page* 88

2a. Halland House from T. W. Horsfield, *History, antiquities and topography of the county of Sussex* (1835). *facing page* 89

2b. Turner's house in East Hoathly. Photograph taken in 1984 by J. Meakin. *facing page* 89

3. The diary, vol. 14, pp. 6–7. *Thomas Turner papers, Yale University Library. facing page* 120

4. East Hoathly overseers' account book, 1762–79: *E.S.R.O. PAR 378/31/1/1. facing page* 121

5. The bastardy bond, written by Thomas Turner on 9 February 1758, by which Thomas Osborne idemnified the parish of East Hoathly against any charge which might arise from the child of Mary Hubbard: *E.S.R.O. PAR 378. facing page* 216

6. The settlement certificate which Thomas Turner's friend James Marchant was obliged to procure from the authorities at Ticehurst: *E.S.R.O. PAR 378. facing page* 217

7. Thomas Turner in account with the East Hoathly overseers for goods supplied from his shop, 1764–5: *E.S.R.O. PAR 378. facing page* 248

8. Title page and illustration of the slide rule from Charles Leadbetter, *The royal gauger . . .* (1755). *Bodleian Library, Oxford. facing page* 249

Abbreviations

Blencowe and Lower	R. W. Blencowe and M. A. Lower, 'Extracts from the diary of a Sussex tradesman, a hundred years ago', *S.A.C.*, xi (1859), pp. 179–220.
Burn	R. Burn, *The Justice of the Peace and Parish Officer* (1755).
D.N.B.	The Dictionary of National Biography.
E.S.R.O.	East Sussex Record Office, Lewes.
Jennings	Thomas Turner, *The Diary of a Georgian Shopkeeper.* A selection by R. W. Blencowe and M. A. Lower, with a preface by Florence Maris Turner. Second edition. Edited with a new introduction by G. H. Jennings (Oxford, 1979).
O.E.D.	Oxford English Dictionary.
S.A.C.	Sussex Archaeological Collections.
S.C.M.	Sussex County Magazine.
S.N.Q.	Sussex Notes and Queries.
S.R.S.	Sussex Record Society.
Tate	W. E. Tate, *The Parish Chest* (3rd ed. Cambridge, 1969; reprinted 1974).
F. M. Turner	*The Diary of Thomas Turner of East Hoathly (1754–1765)*, edited by Florence Maris Turner, with an introduction by J. B. Priestley (1925).

Editorial Method

THOMAS TURNER was one of the handful of men in East Hoathly to whom others went to have documents written or accounts made up, and his hand was a big and bold one, clear and easy to read. His spelling, however, and his use of capital letters was inconsistent, and some of the conventions he employed for abbreviations are confusing to the modern reader. It was therefore decided not to preserve these oddities and inconsistencies in the transcript. There seemed little virtue in retaining, for instance, every variant (and there are many) of his spelling of 'turnips', or his method of writing 'twenty minutes past seven o'clock' which was '20 m.p. 7'. The following editorial method has thus been adopted:

Spelling, the use of capital letters, and punctuation have been modernized. Abbreviations, including ampersands, have been extended, except in the case of personal forenames and his use of 'morn' and 'even' for 'morning' and 'evening'. Surnames have been standardized to the spelling which was used most often. Dates have been rendered in the form 'Mon. 22 Sept.' and times in the form '7.20'. Money is expressed as, for instance, £7 13s. 8d., and abbreviations denoting weights and measures are given in their modern form. References to chapter and verse in the books of the Bible are given as, for instance, *Proverbs* 18.7.

Omissions are shown by the use of three points (. . .) whether or not the omission ends a sentence. For most of the period Turner made an entry in the diary every day. The omission sign (. . .) has been employed to conclude a day's entry only where the diarist wrote more on that day than is here printed. If the next entry printed is for several days later it is to be assumed that the entries for the intervening days have been omitted, not that Turner made no entry on those days. Round brackets are the diarist's own: square brackets have been employed to enclose words supplied by the editor.

Glossary

Bagging: See Hop-bagging

Bait: To eat; or to give food and drink to an animal

Bearskin: A shaggy woollen cloth used for overcoats

Boullis: Probably bullace

Brag: A card-game in which players bet on the value of cards in their hand

Bullace: The wild plum

Bumboo or Bumbo: A liquor composed of rum, sugar, water, and nutmeg

Bumper: A cup or glass of wine filled to the brim

Camlet: A fine material of mixed wool and silk

Cerecloth: A cloth smeared or impregnated with wax or some other substance used as a plaster

Champhire: Camphor

Chaps: Chops

Chine: A joint of meat consisting of part of the backbone with the adjoining flesh

Coast: The side of an animal prepared for cooking, particularly the ribs

Collops, Scotch: A savoury dish made of sliced veal and bacon

Court baron, Court leet: See p. 237, n.18

Diuretic: Substance having the property of promoting the evacuation of urine

Electioner: Deputy or substitute for a parish official

Ferret: A stout tape, most commonly of cotton but also of silk

Frock: See Round frock

Fustian: A thick, twilled, cotton cloth usually of a drab colour

Glauber salt: Sulphate of sodium: a purge

Grist: An allowance of corn to be ground

Groat: A fourpenny piece

Groats: Crushed grain, chiefly oats

Groins: Frameworks of timber or masonry run out into the sea to raise a barrier against encroachment and erosion

Haslet: That part of an animal's entrails which is roasted, otherwise 'pig's fry'

Headborough: Constable or deputy constable

Heal: To re-roof a house with tiles or slates

Heriot: The payment due to the lord of a manor of the best live beast or dead chattel of a deceased tenant; often commuted into a money payment

Hogshead: A cask for beer or cider normally of 51 gallons but with local variations

Hop-bagging: A coarse woven fabric from which bags or pockets for hops were made, otherwise called poking

Huff: To scold or chide in a hectoring manner

Light pudding: Not a soufflé, but a pudding made from lights

Main of cocks: A match involving teams of fighting-cocks

Manna: Juice abstracted from the bark of the manna-ash, used as a laxative

Moul, mool: To crumble

N.S.: New Style; see p.7 n.6

Nantz: Brandy; a corruption of the French place-name Nantes

Orange shrub: An alcoholic drink made with orange juice, sugar and rum

Out-ask: To publish the banns of marriage for the third and last time

Pass: See p. 42 n.21

Patten: A shoe consisting of a wooden sole with a leather strap passing over the instep, the sole mounted on an oval iron ring raising the wearer an inch or so from the ground

Perihelion: The point in the orbit of a heavenly body at which it is nearest to the sun

Pipe: A cask of two hogsheads or four barrels in volume

Pluck: The liver, lungs and heart of an animal

Pocket: A bag or sack for hops equal to about 168 lb.

Poking: See Hop-bagging

Pond pudding: Otherwise known as 'black-eyed Susan'; a roly-poly pudding made with currants

Post a day book: See p. 5 n.3

Puff: Very light pastry, cake or confectionery

Rap: To barter

Ratio: Rasher

Regrating: Buying up a commodity in order to sell it again at a profit in the same market

Round frock: The Sussex labourer's smock; see p. 5 n.4

Sal. glabuler: See Glauber salt

Scotch collops: See Collops

Seton: A thread or tape drawn through a fold of skin in order to maintain an opening for discharges

Sew: See Sue

Shag: Worsted cloth with a velvet nap on one side

Shalloon: A closely-woven woollen material used chiefly for linings

Shamy: Chamois leather

Shrub: See Orange shrub

Sue: Otherwise Sew; to drain

Surcingle: A girth for a horse, especially that which keeps a pack in place

Susi: Otherwise Soosy or Soocey; a striped mixed fabric of silk and cotton

Sweetheart: A sort of sugar cake in the shape of a heart

Tod: A unit of weight for wool equivalent to 32 lb.

Traverse: A shed or lean-to adjoining a blacksmith's shop where horses were shod

Introduction

I

THOMAS TURNER'S diary covers a little over eleven years in the life of a man who kept a shop in the Sussex village of East Hoathly—a village some seven miles north-east of Lewes. Lying on the edge of the Sussex Weald, it is a village where the houses are clustered around a sharp bend in the modern A22, the road which runs down from East Grinstead in the north-west to Eastbourne in the south-east. When the first surviving volume of the diary begins in February 1754 Turner was a newly-married 24-year-old recently arrived in the village. When he stopped writing his diary in 1765 he was thirty-five and one of the pillars of the community. His bride of eleven years before was dead and so was their only child. He had come through times of worry and depression to acquire maturity and a modest status—and he was about to embark on a fruitful second marriage. The diary is that of a man in the second of three generations of shopkeepers—mercers, drapers, grocers, the terms were interchangeable in his day: his father had kept a shop and two of his sons were to keep shops. The family's fortune was not solely dependent on trade, however: they owned land and acquired more. Turner was able in later years to buy both his shop and the principal public house in the village.[1] In the language of the economists the Turner family during the eighteenth century was 'upwardly mobile'.

The village of East Hoathly in Turner's time was not without its own importance, for just within its western boundary stood Halland House,[2] the Sussex mansion of Thomas Pelham-Holles, first Duke of

---

[1] Thomas Turner's father owned land at Chiddingstone, Kent, which passed after his mother's death to his brothers and himself: see below page 301, note 26. Turner himself bought his shop on 1 October 1766 and the public house on 28 February 1772: E.S.R.O., S.A.S. A 663–4 (Laughton manor court books). His will, made on 14 January and proved on 28 September 1793, directed that all his properties were to be sold or disposed of for the benefit of his wife and three surviving sons, Philip, Michael, and Frederick: E.S.R.O., Lewes archdeaconry wills register A 66, p. 728. *The King's Head* was sold to William Verral, a brewer, of Southover near Lewes, in October 1793, while the house and shop passed to Philip Turner in 1798: E.S.R.O., S.A.S. A 666–7 (Laughton manor court books).

[2] See p. 1, note 6, and plate 2.

Newcastle, who for almost eight of the diary years was First Lord of the Treasury. This brought to the parish annually during the summer and particularly during the local race-meeting in August many people of influence on the local and national scene. As the nearest shopkeeper Turner was interested in these visitors often rather more as consumers than as men of power, but, none the less, the Seven Years' War began and ended while he was writing his diary and many of those who controlled the nation's destiny during it passed before Turner's eyes in his own home village. The diary, however, is not about the lives and achievements of the powerful and the articulate: such people appear in its pages simply as names, moving through East Hoathly on their way elsewhere or staying overnight at Halland. One of the most influential, Henry Pelham, Member of Parliament for nearly forty years and First Lord of the Treasury for eleven, appears only as a corpse on his way to the family vault at Laughton.[3] The diary is about those whose livings were much harder won. It chronicles the everyday business of surviving, of getting and spending, of eating and drinking, of work and play, and of life and death, amongst the ordinary people of the mid-eighteenth century; and it was written by a man who was no fool. Thomas Turner's experience was limited and his preoccupations were parochial, but he possessed both the ability and the wit to provide one of the very few intimate pictures of life at parish level in eighteenth-century England.

II

Thomas Turner was born on 9 June 1729 at Groombridge in the parish of Speldhurst only just over the border from Sussex in Kent.[4] In 1714 his father, John Turner (then described as 'yeoman'), and his first wife were said to have been from Withyham just on the Sussex side of that border.[5] Thomas Turner, however, was the eldest son of his father's second marriage—to a girl from Rotherfield—and when Thomas was not quite six, in June 1735, the family moved southwards and deeper into Sussex to take a shop at Framfield. In his 'Notes on family history'[6] Thomas Turner tells us where his brothers were apprenticed and where his sister was at school, but records nothing of his own education. It would seem probable that he went to school in

---

[3] See p. 1, note 5.

[4] For Turner's pedigree see Appendix A. He was baptized at Groombridge on 20 June 1729: bishop's transcripts in the Kent Archives Office.

[5] E. H. W. Dunkin, *Calendar of Sussex marriage licences* (S.R.S., VI, 1906), p. 210.

[6] Printed in *F. M. Turner*, pp. 105–12, and *Jennings*, pp. 79–84.

Lewes where he would have learnt to write the clear hand of the diary and would have acquired the ability to reckon and cast accounts. His subsequent easy familiarity with Lewes merchants and tradesmen together with his aptitude for commerce may also have been acquired during an apprenticeship in that town, though his business sense was perhaps inherited from his father. He was certainly very fond of his father, writing about his death in May 1752: 'In him I lost the best of parents, he was to me a parent, friend and brother.'[7]

Eighteen months before his father's death, Thomas set up in business for himself some three miles away from Framfield in East Hoathly.[8] He was just twenty-one when he took over the premises of a recently-deceased mercer, Francis Weller, in 1750. It appears from a reading of his father's will[9] that Thomas was eventually to run the Framfield shop, subject to his mother's life interest in it, but he did not do so—at least during the diary period—for following his mother's death on 1 April 1759, the business was run by Thomas's younger brother Moses. Seventeen months after his father's death and three years after moving to East Hoathly Thomas married. His bride was the daughter of a Hartfield farming couple, Samuel and Ann Slater. Her name was Margaret, though Thomas called her 'Peggy'. They were married in Lewes where she was then living on 15 October 1753.[10] Thomas was twenty-four and Peggy was not quite twenty-one. How the marriage came about the diary does not make clear, but what is clear is that by the time the diary begins—just five months after the wedding—the young shopkeeper was already discovering that his wife, by that time pregnant, was sickly and capricious, and, perhaps because of the nature of her illness, irritable and quarrelsome. He had also discovered that his mother-in-law was a difficult woman with whom he found it impossible to see eye to eye. The child of this early pregnancy, a boy, lived for only five months. There were no more children and, one suspects, little physical love in the remaining years of the marriage. The domestic atmosphere seems to have been curdled, and running through the first seven years of the diary is the story of this strained relationship and of Thomas's reactions to his wife's steadily declining health. Visits from the medical men became so frequent that people

[7] *Jennings*, p. 80.

[8] *Jennings*, p. 79.

[9] E.S.R.O., South Malling wills register D 9, p. 233. The will is dated 1 May 1752 but was not proved until 1777 when the will of his widow (who had died in 1759) was also submitted for probate: ibid., p. 236.

[10] Lewes, All Saints parish register: E.S.R.O. PAR 410/1/1/5. The bride's surname in the register is spelt 'Slaughter'.

began to talk and, when after an extremely distressing and painful final illness Peggy Turner died on 23 June 1761, there were those (Ann Slater among them) who spread rumours that Thomas had hastened her death by conniving at a savage gynaecological operation.[11]

Following his wife's death Thomas was a lonely man, and in his loneliness he seems to have forgotten the sentiments which he had previously committed to paper. He began to fill his diary with expressions of a regard for his deceased wife which he may have wished he had felt during her life. But if the shopkeeper was lonely, the shop flourished. As is the way with many traders Turner would not admit this. More often than not his complaint was that trade was going downhill, but by 1766 he had so prospered that he was able to buy his shop from the Wellers.

On 4 February 1763 he wrote, 'I doubt my trade will hardly answer the expenses of a family which might . . . be the consequence of marriage',[12] but it is clear from his comments that the East Hoathly villagers did not expect him to remain a widower for long, and the diary ends on 31 July 1765 with his second marriage—to Mary ('Molly') Hicks of Chiddingly, a servant to one of the local Justices of the Peace.[13] Perhaps it is an indication of a new-found happiness that the diary does not continue after his marriage. The rest of Thomas's life was spent with Molly, and seven children were born to them: one daughter and six sons, two of whom (both christened Frederick) died in infancy. Two of the sons, Philip and Michael, became mercers.[14]

Thomas Turner lived for more than a quarter of a century after ending his diary. He died early in 1793: the *Sussex Weekly Advertiser* for Monday 11 February records that 'last Thursday died, at East Hoathly, Mr Thomas Turner, many years a shopkeeper at that place'.[15] His gravestone stands in East Hoathly churchyard to the north of the chancel, and the inscription on it reads: 'In Memory of Thomas Turner of This Parish Draper Who Died February 6th 1793 Aged 63 years'. Molly survived him by fourteen years. Her burial is recorded in the East Hoathly parish register on 1 December 1807: she was in her seventy-third year.[16]

---

[11] See, for example, pp. 230–31, 236–7.   [12] See p. 265.

[13] The marriage took place on 19 June 1765: East Hoathly parish register, E.S.R.O. PAR 378/1/1/4.

[14] See Appendix A, and L. F. Salzman, 'Philip Turner', *S.N.Q.*, XVI (1963), no. 2, p. 37.

[15] He also called himself simply 'shopkeeper' in his will. E.S.R.O., Lewes archdeaconry wills register A 66, p. 728.

[16] E.S.R.O. PAR 378/1/1/3.

III

Those are the bare details of Thomas Turner's life. What makes this man different from hundreds of other village shopkeepers all over England was that he kept a diary and thus provides us with a great deal of information about his friends and contemporaries and their way of life. The diary also tells us a good deal about the character and outlook on life of the diarist himself and this, together with any clues which can be gleaned as to his reasons for keeping the diary, must be borne in mind when we come to evaluate his notes on the world around him.

To keep a village shop in mid-eighteenth-century England was literally to be a jack of all trades. The first editors of the diary defined the mercer as 'grocer, draper, haberdasher, hatter, clothier, druggist, ironmonger, stationer, glover, undertaker, and what not'.[17] Dean K. Worcester put it thus:

Turner dealt in everything that passed through East Hoathly, and most things were to be found at one time or another in his shop . . . Every kind of foodstuff and every useful commodity he tried to keep in stock. He received flour from the millers and cloth from the weavers, nails from the blacksmith, and salt from the fishermen. Some of the goods he had were brought to him, some he had to go and fetch, some he bought in lots and processed himself. He was, furthermore, a purchasing agent for the community, and spent much time attending to errands for his customers . . . He occasionally turned a penny at barbering . . . He manufactured and distributed favours for funerals . . . He weighed, packed, and arranged for the transportation of market produce, wheat, potatoes, cabbages, and all kinds of risk items like wool and hops . . . and his shop was constantly full of odds and ends of merchandise which were awaiting a suitable price or the convenience of a low-priced carter. Anything anyone lacked he could find somewhere, and if time was no object he would be able to make arrangements for any problem that arose.[18]

The diary furnishes details of all these activities and, in doing so, provides a rare first-hand account of a man who supplied virtually every retail item to a small community and also acted as its local expert on money matters.[19] Further, since Turner wrote at a time when there is little evidence of the way of death of middle-rank England, his activities as an undertaker are of especial interest. Time and again he provides information on the organization, extent and cost of burying the dead, together with the ceremonial and the accompanying

---

[17] *Blencowe and Lower*, p. 183.
[18] *The life and times of Thomas Turner of East Hoathly* (1948), p. 53.
[19] For Turner's financial dealings see Appendix D.

distribution of gloves, bands, rings, and other favours. At the funeral of the 28-year-old spinster Alice Stevens, for instance, in January 1757 the cortège must have been an impressive sight as the coffin was carried by bearers some two miles from Framfield to Buxted church followed no doubt by many if not all of the ninety mourners whose names the diary records as having received memorial favours.[20]

Even with all these strings to his bow Turner still found the restrictions of trade irksome, and the uncertainty worrying. He constantly sought to extend his activities into other spheres, fearing that his shop might fail leaving him without a source of income. This was certainly the thinking, for instance, behind his decision to ask Christopher Mason of Eastbourne to teach him surveying and land-measuring in the summer of 1758.[21] For a year between April 1755 and May 1756 he kept the village school, and even after he had handed it over to Francis Elless, he never lost the desire to instruct others—indeed he taught Elless how to do 'the contracted division of decimals',[22] and to others he imparted his knowledge of arithmetic and the use of the slide-rule.[23] He was a conscientious parish officer: during the diary years he was churchwarden in 1757–8, 1764–5 and 1765–6, overseer of the poor in 1756–7, 1762–3, 1763–4 and 1764–5, and surveyor of the highways in 1765. In addition he was collector of the window and land taxes from 1760 until 1766. In almost every year he was called upon to write up the accounts of the other parish officers, and his skills in both writing and casting accounts were constantly in demand by those who needed to have bonds drafted, wills written, inventories drawn up, or charity moneys distributed. As a reliable and competent scribe his advice was frequently sought almost in the capacity of attorney. Not only did he write wills, he also drafted them, though he was aware that his abilities in that sphere were limited. When called in by a dying William Piper in April 1761 and asked to make his will, Turner wrote in his diary: 'Finding it required some person of more knowledge than myself to make it according to form and the testator's meaning, I declined making it, for fear I should make some mistake that might bring trouble into the family, and [cause] the effects of the testator to be thrown away amongst the lawyers, or the testator's intentions frustrated.'[24] The following day Turner called in an attorney, though he did assist with advice and allowed his name

---

[20] See pp. 77–9.   [21] See p. 153.   [22] See p. 43.
[23] See, for example, pp. 136, 137, 246.
[24] See pp. 222–3.

to be inserted as a trustee. In the case of Mrs Mary Virgoe, however, his role seems to have been that of local advisor: he dealt with her creditors, with settling debts left by her late husband, with the insurance of her property, and with the sale of one house and the mortgage on another; he also drafted her will and composed a petition for her.[25] Elsewhere he drew up a flowery, begging petition on behalf of two ex-smugglers, wrote another to the Duke of Newcastle asking him to exercise his patronage in favour of a friend's son, and composed a third for a man whose pig had drowned after getting trapped under the January ice in 1757. He also gave advice to a female servant as to her liability, when her ex-employer claimed that she owed him money which, she held, had not been loaned but had been given to her in return for sexual favours.[26]

Undoubtedly some of this advice and help was given in return for payment, but much of it was not. Turner was a charitable man, tender-hearted towards those less fortunate than himself and normally prepared to put himself out, or put his hand in his pocket, for their benefit. By contrast, some of his sharpest criticism was reserved for those who were in a position to recompense him for his advice and help but who, through niggardliness, did not do so. The main target in this respect was William Piper, but there were others: the rector, the Revd Thomas Porter, was 'abstemious at home . . . to a degree of mean-spiritedness, or if there can be a worse name found it deserves that';[27] the grasping attitude of Dr Snelling of Alfriston, who never took account of Turner's hospitality, provoked the comment, 'Oh, thou blackest of fiends, ingratitude, what an odious colour and appearance dost thou make!'; and when another medical man, Turner's cousin Charles Hill, whom he was visiting, failed to cater for his needs in the way in which Turner was accustomed to cater for those of his own guests, he railed 'Oh, ingratitude, ingratitude! thou common but hateful vice, a vice which in my opinion clouds all our other virtues, and I think no man guilty of it in a higher degree than Mr Hill.'[28]

These occasional outbursts of petulance, however, are not typical. Turner bore few grudges, though in general he seems to have regarded himself as a man of good nature in a world where others were bent on doing him down—as, indeed, in some cases they certainly seem to have been. Faced with what he interpreted as his mother-in-

---

[25] See pp. 2, 16, 17, 28, 29, 37, 40, 43–4, 123–4.
[26] See pp. 122, 193, 79–80, 33–4.
[27] See p. 139.
[28] See pp. 20, 291.

law's 'fantastical, odd, capricious humour', and her constant scornful-
ness towards him,[29] or situations in which his own mother and other
relations seem to have expected him to cheat them or, worse, set out to
outwit him,[30] he expressed either outright bafflement or an inadequacy
which he could only blame on 'that easy temper of mine (that is, easy to
be imposed upon)'.[31] He found it difficult to insist on prompt payment
when money was owed to him and suffered agonies of conscience
when he was driven to distrain on someone's belongings.[32] He was
anxious not to give unnecessary offence and tried to show understanding
when others offended him (even when they suggested that he had
hastened his wife's death). He was unable to agree with Joseph Fuller's
attitude that it was 'a piece of wit and a sharp look-out for a man to
serve himself when he can and his neighbour next', and was depressed
and hurt that Fuller, on whom he had relied, had swindled him over
the purchase of a horse.[33]

Turner was also a man anxious to broaden his mental horizons. He
visited beauty spots and places of interest such as the house and
ornamental garden which George Medley was completing at Buxted
Park, the town and fort at Seaford where he took his brother to prevent
him from going sea-bathing at Brighton and getting into bad company,
the abbey at Battle, and the Despencer house at Mereworth Place
where he particularly admired the paintings and thought the church
'modern-built and excessively handsome, but small'.[34] He noted any
curious events (such as the predicted reappearance of Halley's Comet
in 1758), facts (such as the complete list of the titles and offices of
George, third Earl of Cholmondeley, who attended morning service at
East Hoathly on 10 July 1757), or recipes (such as that for the Duke of
Portland's gout powder).[35] He went to see curiosities when he could:
twice he went over to Halland to see turtles, and once to look at a
strange breed of sheep; he noted the dimensions of an enormous hog
which he saw at Framfield Fair; he was baffled by an 'electrical
machine' which he saw at Jones's public house; he took his sister and
his sister-in-law to see a 'microcosm' where the whole solar system
moved by clockwork; he paid one penny at Uckfield Fair to see 'a man
perform several curious performances on the slack wire, and which I
actually think were very curious'; he was disappointed in the annular

---

[29] See, for example, pp. 164, 80.   [30] See pp. 129, 97–8.   [31] See p. 155.
[32] This was certainly so in the case of Thomas Darby: 'There is something in the
thought of distressing a fellow creature that my mind as it were recoils at.' See p. 252
and, for two other cases, p. 68.
[33] See pp. 102–3.
[34] See pp. 30–1, 170, 43–4, 291.   [35] See pp. 103–5.

eclipse of the sun on 1 April 1764; he was intrigued by the activities of
the Whitesmith mountebank, though it was his wife who fell for the
sales talk; and during his time as a schoolmaster he entertained his
scholars by hiring a fire-eater for 12*d*. to perform for them in the
schoolroom.[36]

He read a great deal. A list of the books which he mentioned as
having read is given in Appendix D. The list contains nothing very
startling, though the range is quite wide and the number very large. It
comes, perhaps, as a surprise to realize that a country shopkeeper,
even one with Turner's inquiring mind, read so many books over an
eleven-year period. Reading was clearly a normal activity in the Turner
house. If Thomas was busy bagging up nails or cutting out Sussex
round frocks, Peggy would read to him: if Thomas was reading and a
friend called then the visitor would be treated to a part of Edward
Young's *Night Thoughts*, or to a few sermons whose style and piety he
admired. He was never happier than when with his books. 'Reading
and study', he wrote in March 1758, 'would in a manner be both meat
and drink to me, were my circumstances but independent'; indeed 'I
believe by a too eager thirst after knowledge I have oftentimes . . . been
at too great an expense in buying books and spending rather too much
time in reading, for it seems to be the only diversion that I have any
appetite for.'[37]

The diary, for all the insights which his reading gave him, is much
more that of a reporter than that of a critical analyst, and the picture of
the diarist which emerges is of a basically insecure man, and one with a
somewhat pessimistic attitude to life. When he was busy and trade was
good he was happy, but when business was slack his confidence was
shaken and he set his sights lower: 'Not that I want to get an estate; no,
if it will please the Supreme Being to bless me with only enough to pay
everyone their own and to maintain my family in an indifferent
manner, I am satisfied.'[38] At times he would almost despair of his
ability to 'make things keep in equilibrium', vowing that 'if it be my
fortune to be poor and low in the world . . . I will endeavour to meet my
fortune with pleasure'; and at other times after a particularly heavy and
confused period spent on parish business he would want to throw it all
aside, praying for 'a retired recluse life, making those about me happy
if it were in my power'.[39] Two principal obstacles he considered stood
in the way of a thriving business and a happy home life: his inability to

[36] See pp. 12, 107, 173–4, 102, 223–4, 161, 183, 288, 11, 208, 36.
[37] See p. 143.   [38] See p. 35.   [39] See pp. 82, 95.

live in perpetual harmony with his wife, and his tendency to fall an easy prey to drink.

It is clear from the diary that Peggy Turner was a woman with a will of her own who was capable of being difficult and obstructive, but it is clear too that she was for many years very ill. Both her strong will and her sickness brought Thomas Turner great unhappiness and, in particular, his reactions to her continual ill-health range from sympathy, through baffled attempts to understand and to do something about it, to downright irritation and anger usually followed by worried guilt. This being so, the abrupt change in the diarist's attitude to his wife which comes about immediately upon her death is not easy for the modern reader to accept at its face value. During her lifetime Peggy Turner is characterized by 'a contrariness or at least spitefulness of temper', and as a woman who caused 'disturbances' of which 'there is no hope of their end, as undoubtedly there is not so long as we both live'.[40] Yet after her death she was an 'inestimable treasure', 'a most engaging and agreeable companion' whose 'smiles of approbation and all the other endearments of conjugal love and affection' the diarist looked back upon as a widower. 'No man ever loved a woman better than I did my wife', he wrote.[41] It is more than likely that Turner could be just as difficult as his wife, however, getting unreasonably cross when she put her pleasure before his business or put her mother's interests before those of her husband's;[42] but lacking a close friend with whom he could discuss the ups and downs of married life, Turner took refuge in his private notebooks in which he could overstate his case without fear of contradiction—and then feel guilty about it. Wherever the truth lies, it was evidently not a completely harmonious marriage, and as early as New Year's Day 1756 Turner wrote, 'I have almost made . . . a resolution to make a separation, I mean by settling my affairs and parting in friendship, but is this that for what I married?'[43] Something similar was going through his mind a month later, and in April 1758 he noted again: 'I have oftentimes thought time and patience would put an end to all the discords subsisting between us . . . but now I despond of their ever ceasing till death, or at least till a separation . . . '[44] The discord, he claimed, affected his business ('it is impossible for my trade to be carried on to advantage amidst such trouble') and the 'family misunderstandings' lay behind much of his

[40] See pp. 113, 120.
[41] See pp. 217, 247, 235, 237.
[42] See pp. 106, 164–5.
[43] See p. 21.   [44] See p. 147.

drinking.[45] For though he did not spend as much time in the public house as did some of his neighbours he was often the worse for drink and this increased his feelings of shame and guilt.

Guilt and remorse were emotions that Turner often felt, both in a general way and in particular instances. He regarded the age in which he lived as a degenerate one, and most of his favourite preachers were those who inveighed against the declining standards which he saw around him. 'Dissoluteness of manners, a spirit of effeminacy and self-interest, together with an intolerable share of pride and luxury, seem almost to over-spread the whole face of this kingdom', he wrote at one point;[46] and on many occasions he saw as prime symptoms of this degeneracy the excessive consumption of liquor with its concomitant idleness. He reflected particularly about this in a purple passage written after the drunken celebrations at Halland for the victory of Prince Ferdinand of Brunswick over the French at Crefeld in 1758.[47] When he himself got drunk (despite his efforts to draw up rules to prevent it) he filled his diary with self-criticism. He saw himself as 'prodigiously silly and apish when I am in liquor, having always, for a great while after, a sting of conscience for the same', and he often meditated on the fact that he seemed to be rushing 'impetuous and headlong into destruction (by drunkenness)'.[48] This guilt which he felt in spending his time unprofitably was also transferred on occasion to those activities which accompanied drinking, such as dancing (or 'jumping about' as he put it scornfully), party games ('behaving more like mad people'), or even playing cards, which, when he had lost 3s., he considered 'quite inconsistent with that which is right'.[49] But if Turner appears to us to have taken his pleasures sadly, it is to be remembered that the pursuit of pleasure was in his day by no means universally recognised as a desirable end in the business of living, nor did he regard his diary as a place in which to record the amusing, light, or humorous incidents in his life.[50] With all his troubles, his worries, his pangs of conscience, his self-pity, and his occasional censoriousness, the impression given by the diary is that there was little that was evil about Thomas Turner, and that in the end this conventional, conscientious, and basically kind man simply wanted life to run

---

[45] See pp. 146–7, 161–2. [46] See p. 125.

[47] See pp. 155–7.

[48] See pp. 24, 76. A good example of the rules which Turner tried to impose on himself is on pp. 26–7.

[49] See pp. 138, 141–3, 172, 193.

[50] *Jennings*, p. x, is therefore less than fair to Turner in referring to him as 'a man very deficient in humour'.

smoothly, undisrupted by those 'good friends, neighbours, acquaintances, intimates, gossips, lovers, haters, foes, farters, friskers, cuckolds, and all other sorts of Christians of what name or denomination soever' who set out to prevent it.[51]

Nowhere did Turner state his reasons for keeping a diary, though more than once he wrote in it in order either to justify his actions 'if any person should by accident or curiosity peruse my several memorandums',[52] or to record for the future the misdoings of others. This certainly seems to have been the case, for instance, in his condemnation in 1756 of 'many of the richest and leading men of our parish' for 'endeavouring to pull down the price of . . . poor men's wages . . . by bringing in many poor into the parish from other parishes'. Having noted this Turner wrote: 'May the annals of future times never record so much barbarity!'[53] In this respect the diary served as a record of his attitudes and intentions and, as often as not, of his contrition or forgiveness. If anyone were to judge him this was to be his defence and his version of the truth. 'Let not this, my readers,' he wrote on 8 October 1763, 'be a matter of speculation.' Writing in his diary was also a release, allowing him in the evening to unwind, to confess, to lament, and to strengthen his resolve. Principally, however, he seems to have written up his journal in the same way that he regularly 'posted' his day book, as a means of keeping track of the many matters parochial, commercial, and personal of which a record needed to be kept somewhere for future reference. It is fortunate that he did, for from the diary emerges much more than a portrait of the writer: what we get is a portrait of a society.

IV

The village of East Hoathly whose life Thomas Turner's diary records was an agricultural village in an agricultural county, a land of sheep, of grain, of orchards, hop gardens and woodland.[54] The population of the village in Turner's time can only be roughly estimated: in the 1821 census the population was 510, in 1811 it was 468, in 1801 it was 395.[55] If we were to work on an estimate for the mid-eighteenth

---

[51] See p. 259.   [52] See p. 83.   [53] See pp. 67–8.

[54] The Revd Arthur Young's *General view of the agriculture of the county of Sussex* was published in 1793, the year of Turner's death. It was the first of a series of such county surveys published by the Board of Agriculture.

[55] Copies of the returns are with Series III of the Thomas Turner papers at Yale University Library.

century of around 350 we would not be far wrong. The 1811 census enumerated 66 inhabited houses: we can infer a smaller number than this for 1750 but not substantially smaller. The tithe map,[56] made in 1839, depicts much the same village that Turner knew. The centre of the village is shown in Plate 1. The road from East Grinstead came in from the west through the area known to Turner as Nursery Quarter (now the village of Halland) and past Whyly—the house occupied through most of the diary years by Jeremiah French. On arriving in the village of East Hoathly, the traveller on the road towards Eastbourne would have found himself at a crossroads with his road swinging right into The Street. To the left and ahead of him at the crossing were smaller roads leading to Hawkhurst Common and Chiddingly. At the crossing and in The Street most of the houses were clustered. Here were the public houses, one of which—*The King's Head* (Burges's house, later owned by Turner himself)—is still there facing the traveller as he bears right. Two hundred yards down The Street was the church with Mr Porter's rectory next to it, and a third of the way between the sharp turn and the church and on the same side of the road stood Turner's house. It still stands there, now embellished with a commemorative plaque, though the pond which lay opposite its front door in Turner's time is now filled in and built over.[57] There were some outlying properties such as old John Vine's farm at Coldharbour, but it was a tight little community and one in which everyone knew everyone else and knew their business too.

Many of those who made their living from the land owned their own small farms; others were tenants: but there were only four who had extensive property holdings. The principal landowner was the Duke of Newcastle, Lord of the Manor of Laughton in which the village lay. As a non-resident landlord, his estates were managed, not always efficiently, by his agent and steward Christopher Coates, who was styled 'gentleman' and whose daughter was married to the rector. Coates was replaced in 1765 by a much more ruthless administrator in the person of Abraham Baley, whose first job was to tighten up on some of Coates's financial laxities. The other three substantial occupiers were Jeremiah French of Whyly, the rector—the Revd Thomas Porter—whose continuous acquisition of further parcels of

---

[56] E.S.R.O., TD/E 48.

[57] A small plan of part of the centre of East Hoathly, including Turner's house, drawn in the early nineteenth century, is contained in a volume of plans by Figg and Son of Lewes: E.S.R.O., Acc. 2327/12. The plaque affixed to the house, and illustrated in *Country Life* 12 October 1978, gives the wrong date for Turner's death.

land the diary records, and Mrs Elizabeth Browne. If the majority of
the parishioners derived their living directly from the land, however,
there was also a broad scatter of other trades. The village supported
two and sometimes three public houses, though Jones's house failed in
1764; there was Turner's general store; there were Thomas Davy and
Robert Hook the shoemakers, Charles Diggens and James Marchant
the tailors, and the two Fuller families—one the butchers, the other
the chandlers and soapboilers; there was a succession of schoolmasters
and of excise officers, a brickmaker, the postman, the carrier, young
John Vine the carpenter and builder, the miller, the gamekeeper, and
the blacksmith.

It was a village, too, with its characters, and many of them come alive
in the pages of the diary. Notes on fifty of the principal figures are given
in Appendix B.[58] As might be expected from the diary of a man like
Turner, it is not the writer's best friends who are the most vividly
depicted. Their characters and foibles he took for granted, and though
on occasion an odd comment helps the reader to grasp their attraction
for Turner, he was not normally at pains to set down a portrait of them
for posterity unless he felt that his friendship for them might somehow
be held against him. He was aware, for instance, that there were those
who might think it 'somewhat odd and profuse or extravagant in me to
entertain' so often a social inferior like Thomas Davy the shoemaker,
and he therefore justified his friendship by noting: 'First he is a very
sober man and one who has read a great deal, by which I oftentimes
learn something. And then he is a man that is always ready and willing
to do anything for me or to go anywhere for me and will never take
anything for so doing; so that it is not altogether for the sake of
company, but for the benefit of improvement and out of gratitude.'[59]
He felt in the same way that he had twice to justify his friendship with
the odd-job man Samuel Jenner since there were those who might
think it strange that he spent so much time at Turner's house.[60] But
few details or opinions were given on the diarist's other constant
companions such as John Long, James Marchant, Lawrence Thornton,
and Mary Martin, though of his friend and successor as schoolmaster,
Francis Elless, he did note that he was 'sober and virtuous, and a man
of a great deal of good sense, and endued with good nature. He has
improved his natural parts with a great deal of useful learning.'[61]

By contrast those with whom the diarist crossed swords, whom he

---

[58] Dean K. Worcester. *The life and times of Thomas Turner of East Hoathly* (1948),
ch. II 'People', also provides character sketches of some of the main inhabitants.
[59] See p. 83.   [60] See pp. 287–8.   [61] See p. 305.

held in contempt, or of whom he went in some awe stand out. Whether Jeremiah French was really such an ogre as Turner makes him sound it is difficult to judge, but the man who emerges from these pages is a daunting figure indeed. Swaggering, grasping, quarrelsome, and loud-mouthed, he was, in Turner's opinion, 'quite stupid through drinking'; he 'would willingly never be without a dram of gin in his hand' even at a funeral; and when he died in September 1763, Turner noted: 'If it was possible to make any estimate of the quantity he drank for several years past, I should think he could not drink less on a moderate computation than 20 gallons a year.'[62] As to his manners, he was one of those whose 'whole discourse seems turned to obscenity, oaths, gaming and hunting'.[63] He was overbearing and always in the right, enforcing his views at vestry meetings in a loud voice 'so the noise of his clamour with the hoarse and grating sound of his huge big oaths almost deafens the ears of any of his audience. But then the best remedy to bring him into good humour and change his perverseness into obscenity and raillery is to give him 2 or 3 drams of old English gin.' He was evidently overbearing, too, in matters of parochial administration, being one of those people with 'the greatest skill imaginable in foretelling and judging right of things when they are past'.[64] More than once Turner complained that it was French who was behind the attempt to keep the paupers of the parish in penury by bringing 'certificate men' in from other parishes to work at low rates of pay; and throughout 1757 when Turner was first overseer of the poor and then churchwarden, French perpetually interfered in the administration of the poor law even though he held no parish office. Turner found it impossible to stand up to him, for though they attended the same meetings and the same parties, he was not French's social equal and obviously regarded him with a mixture of fear and contempt.

The picture which Turner provides for us of the rector, the Revd Thomas Porter, is much affected by Turner's respect for the cloth. On the whole he approved of Mr Porter's sermons, though occasionally he complained of having heard them too often before;[65] but most of what we learn of this clergyman from the diary, apart from the regular births and deaths of his children, concerns the worldly side of his activities. Turner was not slow to point out that the sentiments put forward from the pulpit did not always chime with the rector's actions in the village. His parties could be vigorous affairs from which the revellers rolled home drunk at 5 o'clock in the morning, and often what Porter

---

[62] See pp. 238, 49, 278–9.   [63] See p. 126.
[64] See p. 148.   [65] See p. 271.

considered 'innocent mirth' Turner thought 'abounds too much with
libertinism'.[66] At other people's parties it was often the rector who led
the drinking and the singing and obliged Turner to 'make myself a
beast for fashion sake',[67] and it was Mrs Porter who headed the
revellers in the well-known incident in which after such a party on
23 February 1758 Turner was dragged from his bed at 6 o'clock in the
morning and made to dance about wearing his wife's petticoat.[68] Jollity
and camaraderie at parties was one thing, but normal daily contact was
another: the rector, like Jeremiah French, was Turner's social
superior, on one occasion giving him a wigging apparently for the way
he looked at him during a dispute over a cheese, and on another
pointedly not offering him a cup of coffee when he called even though
he was himself drinking one at the time.[69] To add to this, Turner
always felt that the rector's wife treated him with scorn as if he were an
'abject slave', and she was inclined to argue with him about his
prices.[70] The Porters were clearly wealthy and getting wealthier,
acquiring land and property wherever they could: from the Pipers,
from the bankrupt John Vine, and from Joseph Burges in a transaction
whose fairness Turner found very difficult to justify.[71] It is not
altogether surprising that the rector's daughter, Harriet, who subsequently married the Archdeacon of Lewes, became one of the
wealthiest landowners in East Hoathly.

Turner's diary provides sketches of many other characters: the
violent excise officer, George Banister, who was apt to lay about him
with a horsewhip if he felt insulted and who shared confiscated brandy
with his neighbours; the feckless, argumentative rogue Peter Adams,
who 'seems to glory in and give encouragement to crimes of the
deepest dye, and his chiefest discourse consists in obscene words and
oaths'; the butcher, Joseph Fuller senior, who swindled Turner over
the purchase of a horse, and whose wife was 'afflicted' with a 'volubility
of tongue' which made her behaviour, in Turner's eyes, 'obstreperous
and masculine'; the waspish mother-in-law Ann Slater, whose idea of
a conversation with her son-in-law was 'to level a sharp satirical
sentence or two at me'; Samuel Beckett, the publican and post-chaise
operator from Eastbourne, who seems to have been responsible for a
nasty traffic accident on 30 June 1764 which led to the death of one of

---

[66] See pp. 23, 141. [67] See p. 141.
[68] See p. 138–9. [69] See pp. 155, 139.
[70] See p. 98. 'If Mrs. Porter is neither Turk nor infidel, I am sure her behaviour is
not Christian, or at least not like that of a clergyman's wife.'
[71] See pp. 221–2 for Turner's doubts over the Burges affair.

Mrs French's wagon horses; the poor suicide Elizabeth Elless; the troublesome labourer George Hyland and old Thomas Osborne who caused the parish such problems when they fathered illegitimate children; Richard Parkes, the Ringmer husbandman whose marriage was publicly objected to in October 1757 when the banns were called for the second time; and a host of traders and dealers who called at Turner's shop.[72]

One other inhabitant of East Hoathly stands out in very sharp relief because of the diarist's dislike for his principal characteristic—his meanness. This was William Piper, a none-too-well-off farmer whose wife Turner buried in March 1755 but who before the year was out had married his servant. When Piper himself died on 22 April 1761 Turner was deprived of a permanent grouse, and the reader of a character brought to life by Turner's irritation. His niggardliness 'deserves very justly to be ridiculed'; his parties were informed by 'a mean and stingy spirit' and at one of them 'there was not any liquor I suppose worth drinking, though I tasted only some small beer and that came like drops of blood', while at another the guests paid him back for his meanness by 'emptying as many of his old rusty bottles as they possibly could'.[73] Turner considered him 'a wretch who sets his whole delight in nothing but money and knows not the real use of it farther than it is conducive to the adding of store to store'.[74] He was niggardly with his feelings too. When at the tithe feast on 5 November 1756 the news was brought to him that his new wife was safely delivered of a daughter, he simply stayed on so as not to forgo the rector's hospitality, 'saying he should get home soon enough to kiss the old woman, for they were no starters';[75] and at the christening party for the second child of this marriage (for whom the diarist stood godfather) Piper's main contribution was to complain of 'the expense of the christening and the too great quantity of liquor the women drank when his wife was bought abed'.[76] Turner took a particular delight in his discomfiture when Thomas Fuller, during a heated argument, told him just how mean other people considered him to be, upon which he got 'so angry that he cried and bellowed about like a great calf'.[77]

---

[72] See pp. 285 (Banister), 71 (Adams), 102–3 (Fuller), 73 (Fuller's wife), 80 (Ann Slater), 297–8 (Beckett), 50–4 (Elless), 85–91 (Hyland), 125, 133–5 (Osborne), 116 (Parkes).

[73] See pp. 176, 24, 72.

[74] See p. 50.

[75] See p. 71.

[76] See p. 176.

[77] See p. 76.

V

The real value of the diary, however, lies not in the picture it gives of the literate village grocer who wrote it nor in the East Hoathly characters, some drawn larger than life and some to whom justice is almost certainly not done, but in the insight it gives the modern reader into the life of an English village in the central years of the eighteenth century. For this one village in East Sussex, the shopkeeper's diary puts the flesh on those dry bones which survive for almost all English villages—the documents which were preserved in the parish chest: parochial registers; the account books of the churchwardens and the overseers of the poor; clutches of filled-out printed forms produced in connection with the control of the poor—settlement certificates, removal orders, and bastardy bonds; the laconic entries of the vestry minutes. In East Hoathly we can see the machinery which produced this documentation actually at work and we can see how personalities affected its operation.

The vestry meetings, for instance, which elected the parish officers and dealt with parish problems are seen not to have been peaceable affairs. It is known that in many parishes, perhaps in most, there were those whose aim was not so much to tackle the problems of the weak and the needy, but to keep the rates down. Through this diary we can see, for one parish, greed at war with charity. Swearing, quarrelling, and bickering preceded most decisions (indeed Turner felt the need to comment when the parish meeting was not a stormy one) and when it came to assessing a poor-rate all kinds of 'artifice and deceit, cunning and knavery'[78] were used in order to conceal a ratepayer's true liability.

Through this diary we are given a view of how the smooth running of local administration was continually hampered by the clash of personality, and by other unquantifiable influences such as the weather and the absence of easy communication over even quite small distances. Without a document such as this, it is easy to forget that Thomas Turner and his neighbours simply spent a great deal of their time getting from one place to another. The case of Ann Durrant and George Hyland which occupied Turner as overseer of the poor between 18 February and 1 March 1757 serves as a case in point.[79] On the face of it it was a straightforward matter. Ann Durrant, an East Hoathly parishioner living in Laughton, was pregnant and had sworn the child upon George Hyland, a parishioner of Laughton. The task facing Turner was simply to get the two people married and settled in

[78] See p. 267.   [79] See pp. 85–91.

Laughton parish so that Ann Durrant would not become an unmarried parishioner of East Hoathly with an illegitimate child entitled to support from the parish rates. Yet at every turn problems, expense, and travel were concerned: in getting co-operation from the headborough of a neighbouring parish; in meeting the demands of the putative father, who was determined to get what he could out of his predicament; in keeping the man under some sort of surveillance until the matter could be settled (often, unfortunately, in a public house where the drink was too much of a temptation for the guards); in getting a marriage licence from Lewes when Turner did not have his own horse and was wrongly informed as to the documentation needed, so that two journeys were necessary; and finally in getting the marriage paid for and performed at Laughton when the parson could not be found. The sheer burden on Turner as a parish officer both in terms of time spent and of general wear and tear on the nerves is graphically described. It is small wonder that he complained that it was affecting his trade; and yet it was only one of many such cases, some of them unusual such as the suspected suicide of the unfortunate Elizabeth Elless when a post mortem had to be arranged in a hurry,[80] occupying Turner's whole time for three days, and others seemingly never-ending such as the saga of Thomas Daw and the attempts to provide him with the means to earn his living as a blacksmith which dragged on during the year in which Turner was churchwarden in 1757–8. Throughout its entire length the diary illustrates how the perpetual problems of poverty, vagrancy, and bastardy put continuous pressure on a conscientious parish officer, occupying as he did the lowest level of English local government.

With poverty came death, though a diary kept by a man who earned some of his income from arranging funerals and who was pessimistic by nature may over-emphasize the preoccupation with poverty and death on the part of those in Turner's level of society. Death was, none the less, never far away in an age when the expectation of life was short, and much of Turner's reading was in contemplation of it. To some it came as a relief, as was evidently the case with Peggy Turner; but to others it came suddenly, as it did to one of the poorest people mentioned in the diary, Matthew Lewer, a lath-cleaver, who was found dead in the woods one November night in 1764 and at whose inquest Turner was foreman of the jury, or to one of Turner's special friends, John Long, who died in the same month following a

[80] See pp. 50–4.

smallpox inoculation.[81] Turner himself was in the social class which could afford to join a benefit club (in his case the Mayfield Friendly Society) to provide help in times of sickness and death. It may be, too, that he could afford to insure his property against unforeseen disaster. Most people were not in this position, however, and charitable giving is shown by the diary to have played a considerable part in life in the parish, whether it was by helping the poverty-stricken traveller on his way with a small coin, or taking up a collection throughout the parish by means of a church brief. On one such occasion on 18 March 1765 Turner accompanied the rector round the village collecting for those who had suffered losses in a disastrous hailstorm in August 1763. The list of donors which he copied up into his diary is an indication of the range of this generosity: from the rector's 5*s*. to George Banister's 1*d*.[82]

The shopkeeper's pen recorded not only events and attitudes but also the general underlying annual rhythm of local life as it affected him. The regular cycle of winter and summer, of seed-time and harvest, of sheep-shearing and hop-gathering, of tax-collecting, of office-holding, of open days at Halland House, of race-meetings at Lewes, of the distribution of parish charity moneys and of beating the parish bounds, is portrayed as forming the very framework of life. Set in this framework are activities special to East Hoathly or East Sussex but nevertheless part of the annual rhythm. The regularity of the round of New Year parties is a case in point. Christmas Day had little significance, though Turner tended to have a group of regular visitors on that day every year. Each year, however, the middling ranks of East Hoathly society, the Fullers, the Porters, the Pipers, the Frenches, the Turners, the Coateses, and the widow Atkins, entertained one another to tea, an evening meal, and games throughout January and early February and in one case, in 1758, until 14 March. The diarist was a regular, if rather censorious, attender at these gatherings except in the year of his wife's final illness, and always expressed relief when the party season—'the old frantic sports' as he called it—was at an end. In 1758 it was his party which ended the exceptionally long season, and afterwards he wrote, 'Now I hope all revelling for this season is over, and may I never more be discomposed with too much drink or the noise of an obstreperous multitude.'[83] In the summer cricket played a large part in the young men's lives. Turner played himself, but also went to watch. Matches were evidently recognized meeting-places: on

---

[81] See pp. 307–8.
[82] See pp. 316. For the system of briefs see p. 158, note 37.   [83] See p. 142.

at least one occasion it was hoped that a quorum of the vestry would be able to deal with parish business during a lull in the game between East Hoathly and Framfield.[84] The matches, too, were not just social and sporting occasions: they provided a chance to lay bets and win money. Betting, indeed, is shown to be very much part of everyday life amongst Turner's friends. They bet not only on cricket and on horse-races, but also on cock-fights, on who could run the fastest, on the length of a path, on whether a horse could travel a given distance in a given time, on the likelihood of a friend's marriage before the following Christmas, and even on one occasion on whether Jeremiah French could sit through a whole sermon in Hellingly church and then come out and repeat the preacher's text together with chapter and verse.[85]

Turner was always keen to note what food he had consumed during the day and the importance of the diary as a source of information on the standard of food which someone in his position might hope to provide has already been mentioned.[86] It is also important in showing the range of people with whom Turner would share his table. Those who formed the charmed circle of the annual parties were far from being the only ones with whom Turner sat down. His mealtime companions ranged from visiting suppliers and merchants to chance acquaintances, from a casual customer in the shop to the man who helped Turner to pick his apples, from the local carrier to old servants come back to visit him, from wealthy friends such as Thomas Tipper of Newhaven to the woman who did his washing.

If Turner's preoccupations were for the most part parochial, his horizons were not completely limited by the parish boundary. He travelled to neighbouring towns, especially to Lewes and Tunbridge Wells, quite regularly, and occasionally went further afield to Eastbourne, to Maidstone, or to Brighton. Sometimes, though rarely, these trips were for pleasure. Travelling, for Turner at least, was not a pleasurable affair and it was normally business which took him further from home. The diary records only one longer journey. This was the three-day trip to London in March 1759, and if the diary is a faithful record of the time spent there, pleasure played little part in it. He stayed in and around Southwark, settling accounts, buying goods, and exchanging bills, and he returned home as quickly as he could to his normal everyday round.[87]

Beyond this everyday round what else did the diary record? National

[84] See p. 41–2.
[85] See pp. 292, 104, 109, 55, 206, 60, 67.
[86] See above p. viii.   [87] See pp. 177–9.

and international events were bound to intrude into the life of a village some twenty miles from the south coast at a time when the nation was engaged in armed conflict on the continent, in America, in India, and in the West Indies. This was the Seven Years War and Turner, an avid reader of the newspapers, kept abreast of it. When the struggle was not going well the village observed the decreed days of fasting and went to church to pray for victory, and when successes came, thanksgiving services were held and the village celebrated, fires were lit, and toasts were drunk. Turner's reflections at these times were moral, conventional, and, one suspects, sometimes second-hand. On occasion, however, the war came closer: on 7 July 1759 he noted with somewhat muted alarm the 'disagreeable news of the French being landed at Dover'. Happily this report proved false, and two days later the news of Rodney's attack on Le Havre provided 'a sudden transition from sorrow to joy'.[88] Turner's cousin, Charles Hill, was a surgeon on board the vessel on which Admiral Byng was executed, and was able to deliver a first-hand account of his appearance and demeanour. Turner, in common with most of his contemporaries, had seen the loss of Minorca as a terrible disaster, but when it came to the execution of Byng he thought the sentence neither just nor prudent and he saw Byng as having been sacrificed to a 'clamorous and enraged populace'.[89] When peace eventually came with the Treaty of Paris in 1763 East Hoathly did not celebrate. 'I think almost every individual seems to be dissatisfied with the peace', Turner wrote, 'thinking it an ignominious and inglorious one.'[90] Rarely, however, are national matters or personalities mentioned in Turner's memoranda. He approved of the 'spirit of liberty' expressed in *The North Briton*, which he read in July 1763, and mentioned that Wilkes had been released from the Tower; he recorded the births of royal children, and the death of the monarch in October 1760. Other events, however, were entered in the diary only if there was some local significance and then normally only if they affected the diarist's own interests. Thus the imminent arrival of the Duke of Cumberland to view the South Coast was noted, but the actual view and its progress was then ignored.[91] The Lewes election of 6 April 1761 figured because of the great concourse of people it brought to Halland, which Turner resented since it 'puts me out of that regular way of life which I am so fond of ', with the risk that 'when reason is in a flustering state the passions are apt to ride triumphant'.[92] The great Lisbon earthquake was mentioned only because Turner was

---

moved by a powerful sermon which the rector preached on the ensuing public fast-day.[93]

The great events of the day, like the great works of literature of the period, are not reported and analysed in the diary because Turner did not keep a diary in order to record such matters. He kept it to ensure that his actions were rightfully remembered and his books straight. His was a little world, to be sure, but the diary's first editors and those who have subsequently reproduced their work have been neither fair to Thomas Turner, nor as helpful as they might have been to students of eighteenth-century history, in the selection which they made from it. Blencowe and Lower wished primarily to show how advanced the Sussex society of 1859 was when compared to that of 1759. Their extracts from the diary emphasized the parties, the drunkenness, and the curiosities of what they regarded as a backward age, with the result that Turner was portrayed as a naïve nonentity, a figure of fun surrounded by boorish rustics in a very peculiar village. This present edition, with its more detailed extracts, aims to do greater justice to East Hoathly and to its inhabitants, but above all it aims to accord to Thomas Turner his rightful place as an important observer and recorder of eighteenth-century life.

[93] See p. 25.

1754

Thurs. 21 Feb.[1] I had the gardener at work to plant beans etc.; went to Wm. Rice's to inform him of Mr. Chatfield's intention to repair Chantler's. Came home and found Mr. Porter's servants at cards and went to bed drunk.

Mon. 25 Feb. John Richards called at my house and told me of the malicious report at Lewes of my wife's and my not agreeing. I then hired Elphick's horse and went to Lewes. Found Mr. Madgwick of that opinion. Came home in the even and drank tea with my wife with a great deal of pleasure to both. Spent 10*d.* at Lewes.

Thurs. 28 Feb. Killed an hog. Had it of Mr. French. Weighed 30 stone at 2*s.* 2*d.* [the] stone.[2] Cut him out in the even and sat up all night with Mr. French, Tho. Fuller and Mr. Hutson at cards and lost 2*d.*

Weds. 6 Mar. Dr. Russell came to my wife and declared her very dangerous.[3]

Sat. 9 Mar. My wife continues worse and nothing more material.

Sun. 10 Mar. Mr. Tucker and Moses came to see me and dined with me on a pig sent me from Hartfield.[4] Heard of the death of the Rt. Hon. Hen. Pelham who died on Wednesday the 6th instant about 6 o'clock in the morn.[5] My mother came and spent the afternoon with us. Not at church all day.

Weds. 13 Mar. Set out on foot in order to go to Inchreed in Rotherfield where there was a sale, but was obliged to return by reason of the continual snowing, and the snow in most places was upwards of 16 inches deep. About 3.20 Mr. Pelham was brought to Halland[6] (from

[1] The first entry in the first surviving volume of the diary.

[2] The stone weight when applied to meat was equal to 8 lbs.

[3] Margaret Turner was in the fourth month of her pregnancy.

[4] Hartfield was the home of Turner's parents-in-law. Moses was his younger brother.

[5] For Henry Pelham (1695?–1754), see *D.N.B.* His death on 6 March was said to have been 'from an attack of erysipelas . . . brought on by immoderate eating and want of exercise'.

[6] Halland, the principal seat of the Duke of Newcastle, stood on the western boundary of East Hoathly parish. The boundary between East Hoathly and Laughton parishes was said to run through its front entrance. An Elizabethan mansion (see Plate 2), it was, soon after this diary ends, abandoned by the Pelham family. Following the death of Thomas, Duke of Newcastle in 1768, his heir, Sir Thomas Pelham, dismantled it in order to use materials from it to beautify his house at Stanmer. The property gradually dwindled into a farm house. Viscountess Wolseley, *Some of the smaller manor houses of Sussex* (1925), pp. 97–115; Frederick Jones, 'Random notes on Halland Park', *S.C.M.* 4 (1930), pp. 373–80.

Kidbrook, they lodging there last night) in a hearse being attended by two mourning coaches and six persons on horseback in cloaks.

Thurs. 14 Mar. Continued snowing all night and now the snow falls a great pace. Continued snowing all day. About 10.30 in the forenoon Mr. Pelham was carried from Halland and interred in the family vault at Laughton about 12.15.

Thurs. 21 Mar. Tho. Butler's wife buried. Paid Tho. Burfield for beehives £1 8s. Went with Jo. Durrant to Mr. Will. Piper's in the even concerning Tho. Ling.

Fri. 22 Mar. About 10 o'clock Mr. French called on me and brought me a warrant to serve on Tho. Ling. We immediately went to Thunderswood, took him up, carried him to Uckfield where he swore himself on East Hoathly parish before George Courthope Esq. I spent 4d. at Halland's. Came home with Mr. French and stayed there some time. Came home and went to bed drunk.

Sat. 23 Mar. . . . Engaged myself to assist Mr. Jos. Burges in getting Mrs Virgoe's debts in.

Mon. 25 Mar. . . . Mary Martin came to live with me at 30s. a year. Read *The Conscious Lovers* in the even.

Thurs. 28 Mar. . . . Paid Mr. Tho. Harman[7] for pipes as under:

12 gross at 14d.		14s. 0d.
6 *do.*	18d.	9s. 0d.
4 *do.*	2s.	8s. 0d.
1½ *do.*	13d.	1s. 6d.
		£1 12s. 6d. . . .

Fri. 29 Mar. . . . Went in the even to shave Tho. Fuller. Paid Jos. Fuller for a shoulder [of] mutton 18d.

Tues. 9 Apr. . . . Went to Chailey and agreed with Mr. Beard to take Richd. Turner as an apprentice for seven years to come,[8] from the first day of last January, and to be paid in hand £10, but it is agreed for the said Mr. Beard to pay Richd. Turner 2s. 6d. on every quarter-day in the first year of his time, 5s. on every quarter-day in the second year, 7s. 6d. a quarter in the third year, 10s. a quarter in the fourth year, 12s. 6d. a quarter in the fifth year, and 15s. a quarter for the sixth year and 17s. 6d. a quarter for the 7th and last year. But the said Geo. Beard is, after the expiration of the 7 years, to be paid £1 4s. if he, the said Geo.

---

[7] Pipemaker of Lewes. See D. R. Atkinson, 'Sussex Pipemakers', *S.N.Q.*, XVI, p. 170; and W. K. Rector, 'Pipemakers of Lewes in the 18th and 19th centuries', *S.N.Q.*, XV, pp. 315–17.

[8] Richard Turner was the diarist's youngest brother. At this date he was twelve years old. For George Beard see Appendix B.

Beard, thinks proper to demand it. I came home in the even. Spent 5*d.*

Good Friday, 12 Apr. Bought half a hog of Will. Piper, about 9 stone at 2*s.* 2*d.* per stone . . . Mouled up 140 papers of tobacco.[9] Busy all day.

Weds. 26 June.[10] This day made an end of instructing Miss Day. Read part of *The Spectator*; prodigiously admire the beauties pointed out in the eighth book of Milton's *Paradise Lost* by *The Spectator*'s criticism wherein is beautifully expressed Adam's conference with the Almighty, and likewise his distress on losing sight of the phantom in his dream, and his joy in finding it a real creature when awake.[11]

Tues. 30 July. At home all day. My wife, self, Ann Slater, and maid very busy a-making of bolster and pillow-ticks and a bed bottom for Jos. Fuller.

Sun. 4 Aug. Very bad with a swelled face. Sam. Slater came for Nanny; he breakfasted and dined with us. I received a letter which informed me of my mother's cruelty against my wife and self. Freely forgave her and am willing to serve her to the utmost of my power, but came to a resolution to keep Will. Cripps no longer. Sam. and Ann Slater went from Hoathly after dinner. I at home all day. Read part of Hervey's *Meditations*.

Thurs. 8 Aug. Called up in the morn to John Vine Jr. Sold him a frock and two waistcoats. Agreed with Dame Smith to nurse my wife at 2*s.* 6*d.* per week. Paid 2*d.* for a quarter of gooseberries. At home all day.

Thurs. 15 Aug. Went in the morn to Mr. John Cornwell's. Bought his and George Cornwell's wool at 6*d.* per pound for my mother. Called on Mr. Peckham; bid him 18*s.* per tod, being the price he asked me, but at the same time gave him the opportunity of selling it till Tuesday next, provided he can sell it for more. In the afternoon went to Mr. Fran. Hicks's; bid him 6½*d.* for his wool, but did not agree for it.

Mon. 19 Aug. My wife safely delivered of a son about half an hour past 7 o'clock in the morn. . .[12]

Sun. 25 Aug. My wife very bad. My mother came to see her. Sent for Dr. Stone to her.[13]

[9] To 'moul' or 'mool' was to crumble. It is clear from the diary entry for 17 Feb. 1757 that each 'paper' contained four ounces.

[10] The volume of the original diary covering the period between 14 April and 25 July is missing. The entry for 26 June is taken from *Blencowe and Lower*, p. 185.

[11] *The Spectator* no. 345, 5 April 1712. The article which so impressed Turner was by Joseph Addison.

[12] Turner's son, Peter, was baptized on 3 September. The godparents were Mrs. Roase, Thomas Scrase and John Madgwick, all shopkeepers in Lewes. Peter died on 16 January 1755. See 'Notes on family history written by Thomas Turner' printed as an appendix to *F.M.Turner* and *Jennings*.

[13] A volume of the original diary ends here. The next volume begins on 1 January 1755.

1755

Sat. 11 Jan. At home all day a-writing. In the even read *The Universal Magazine*[1] for December; think the following observations worth notice:

1st. He who is vexed at a reproach may thereby know that he would be proud if he were commended.

2dly. Pride upon the account of preferment shows that it is not deserved, for he who values himself upon his outward character acknowledges thereby that he wants intrinsic worth. But the greatest men are valued more for their abilities than for their fortunes; and if virtue were esteemed above all things, no favour or advancement would change men in their temper.

3rdly. We should often blush for our best actions if we saw all the motives upon which they were grounded.

4thly. It is idleness which induces men to be guilty of bad actions, but whatsoever art is able to busy the minds of men with a constant course of innocent labour will certainly have its effect in composing and purifying their thoughts, surer than all the precepts of the moralists.

5thly. There can be no arguing with an obstinate man, for when he has once contradicted you all conviction is excluded from his mind. None but manly souls can acknowledge themselves mistaken and forsake an error when they find themselves in the wrong.

6thly. We should not measure men by Sundays without regarding what they do all the week after, for devotion does not necessarily make men virtuous.

Tues. 14 Jan. My wife very bad. Paid Dame Vinal for washing and paid for odd things together 12*d.* Paid Jos. Haines in goods 6*s.* for 4 chairs. Gave Tho. Mepham 6*d.* for going to Dr. Stone's. At home all day. In the even read the 9th book of *Paradise Lost.*

Weds. 15 Jan. At home all day. My wife and little boy both very bad. Read the 10th book of *Paradise Lost* in the even.

Thurs. 16 Jan. This morning about 1 o'clock I had the misfortune to lose my little boy Peter, aged 21 weeks, 3 days. Paid for flour and other small things 16*d.* At home all day. In the even read the 11th and

---

[1] The six observations are extracted from 33 similar 'moral considerations' communicated to the magazine by 'H.H.'; *The universal magazine of knowledge and pleasure*, XV (Dec. 1754), pp. 249–51.

12th books of *Paradise Regained*, which I think is much inferior for the sublimity of style to *Paradise Lost*.

Fri. 17 Jan.   Went to Framfield concerning the burying of little Peter. Met at Framfield Messrs. Grant, Barlow and Wigginton's man. My mother paid him £30 in full to September last. This day balanced accounts with John Dulake and paid him for the carriage of 2½ tuns salt, 2 bushels sand and 2 cwt. Cheser[2] cheese: 18s., which with 6s. 6d. received of him was in full to this day. Paid Francis Paris for 3 tubs brandy 49s. 6d., but none for myself. Supped at Mr. Porter's, who informed me that he had paid to Messrs. Margesson and Collison on my account £35. Received of Tho. Diplock £27, for which sum I am to give him a draft to send to his landlord.

Sat. 18 Jan.   After dinner went to Framfield and buried my little boy. Mr. Collison came there to me and came home with me. Paid him in cash £5 12s. and by balance of cash account £40 14s., which I am to draw on him for, payable to Tho. Diplock. Mr. Collison spent the even and lodged at my house. Paid for meat 2s. 1d.

Weds. 22 Jan.   Paid Fran. Paris for a tub of brandy 16s. 6d. for self . . .

Thurs. 30 Jan.   At home all day. Paid Mr. Porter 2s. for 2 pairs gloves. In the even went into Master Durrant's to teach Thomas.

Sun. 2 Feb.   . . . In the afternoon at church; sat in Mrs. Browne's seat. The text: *Psalm* 33.8. This day John Cayley informed me that my wife and I might sit in Mrs. Browne's seat for 12d. per year each.

Thurs. 13 Feb.   At home all day; posted my day book[3] and cut out round frocks.[4] In the even began Tournefort's *Voyage into the Levant*. Read his *Life* and the *Eulogium* on it by M. Fontenelle. Memorandums on his life: He was born at Aix; first travels to Provence, Savoy and Dauphine; then Montpellier and Spain; he was robbed on the Pyrenean hills, traversed Catalonia. In a house where he lodged near Perpignan it fell in the night, by which accident he miraculously escaped with his life, being some time buried in the ruins. Returned to Montpellier, then to Orange and Aix. He likewise travelled to the Alps, Spain, Portugal, England and Holland—all in the search of botany, chemistry, medicine and anatomy, all of which he seemed to excel in, and in particular in botany, in which he seems never to have been

[2]   Cheshire.

[3]   Turner is here transferring entries made in his day book (which were in consecutive order) into his ledger where the entries were arranged by client.

[4]   The round frock was the loose outer garment worn by men, elsewhere called a smock. See 'The Sussex smock or round-frock' by one who wears it (not in public), *S.C.M.*, 11 (1937), p. 371.

excelled, it being his greatest talent. He renewed and, which is more, demonstrated a system of the vegetative life of stones. He also found several surprising particulars relating to the formation of corals, sponges, sea mushrooms, lithophytes and stony plants or others that grow at the bottom of the sea. He also extended his system of vegetation to minerals, and even to metals, rock crystals and precious stones. He also went further and proved that shells vegetate. He was upon the whole a true philosopher, a good geometrician, an attentive anatomist, an exact chemist and a penetrating naturalist. He died at Paris 28 December 1708.

Sat. 15 Feb. . . . Paid Clymson for 22 pairs men's outseamed tan [gloves], 4 pairs men's inseamed *ditto* and 10 pairs boys' outseamed *ditto* 17s. 6d. (to wit) by cash 9s. 6d. and goods 8s. . . .

Tues. 18 Feb. At home all day. Ed. Smith and I balanced accounts, and I paid him 4s. for 8 packs cards. He spent the even with me.

Tues. 25 Feb. . . . Brewed and filled the old barrel, the new being filled about 12 weeks ago . . .

Thurs. 27 Feb. At home all day; wrote out bills. In the Lewes newspaper of the 24th think the following worthy of a memorandum: 'York, Feb.11—At the adjournment of the Quarter Sessions for this city and country thereof, James Monkman of New Buildens, near Sutton-on-the-Forest, yeoman, was convicted of regratings contrary to statute 5 and 6 King Edward VI. The fact was that he bought 6 chickens in open market in the city at 6d. each and immediately after sold them again the same manner at 7½d. each; for which offence he forfeited the value of the goods, received sentence of 2 months' imprisonment and was accordingly committed to the city gaol. By the abovementioned statute the punishment for 2nd offence of forestalling, engrossing or regrating is a forfeiture of double the value of the goods and six months' imprisonment; and for the 3rd offence the offender loses all his goods, must stand in the pillory and be imprisoned during the King's pleasure.' At the even at Master Durrant's instructing Thomas. After supper read part of Tournefort's *Voyage into the Levant.*

Tues. 4 Mar. . . . Dame Piper died.

Weds. 5 Mar. Mr. Tho. Tomsett died . . . Tho. Fuller killed my hog had of Jer. French. Weighed him in the even; weight 19 stone 4 lbs. Paid Mr. French for him £2 1s. 6d. In the afternoon I went to Framfield for to get a sheet for Dame Piper. After supper finished *The Tragedy of Cato.*

Fri. 7 Mar. . . . At home all day, except carrying up the wine to Mrs. Piper's . . .

Sat. 8 Mar. At the funeral of Mrs. Piper. Distributed at the funeral and afterwards 20 pairs men's and women's 2*d*. shamy gloves, 3 pairs youth's and maid's and 8 hatbands.[5] Will. served Mr. Tomsett's and gave away 8 pairs men's white lamb. There was a sermon for Mrs. Piper. The text: *Revelations* 2.10 . . . In the even went down to Whyly to hire Mr. Virgoe's house to teach school in. After supper read part of Tournefort's *Voyage into the Levant*.

Sun. 9 Mar. In the morn Mr. Clinch and myself went down to the late Mr. Tomsett's house and looked over his affairs, but found nothing material. I engaged myself to get in his debts and pay all disbursements . . .

Tues. 11 Mar. . . . Hired Mr. Virgoe's new house at 50*s*. per year, to enter upon it at Lady Day next, N.S.,[6] to make whatever use I like of it as a warehouse; agreed to give him 1*s*. 6*d*. for beans etc. planted. Balanced with Mr. Will. Piper and received of him as under:

My bill on him		£9 18*s*. 11¾*d*.
Received 84 lb butter	£2 2*s*. 0*d*.	
Ditto 1 pair gloves	1*s*. 4*d*.	
Ditto 3 bottles of wine	4*s*. 6*d*.	
		£2 7*s*. 10*d*.
Balance received		£7 11*s*. 1¾*d*.

. . .

Thurs. 13 Mar. In the forenoon posted part of my day book and wrote out some bills. After dinner Jos. Durrant and I went to Mr. Miller's at Burghill. Mr. Miller promised me his son should come to me to school. I received of him 18*d*. due to Mr. T. Tomsett for

[5] Here Turner, like many grocers, mercers and drapers of the time, acted in the role of undertaker for funerals, distributing mourning gifts and favours and getting in the wine for the wake. He was always anxious on hearing of a death to secure the right to 'serve the funeral' and often assisted his mother with funerals in the Framfield area. The supply of mourning gifts of gloves, hatbands and favours when these were distributed on a grand scale (as they were evidently not at this funeral) represented a considerable profit for someone in Turner's position. There is no comprehensive work on the organisation of eighteenth-century funerals, but see Bertram S. Puckle, *Funeral customs, their origin and development* (1926) and below p. 77, note 1.

[6] New Style. Turner was writing less than three years after the adoption of the Gregorian calendar in Great Britain. Chesterfield's Act (24 George II, c. 23) laid down that the dating of the beginning of the year from Lady Day (25 March) should cease; that 1 January 1752 should be the first day of 1752; and that, in order to rectify the calendar, 2 September 1752 should be immediately followed by 14 September. Turner here made it clear that his lease was from *new* Lady Day (i.e. 25 March) not 5 April which still remains as 'Lady Day' for fiscal purposes.

6 weeks' schooling Henry. Came home about 5.30, drunk tea and went into Joseph Durrant's to instruct Thomas. After supper read *The Tragedy of Macbeth*, which I like very well.

Thurs. 27 Mar. In the morn went over to Framfield. Bought of Mr. Stone at Stonebridge 11 pairs parish sheets at 10s. pair. Sold my mother one pair at 10s. 6d., brought home the other 10 pairs . . .

Mon. 7 Apr. Began to teach school[7] . . .

Thurs. 17 Apr. . . . In the even I went down to Mr. French's. Carried some patterns for Mrs. French for a gown . . .

Tues. 6 May. . . . Paid Mrs. Weller £8 in full for one year's rent due the 5th April and paid her £4 in full for one year's interest . . .

Fri. 9 May. Busy all the day in settling Mr. Thomas Tomsett's affairs with his two brother-in-laws and bought of them all his books, paper, forms etc. for £4 12s. and I balanced accounts with them . . . [the] balance of £1 5s. 5d. I paid them and afterwards bought of them some pork and a crock which I gave them 12d. for.

Sat. 10 May. At home all day and busy. Paid Fran. Smith 5s. for 2 gallons gin . . .

Tues. 13 May. At home all day . . . In the even read part of that simple thing called *The West Country Clothier*[8] and, notwithstanding the meanness of the language, I think the character of the midwife and gossips is in some measure painted in their true colours; and the thoughtlessness and extravagance of many women are in some respects justly exposed by its often terminating in the husband's ruin (as in the case of Mr. Wilmot and, I think, the character of Hump. Shewshow). Oft fidelity and honesty is obscured by meanness of birth and the want of education; and I farther think nothing is more ridiculous than to see a waiting-maid made the lady's confidante, either when single or in a married state. If in the first, the young lady stands candidate for ruin, and, if after marriage nuptials, differences must in all probability ensue, for it many times happens that the insinuating expressions of an Abigail fail not of drawing the lady's affections from her husband by detraction though she should be almost the only one for him before.

Mon. 19 May. Tho. Durrant and I went to Hawkhurst. Lay there all night. Went to bed drunk.

Tues. 20 May. Came home about 10 o'clock at night and in liquor, but have set a resolution never to get in liquor again. Spent in all about 8s.

---

[7] In succession to Thomas Tomsett who had died on 5 March.
[8] Not identified.

Mon. 26 May.   Very busy all the forenoon. Charles Diggens came over and took up a coat, waistcoat, and breeches for Mr. Porter, and a coat and waistcoat for Tho. Fuller. In the afternoon went to see a cricket match in the park between Hoathly and Framfield, which turned out a pretty match though it was not played out by reason of not having time, Framfield having 5 to fetch and 3 wickets to go down.[9] Spent 2½d.

Weds. 4 June.   . . . Dame Dallaway all day a-helping my wife quilt. Halland gardener gave me 3 cucumbers and Tho. Fuller, chandler, gave me 2 fine roach . . .

Weds. 11 June.   At home all day. Paid Mary Martin in goods 30s. being in full 1 year's wages, due 25 March last (N.S.). In reading *The History of England* I find that England first took that name under Egbert the 1st monarch of England after the Saxon Heptarchy, anno 801.

Thurs. 12 June.   At home all day. In the afternoon Tho. Cornwell beat me at cricket; lost 6d. Paid Halland gardener 12d. for two gallons of gooseberries. Found in *The History of England* that England was first divided into counties, parishes etc. in King Alfred's reign, about the year 890 . . .

Fri. 20 June.   At home all day. Mr. Humphrey and Mr. Robt. Noakes of Horsham called on me. This day being my birthday, I treated my scholars with about 5 quarts of strong beer. Had an issue cut in my leg. Paid Joseph Fuller for 3¾ lbs of the scrag end of a neck of mutton 11d. This day I was took very ill.

Mon. 30 June.   Received of Will. Vine 5s. 6d. in full to this day. Balanced accounts with Fran. Smith and agreed to pay Mr. James Hutson £2 17s. 10d. and Smith owed me 9s. 9d. which makes £3 7s. 7d. and is full to this day. I likewise received of Fran. Smith £1 10s. 9d. which I am to pay Mr. Hutson with the above £2 17s. 10d. which together makes £4 8s. 7d. and is in full for the money Fran. Smith received of Mr. Pollard for Hutson's hops, with the carriage being deducted.[10]

Tues. 1 July.   This day paid Mr. James Hutson £4 8s. 7d. as I yesterday agreed with Frank Smith to do. Afterwards I balanced

[9] Some of the references to cricket in the diary were published (with many inaccuracies) in 'Diary of Thomas Turner of East Heathly [*sic*], 1729–1793', *The Cricket Quarterly*, IV, 4 (Autumn 1966), pp. 239–42. See also H. F. and A. P. Squire, *Pre-Victorian Sussex cricket* (1951) and, for the wider context, J. Marshall, *Sussex cricket* (1959).

[10] Turner's shop was here being used as a sort of depot or clearing-house for transactions involving cash. Smith, the carrier, had delivered, and received money for, Hutson's hops. He then topped up with cash the sum with which he was in credit with Turner, and Turner paid Hutson.

accounts with Mr. Hutson and received of him 14s. 4d. which with £1 4s I owed him makes £1 18s. 4d. and is in full to this day, Joseph's schooling excepted. Received of Robt. Hook 4s. for 4 bottles of brandy. Paid Jos. Fuller 2d. for a piece of mutton for broth. Received of Rippington's boys 6d. for 1 week's schooling due a-Saturday last.

Sat. 5 July. At home all day, busy ... Mr. French, Piper, Fuller and Durrant here in the even to get me to make the land and window tax books against Monday morning.[11] This day I heard of my late servant George Beard of Chailey being married to his servant ...

Mon. 7 July. ... Paid Halland gardener 17d. in full for cucumbers sold for him.

Thurs. 10 July. At home all [day]. Paid Tho. Balkham for William Funnell in goods 4s. for 2 hundred of bricks received this day by Tho. Balkham's wagon. This day Tho. Tester's wife brought me 2 carp for a present; gave her about 1 lb lump sugar and 6 fish hooks, value about 9d. Paid for bread and milk 1d.

Sat. 12 July. This day sold Mr. Blackwell of Hawkhurst[12] all my white rags at 27s. per cwt., and bought of him 12 couple of white pound and half-pound paper, 8 reams of midd[ling] hand *ditto* at 4s., 2 reams of brown cap and 1 ream of 2 lb. *ditto* at 4s. 3d. ...

Mon. 14 July. ... This day was played in the Broyle a cricket match, Mayfield against Ringmer, Framfield and St. John's, which was won by Mayfield with ease ...

Tues. 15 July. ... Paid Master Elphick in cash 3s. 6d. and in goods 6d. which together is 4s. for bringing my hay from Framfield. Gave two of my scholars, for helping in with the hay, gingerbread value 2d. ...

Tues. 22 July. At home all day. Gave over teaching school for this week, being very ill ...

Weds. 23 July. At home all day. Paid for bread 3½d., for butter 3d., for a sheep's heart 1½d. Paid John Jenner at Hailsham, hatter, £1 16s. (as under) for 3 hats at 18d., 3s., 3s. 6d. and 4s. Received this day (to wit):

[11] Here the collectors of the land and window tax were prevailing upon Turner to use his accounting and writing skills to write up the collectors' books, presumably for transmission to the justices the following Monday. Turner always complained of this extra chore for which no one paid him. He himself was the collector from at least 1760 onwards.

[12] Edward Blackwell made paper at Ockley Mill, Hawkhurst, Kent. See A. H. Shorter, *Paper mills and paper makers in England, 1495–1800* (1957), p. 197.

By balance of account	5s. 9½d.
By goods	18s. 2½d.
By cash	12s. 0d.
	£1 16s. 0d.

Gave him orders for 3 boys' hats at 13*d*. and 22*d*. and 6 men's at 2*s*. and 2*s*. 6*d*. . . .

Thurs. 24 July. . . . Except going down to Halland, at home all day. Mr. Coates informed me of the Duke's being down soon.

Fri. 25 July. . . . This day the parish was confirmed at Lewes by the Rev. Mr. William Ashburnham, Bishop of Chichester; my maid went.

Sun. 27 July. At home all day. Not at church all day. Read part of Boyle's lectures[13] and Smart's poem on eternity and immensity. This day I turned my horse out to keep at Mr. French's. Paid for 1 lb. cherries 2*d*.

Weds. 30 July. . . . Left off school at 3.40. About 5 minutes before 4 my [wife] went to Whitesmith to see the mountebank. Bought a packet which cost her 12*d*. She came home about 10.45[14] . . .

Fri. 1 Aug. . . . In the even went down to Halland to look upon Mr. Coates's wool, which was all West-Country, and because I would not give him 7*d*. per pound for it he huffed me prodigiously.

Mon. 4 Aug. At home all day. In the morn about 6 o'clock Char. Diggens and Rob. Rice called me up to take up a coat and waistcoat for Rob. Rice. Charles Diggens took up a settoo[15] coat for me . . .

Fri. 8 Aug. At Halland at about 7 times; otherwise at home all day. Nothing more of moment, but I find I sell but little to Halland on the account I would not give Mr. Coates 7*d*. per pound for his West Country wool on the 1st instant, which I should have lost not less than 40*s*. by, had I bought it at that rate and fulfilled my other engagements. Paid for 4 doz. of brooms 4*s*.

Sat. 9 Aug. At home all day except at Halland 6 times . . . This day the Duke of Newcastle came to Halland about 7 o'clock in the evening . . .

Sun. 10 Aug. This day the public day was at Halland. The 2 judges

[13] Turner seems here to be referring to William Derham, *Physico-Theology* which he was reading again on 7 September and 25 December. See Appendix D.

[14] The mountebank, or travelling purveyor of patent and quack medicines, visited Whitesmith weekly and used entertainment as part of his sales technique. Margaret Turner evidently fell for this sales talk, though the diarist (see below, p. 208) was himself very contemptuous of it.

[15] Meaning unidentified: *O.E.D.* has a 1657 reference to a 'sithy coat' but the precise meaning is obscure.

came from Lewes and dined there. There was on the whole but a small company. I was there about 4 times. I came home the last time about 9.20 with several of our parishioners, a little matter enlivened by liquor, but no ways drunk. Paid 2*d.* for bread. Not at church all day, neither looked in any book all day except *The Tatler.* My wife after dinner went down to Halland to see the turtle. Our maid went to Framfield; was gone all day. Ed. Smith called to see us in the evening.

Mon. 11 Aug. At home all day. Nothing more of moment. Paid for bread ½*d.* Mr. Miller sent me for a present 4 pikes and one carp. This day the assizes at Lewes and only one prisoner. Gave a couple of travellers 1*d.* . . .

Tues. 12 Aug. . . . At home all day. Paid for bread ½*d.* In the even read part of the 4th volume of *The Tatler,* in which I find some very agreeable stories, in particular one wherein a beautiful and virtuous young lady is ruined by a young debauchee and a sordid parent, his father.[16]

Weds. 13 Aug. Called up about 6 o'clock. Wrote to Mr. Tho. Friend to let him know my wool will be ready for sale about the middle of next week, and gave him the preference before another. Sent by Bonnick's carter a crock of butter which weighed 39 lbs. to Mr. John Wilson, which I had from Mrs. Cayley. About 8.30 I was sent for down to Halland to bring my bill. Went down with it and received of Mr. John Greening £13 17*s.* in full on account of the Duke of Newcastle to this day. Heard from the assize by Mr. French. Two cast[17] at the last assize now to be transported for 14 years and the boy tried this time, acquitted . . .

Fri. 15 Aug. This day the King's Plate was run for at Lewes;[18] only one horse. Mr. Tho. Chittenden of Hawkhurst called and dined with me. I paid him 38*s.* 6*d.* for 11 yards bearskin, which he brought today. After schooltime went over to Framfield . . . Came home about 9.20. My mother and I had some words, but really I cannot well tell for what.

Sat. 16 Aug. The last race-day at Lewes. The plate won by the Lord Portmore's[19] horse Partner.

[16] 'The history of Caelia', *The Tatler,* IV, no. 198, 13–15 July 1710.

[17] i.e. convicted.

[18] Lewes racecourse lay to the west of the town. The three-day race meeting attracted great crowds from all over the south of England. T. W. Horsfield described the racecourse in 1824 as 'justly regarded as one of the finest in England'. *The history and antiquities of Lewes and its vicinity,* I (Lewes, 1824), p. 342.

[19] Charles Colyear, second Earl of Portmore (1700–1785), a great patron of the turf and, according to *D.N.B.,* 'conspicuous in London society by the magnificence of his equipages'.

Mon. 18 Aug. After schooltime went over to Framfield on foot to talk to Chas. Diggens about some clothes for Master Darby's little boy . . .

Fri. 22 Aug. In the morn paid for flour 1¾d. Received of Mr. Sam. Virgoe 10 fleeces of wool, 25 lbs. After schooltime went over to Framfield to carry little Darby's clothes. Found at Master Diggens's a journeyman whom I think to be the best scholar I ever conversed with. He writes very good square text, engrossing, running hand, round, and Italian hands, Rich's and Weston's shorthand;[20] and as far as I am a judge seems to speak Greek and Latin very fluent, and is perfectly well acquainted with all the authors of note, so far as to quote almost any remarkable passage . . .

Mon. 25 Aug. At home all day. Master Diggens's man Joseph Harrison was a-writing for me. Gave him a new *Hudibras*. Paid the post 2d. for bringing a parcel for me from Lewes . . .

Weds. 27 Aug. At home all day, except going to Waldron between schooling. Paid for butter and a pane of glass 4¼d. and bread 1d. In the even read part of the 4th volume of *The Tatler*, which I think the oftener I read the better I like it. I think I never found the vice of drinking so well exploded in my life as in one of the numbers.

Thurs. 28 Aug. Went over to Framfield after dinner to meet with Fran. Smith. Drank tea there at my mother's with Mrs. Diggens. While we was a-drinking of tea Mrs. Beard and Dicky came in. There I found out that Harrison has run away with the 1st volume of *The Tatler* which I, like a good-natured fool, lent him the 26th instant. Came home about 7 o'clock; read several numbers in the 4th volume of *The Tatler*.

Sat. 30 Aug. This morn my wife and I had words about going to Lewes tomorrow. My reason for not going was on account of my owing Mr. Roase some money, and was loath to go till I could pay him the balance. Oh! what a happiness must there be in a married state when there is sincere regard on both sides and each party truly satisfied with each other's merit; but it is impossible for tongue or pen to express the uneasiness that attends the contrary. Paid for bread ½d. At home all day and very busy. Paid Halland for cucumbers 12d. in full.

Sun. 7 Sept. . . . After churchtime Master Piper stayed and smoked a pipe with me. After we had drank tea, my wife and I took a walk. In the

---

[20] Jeremiah Rich (d.1667?) by his *Semography: or short and swift writing . . .* (1642) and *The pen's dexterity* (1659), and James Weston (1688–1751) by his *Stenography compleated* (1727), achieved fame as stenographers whose systems were widely used. Portraits of both are to be found in E.H. Butler, *The story of British shorthand* (London, 1951).

even read Derham's *Sermons at Boyle's Lectures*, wherein I find a man evacuates as much in one day by insensible perspiration as in 14 by stool.

Weds. 10 Sept. At home all day, very bad with an inflammation in my left eye. Paid for bread and butter 4*d.* Mrs. Atkins and Mrs. Hicks drank tea here.

Thurs. 11 Sept. At home all day. Bleeded. Received of Jos. Fuller 15 lbs of beef, 3*s.* 9*d.* not paid for. In a great deal of pain with my eye.

Fri. 12 Sept. At home all day. Took physic. Remained in great pain with my eye. Mr. Miller sent me for a present a very fine roasting pig. Gave the boy 12*d.* Borrowed of Mrs. Virgoe 12*s.* 6*d.* Paid for bread 1*d.*

Sat. 13 Sept. At home all day. Very bad with my eye, and very busy. Paid for bread 1*d.* My wife took bad with the rheumatism.

Sun. 14 Sept. At home all day in very great pain with my eye; took physic . . .

Mon. 15 Sept. At home all day. My eye very bad. Nicholls at work for me. In the even had a blister laid on. Busy all day. My brother Moses came to see me in the afternoon. Gave Richd. Fuller about 1*d.* in pears for writing my London letters etc. . . .

Tues. 16 Sept. At home all day. My blister did not rise at all; my eye much the same . . . In the even another blister laid on. Paid John Saxby for a forehead comb which he got at Lewes for me, 2*d.* . . .

Weds. 17 Sept. At home all day. My blister did not rise scarcely at all. Nicholls at work for me . . .

Thurs. 18 Sept. At home all day. Took physic; my eye continued bad. Mrs. French drank tea here. Paid for a sheep's bell 4*d.* Nicholls at work for me; very busy all day.

Fri. 19 Sept. At home all day. Bleeded in the morn. Paid for 12 pairs pattens and 13 pairs clogs (received this day from Thomas Freeman of Mayfield by his servant) 14*s.* 5*d.* . . .

Mon. 22 Sept. . . . Doctor Snelling called on me and ordered a poultice of conserve of roses, and about 6 gr. of champhire in each poultice, to be laid to my eye, with purging twice a week with sal. glabuler and manna. George Richardson called on me. I gave him 1¼ yds. silk and silk to have a bonnet made . . .

Thurs. 25 Sept. . . . My wife read the 20th and 21st numbers of *The Guardian* to me, which I think extremely good, the first of which shows how indispensable a duty forgiveness is and the last how much mankind must be delighted with the prospect of the happiness of a future state.

Fri. 26 Sept. . . . Paid Mr. Ed. Lashmer for 18 lbs. 6*d.* sugar, 8*s.* 9*d.*

Paid Fran. Smith in cash £14 3s. 6d. which, with the £10 I gave him the 22nd instant, makes £24 3s. 6d. and is in full for money paid in London for me this week (to wit):

To Messrs. Margesson and Collison   £24 0s. 0d.
For 6 alphabets of round hand copies   3s. 6d.

Tho. Davy here in the even. Stayed and went to cribbage. I lost 1 gallon of pears valued at 1½d. Borrowed of Tho. Davy in cash 5s. . . .

Sun. 28 Sept.   Geo. Richardson called and breakfasted with us. I lent him Young's *Night Thoughts*. Dr. Stone came and bleeded me. Sent my maid to Framfield after dinner to borrow some lump sugar, where she heard I had cried off all my customers from coming before 8 o'clock in the morn or 4 in the even. At home all day and not at church.

Thurs. 2 Oct.   . . . About 5 o'clock my brother Moses came and called me to go to Lewes to meet the Manchesterman.[21] Received of Mr. Isaac Hook £1 1s. 0d. in full. Spent the even in part at Mr. Hook's, Mr. Scrase's, and part at *The White Horse*. Paid 2d. for the turnpike. Paid for supper and wine 12d. Went to bed about 12 o'clock.

Fri. 3 Oct.   Got up in the morn about 6 o'clock. Bought my Manchester goods at Mr. Hook's . . . Paid Mr. Will. English 10d. for grinding 5 razors. Met with Mr. Geo. Beard at Lewes today. Paid for my horse 11d. Gave the maid 3d. Gave the ostler 3d. Paid for ⅓ of a pint of wine and water 5d. Paid for 10 whitings 3d. Paid for the turnpike 2d. Came home about 1.40. Received of Will. Sinden ½ [a] bushel [of] wheat. My brother dined with us.

Thurs. 9 Oct.   . . . This day wrote an agreement between John Watford Jr. and Sarah Thunder, wherein she hires part of John Watford's house at 5s. per quarter, to be paid quarterly, and she to pay the window tax. On failure of her paying it quarterly, she is to go out of it at a quarter's notice. N.B.: she is to enter upon it the 11th instant . . .

Sat. 11 Oct.   . . . This day weighed of my wool; to wit, 334 fleeces, 20 tod 20 lbs, which I have sold to Messrs. Reeder, Cocks & Co. at 19s. per tod.[22] Paid for butter 3d. This day I left Mr. Sam. Virgoe's shop and I owe him 1¼ years' rent, 12s. 6d. . . .

Mon. 13 Oct.   Received of Sam. Elphick 6s. 10d. in full for his indentures, which I see Wm. Weller and he cancel today in my kitchen . . .

---

[21] A packman from the north of England, usually from the Manchester area, selling mainly cotton goods.

[22] The tod was a unit of weight used in the wool trade. It was normally equal to 28 lb. but there were local variations and it is clear from entries elsewhere in the diary that Turner reckoned it at 32 lb. See for example below, pp. 62, 108.

Weds. 15 Oct. At home all day. Paid for milk ½*d.* Nothing more of moment; only posted part of my day book. My wife read part of *Clarissa Harlowe* to me in the even as I sat a-posting my book.

Mon. 20 Oct. Got up in the morn about 3 o'clock. Weighed of my rags. Sent them by Smith's cart to Waldron in order to go by Noakes the Wadhurst carrier to Loose,[23] where I am to have 29*s.* 6*d.* per cwt. I am to pay carriage. The weight was 6 cwt. 0 qr. 18 lbs., £9 1*s.* 9½*d.* ...

Sat. 1 Nov. ... My brother Moses came over and dined with us and stayed and kept shop while my wife and I went to Lewes. We set out about 2 o'clock; got there about 3.45 ... We spent the even at Mr. Roase's and lay there. Paid the turnpike 2*d.* I insured Mrs. Mary Virgoe's house in Lewes in the Sun Fire Insurance for £300, which will cost her the annual sum of 9*s.* it being 3*s.* per cent per annum, on the account of its being deemed hazardous insurance, or otherwise it would have been but 2*s.* per cent per annum; the policy and carriage etc. will cost about 8*s.* ...

Weds. 5 Nov. At home all day. Paid for 32 herrings 8*d.* Paid for bread 1*d.* Mr. Madgwick and my brother Moses drank tea here. Gave the boys powder[24] to the value of 3*d.* ...

Sat. 8 Nov. Mr. Minifie called on me again in the morn; I gave him orders for 2 half-pieces of napped rug at 4*s.* 3*d.* Before the Reformation ⅓ of the best benefices in England were appropriated to abbeys and other religious houses, 190 of which were dissolved by Henry VIII, whose yearly revenue amounted [to] £2,653,000 per annum, a great part of which went to Rome, the governors and governesses of several of the richest of them being foreigners residing in Italy. Paid Halland gardener 3*s.* 5*d.* in full. At home all day. A remarkable quantity of rain fell today and in the night. Paid for milk ½*d.* In the evening read Tournefort's *Voyage into the Levant*, where I find the Turks think the dead are relieved by prayer. There is yearly spent in the Seraglio 40,000 beeves, besides the purveyors are to furnish daily 200 muttons, 100 lambs or goats, according to the season, 10 veals, 200 hens, 200 pair of pullets, 100 pair pigeons, 50 green geese.

Sun. 9 Nov. Not at church in the morn on account of my looking for Mr. Tucker while the people were at church. Mrs. Virgoe sent for

---

[23] The rags were going to the paper mills at Loose, near Maidstone in Kent. There was more than one mill at Loose; see A. H. Shorter, op. cit., 193–4, and R. J. Spain, 'The Loose watermills', *Archaeologia Cantiana*, 87 (1972), pp. 43–79, and 88 (1973), pp. 159–86.

[24] Presumably gun-powder for fireworks.

me to acquaint me that she that morn had received a letter from Mr. Dungate, wherein he told her he had orders from Mrs. Edwards to prosecute her unless she paid the money due from her late husband to Mrs. Edwards in 4 weeks time, upon which, after dinner, Mr. Jos. Burges and myself went to talk with Mr. John Burges of Rotherfield about it and came home about 7 o'clock.[25] When I came home I found word sent from Framfield to acquaint me my mother was very ill. I got Master Durrant's man to walk over to Framfield with me, where I found my mother better, but I thought still very dangerously ill. We came home about 10.45.

Mon. 10 Nov.   Paid for bread 1*d*. Gave the post 3*d*. for bringing some parcels. After schooltime Step. Vine and I walked over to Framfield; found my mother better. Gave the boy 3*d*. for going with me. Came home about 6.30 . . .

Weds. 12 Nov. . . . In the even went down to Mr. Porter's to talk to him about Mrs. Virgoe's affairs. Paid Eliz. Mepham by Lucy Mepham 5*s*. for the making of 10 round frocks . . .

Thurs. 13 Nov.   An acre of land was first fixed to be 40 poles in length and 4 in breadth, or 160 square poles, by a statute of 31st of Edward the first, or in the year of 1301. The six ages of the world are as under: 1st from the creation of the world to the deluge, being 1656 years; the 2nd from the deluge to Abraham's coming into the land of promise in 2082, containing 426 years; the 3rd from Abraham's entrance into the promised land to the deliverance of the Hebrews out of Egypt in the year of the world 2523, including 430 years; the 4th from the going out of Egypt to the foundation of Solomon's temple in the year of the world 2992, being 479 years; the 5th from Solomon's laying the foundation of the Temple to the Babylonish captivity in the year of the world 3416, containing 424 years; the 6th from the Babylonish captivity to the birth of Jesus Christ, which happened in the year of the world 4000 and includes 584 years.After dinner was sent for over to Framfield. I found my mother very bad. Came home about 8.10. Paid for bread ½*d*. Gave Richard Fuller 6*d*. for cleaning their horse.

Mon. 17 Nov.   . . . At home all day. Cut out 10 round frocks. Mrs. Virgoe supped with us on a roast pig. A very violent tempest of thunder, lightning, hail and rain between 8.30 and 9.45.

Tues. 18 Nov.   . . . Delivered to Eliz. Mepham 10 round frocks with thread and buttons . . .

[25] Mary Virgoe was the sister of the Burges brothers. See Appendix B.

Tues. 25 Nov.   This night has been a very remarkable windy night and a great quantity of rain . . . At home all day. Molly French, John French and Fanny Weller drank tea with us and we played at brag in the evening. My wife and I lost 1½*d*. . . .

Sat. 29 Nov.   I wrote the banns of marriage for Master Piper to be published in the church tomorrow between himself and his servant Mary Denmall . . .

Sun. 30 Nov.   In the morn I got up and went to Mrs. Day's to invite her and Miss Suky to dinner with us, and breakfasted with her. While they was at church in the forenoon, Sam. and Nanny Slater went home. My brother Moses came to see us while they was at church and he, Mrs. Day and Miss Suky dined with us on a hare, a piece of boiled beef and turnips and a bread pudding. After dinner Charles Diggens came in. Mrs. Day, Miss Suky and my wife went to church and came back and drank tea. Mrs. Day and Miss Suky went home about 5.25. Charles Diggens and my brother Moses stayed till about 7.15 and both went home a little merry. Paid for bread 2*d*.

Weds. 3 Dec.   At home all day . . . Made an entry of my shop and kitchen to retail coffee, tea etc. in, and dated it as tomorrow . . . [26]

Fri. 5 Dec.   At home all day. Dr. Stone cut me an issue on my back and drawed a tooth for my wife. Molly French dined here. Robt. Diggens drank tea with us . . .

Sat. 6 Dec. . . . Paid Eliz. Mepham 5*s*. for making 10 round frocks.

Sun. 7 Dec. . . . Dame Reeve died this morn about 10 o'clock.

Weds. 10 Dec.   About 3 o'clock went down to Mr. French's and borrowed his little horse to go to Lewes upon for wine for Dame Reeve's funeral, having sent for some but it did not come as I expected. I rode him up home to put on my greatcoat etc., and accordingly got up at the block, but by accident, either by touching him with the spur or his taking fright of the dog, he fell a-kicking and running etc. and threw me down near the corner of Mr. Virgoe's stone wall, and hurt my side very much. I sent Tho. Davy for the doctor in the even but he did not come. Paid Dame Vinal for washing 9*d*. Gave a man who brought the wine for me from Lewes by accident (it not being the same I sent for it by) 6*d*.

Thurs. 11 Dec.   At home all day; my side very bad. Dr. Stone came

[26] By the statute 10 George I, c. 10 every retailer dealing in coffee, tea, cocoa or chocolate was required 'before he take any the said goods into his possession [to] make entry in writing of all storehouses, shops, rooms, and other places intended to be used by him at the [excise] office for the division' on pain of forfeiting the goods plus £200. This is what Turner was registering here. See *Burn, sub* excise.

in the morn to see me and examined my side, but said I had no ribs broken. Tom Cornwell brought me 2 carps; I gave him 12*d*. I sent them by Dr. Stone to my mother. Gave Tho. Cornwell 6*d*. for carrying the wine up to Master Reeve's. Dame Reeve was buried today; my wife went to the funeral in the afternoon . . .

Fri. 12 Dec. At home all day; my side very bad . . . This morning I had a cerecloth laid on my side.

Tues. 16 Dec. . . . This day Mr. Will. Piper and his servant Mary Denmall was married at this church . . .

Fri. 19 Dec. At home all day. My boys broke up. Received of Edwd. Russell £3 11*s*. 10½*d*. in full to this day for himself and his servant Richd. Brazer. Received of Mary Brazer 12*d*. for breaking up; *ditto* of Jos. Hutson 6*d*. for breaking up. Nothing more of moment. Paid John French 6*d*. for 1 pack of playing cards . . .

Mon. 22 Dec. At home all day; it being St. Thomas's Day I gave to the poor of this parish about 2*s*. 6*d*., being about 30 in number (giving to each 1*d*. and a draught of beer) . . .

Weds. 24 Dec. At home all day. Received of Jos. Fuller 1 leg mutton, 6½ lbs at 3*d*: 1*s*. 7½*d*. Received of butcher Bonwick 1 lot of beef; the weight I do not know. Paid Will. Harvey 18*d*. for making 3 cotton waistcoats . . . Paid John Jenner of Hailsham . . . 21*s*. . . . in full for hats received of him today . . . 3 hats at 3*s*. 6*d*., 3 *ditto* at 2*s*., 1 *ditto* at 4*s*. 6*d*. Dr. Snelling called on us in his journey to Ticehurst, but did not get off his horse. In the even wrote out Peter Adams's bill amounting to £8 16*s*. 1½*d*. Mr. Tho. Reeve's bill amounting to £1 17*s*. 10¼*d*. Gave a carter 2 oz. tobacco for bringing 2 cheeses from Lewes for me. Gave Jenner's boy 6*d*. for his box.

Thurs. 25 Dec. Being Christmas day my wife and I both at church in the morning, the text *Hebrews* 9.26: 'But now once in the end of the world hath he appeared to put away sin by the sacrifice of himself.' My wife and I stayed the communion; gave, each of us, 6*d*. At home all day. On reading Derham's notes on Boyle's lectures I find he says that Mr. Boyle demonstrates that so slender a wire may be drawn from gold that from one ounce of gold a wire may be drawn 777,600 feet in length or 155 miles and a half. In the even Tho. Davy here and supped with us and stayed till 11 o'clock but drunk nothing, only 1 pint of mild beer. We read Smart's poems on immensity, omniscience and power.

Fri. 26 Dec. About 11 o'clock my brother Will. came to see us and dined with us. After dinner I went down to Jones's (it being a meeting of the parish for choosing of surveyors).[27] The company there was

---

[27] There were two main vestry meetings in the year: one, normally held on Easter

J. French, T. and J. Fuller, Ed. and Rd. Hope, J. Watford and son, T. Reeve, J. Durrant, W. Rice, W. Piper, J. Burges, J. Hutson, J. Cayley and Rob. Hook. The surveyors chosen for the year 1756 were John Cayley and Rob. Hook; the electioners, Jos. Durrant and James Hutson. I stayed there about 2 hours; came home about 5.30. About 6.30 Will. Piper, T. and J. Fuller and P. Adams came in here from the parish meeting. Peter Adams went away directly. Jos. Fuller went away about 7.30 very much in liquor. About 9.30 I went home with Wm. Piper and Tho. Fuller as far as Mr. Piper's.

Sat. 27 Dec. At home all day . . . Gave away to boys for their Christmas box 21*d.* as under:

To James Fuller	1*s.*	0*d.*
To Ben Mott		6*d.*
To John Barnard		3*d.*

. . . Paid Jos. Durrant 2*d.* for 1 peck turnips. My brother Will. went away after breakfast . . .

Sun. 28 Dec. . . . Just as we was drinking tea, Dr. Snelling came in, as did Halland two men. Dr. Snelling cut me a seton and stayed all night. My brother and Halland two men went away about 7.50.

Mon. 29 Dec. Dr. Snelling went away after breakfast. I paid him half a crown for cutting my seton and likewise am to pay John Jones for his horse's hay, oats etc. 18*d.* which together make 4*s.* Oh, could it have been imagined that he could have took anything of me, considering that I paid him £39 for curing my wife, great part of which I paid him before he had it due, and all of it within 5 months after he had performed the cure. I always do and ever did use him after the best manner I was capable of when he was at our house. He was that man that never gave my servants anything, no, not even the meanest trifle that could be. Notwithstanding they always waited on him like as if they were his own servants. Oh, thou blackest of fiends, ingratitude, what an odious colour and appearance dost thou make! Oh, may the most ever-to-be-adored Supreme Excellency that sees and views all our most private and secret actions and even knows our most secret thoughts before we bring them into action guide me with His grace that I may never be guilty of that hateful crime, nor even so much as to indulge an ungrateful thought . . .

Monday at which the churchwarden and the overseers of the poor (together with their deputies or 'electioners') were appointed, and the other in late December at which were chosen the surveyors of the highways and their 'electioners'. For the regulations regarding the appointment and duties of the surveyors see *Tate*, Ch. IX.

Weds. 31 Dec. . . . At home all day . . . About 2.45 Sam. Slater came in from Lewes. He and Tho. Durrant spent the evening with us . . . We did not go to bed until 12.25, but all very sober.

1756

Thurs. 1 Jan. . . . This day my wife and I had a great many words, but for what reason I cannot recount, though doubtless if we could be proper judges of our own actions we should find that we are both but too much to blame and possibly should find all our differences to arise from so trivial a cause that we both might have cause to blush. But oh! was marriage ever designed to make mankind unhappy? No! unless by their own choice. It's made so by both parties being not satisfied with each other's merit. But sure this cannot be my own affair, for I married, if I know my own mind, with nothing in view but entirely to make my wife and self happy, to live in a course of virtue and religion, and to be a mutual help and assistance to each other. I was neither instigated to marry by avarice, ambition, nor lust. No, nor was I prompted to it by anything; only the pure and desirable sake of friendship. Sure, many of my actions must be convincing proofs of love and friendship (to one who once I hoped to be ever dear to); though other of my actions may doubtless in the eye of the world render me not so, but if in my worst actions were the motives traced up to their first origin I doubt not but my sincere love and friendship would there appear in still more stronger lights. Oh, what am I a-going to say I have done? I have almost made as it were a resolution to make a separation, I mean by settling my affairs and parting in friendship, but is this that for what I married? Oh, how are my views frustrated from the prospect of a happy and quiet life to the enjoyment of one that is quite the opposite! Oh, were I but endued with the patience of Socrates; then might I be happy, but as I am not I will endeavour to pacify myself with the cheerful reflection that I am well assured I have done to my utmost to render our union happy, easy, good and comfortable to ourselves and progeny.

Fri. 2 Jan. Sam. Slater went home in the forenoon. I was at home all the morn. In the even all Mr. Porter's servants came to see us, and they and Tho. Davy stayed and played at cards with us; my wife and I lost 7½*d*. I went down to Mr. French's about 4.20 concerning a fat hog I am to have of him tomorrow . . .

Sat. 3 Jan. In the morn I went down to Mr. French's; brought home my hog, and Jos. Fuller Jr. killed him for me. Paid for 10 dozen quills 10*d*. . . .

Mon. 5 Jan. At home all day. In the morn Mr. French, Jo. and Tho. Fuller weighed and cut out my hog. He weighed 26 stone 7¼ lbs at 2*s.* per stone, but I did not pay for him . . . An entertainment at Mr. Porter's today.

Mon. 12 Jan.[1] . . . This day began teaching school again, being the first day since Christmas . . . Read *The Merry Wives of Windsor* wherein I think the genius of the author shows itself in a very conspicuous manner as to humour. But I cannot find in my heart to say I think there is one good moral character . . .

Tues. 13 Jan. . . . Mr. Peter Adams came to balance accounts with me but could not, he being in liquor, but appointed to come tomorrow night. Mrs. Virgoe sat about 2 hours with us. In the even I posted part of my day book. Tonight there was a dance, made, I conject, by the servants, there being all the young people of both sexes in the parish . . .

Weds. 14 Jan. . . . In the even my wife and I put up 120 papers of tobacco. After supper read part of Tournefort's *Voyage into the Levant* wherein I find the following remark: They breed (says he) the finest goats in the world in the Champaign of Angora. They are of a dazzling white, and their hair, which is fine as silk, naturally curled in locks of 8 or 9 inches long, is worked up into the finest stuffs, especially camlet. But they don't suffer these fleeces to be exported because the country people get their living thereby. Their young are degenerate if carried far.

Fri. 16 Jan. At home all the fore part of the day. Mr. Ed. Relfe and Mr. John Aliot of Lewes called on me, both in liquor. They stayed and drank 2 bottles of beer with us. In the even went down to Halland to balance accounts with Mr. Coates, but he was not at home, though he came home while I was there—in liquor—so I came away without balancing accounts. When I came home, I found Mr. John Collison at our house. We balanced accounts, and there remains due to Messrs. Margesson and Collison £17 16*s.* 0*d.* in full . . . Mr. Collison supped with us and stayed all night at our house.

Tues. 20 Jan. At home all day. Mr. T. Earle and Mr. S. Tupper called on me but did not get off their horses . . . Borrowed of Mrs.

---

[1] Between 12 and 16 January Turner confused the month and wrote 'December'.

Mary Virgoe in cash £2 17s. 0d., to wit: 1 36-shilling piece[2] and
1 guinea . . .

Weds. 21 Jan. . . . Halland gardener cut my grape-vine and drank
tea with us. Tho. Davy supped with us and he and I played at cribbage;
I won 1d. I paid him 13d. for my share of *Martin's Magazine*; that is,
1d. per piece over half price, on account of their being to be mine . . .

Fri. 23 Jan. While we was at breakfast, my father Slater came in.
He stayed and breakfasted with us and also dined with us, and after
dinner he went to Laughton to bury Mrs. Shoesmith, who was brought
from Hartfield to Laughton yesterday by men. I tied up 8 hatbands for
him . . .

Sat. 24 Jan. My father Slater went away after dinner. Lent him a
book of plays. In it was *The Revenge, Journey to London, Careless
Husband, Way of the World, Love for Love,* and *The Busy Body.*[3] Paid for a
bean net 2d. . . .

Sun. 25 Jan. . . . In the afternoon the proclamation for a general
fast on the 6th of Feb. next was read on account of the dreadful
earthquake which happened at Lisbon and many other places on the
1st day of Nov. last, and several times since.[4] Not out of doors all day
except to church. This day John Jones's wife and Mr. Peter Adams's
daughter died. Paid for milk ½d.

Mon. 26 Jan. This day paid Mrs. Virgoe the £2 17s. 0d. I
borrowed of her the 20th instant. Paid for 6 bean nets 12d. The wife of
Richd. Heath was buried today. Sent my maid over to Framfield for a
shroud for Mr. Adams's daughter. Martha Mepham dined here. My
wife went down to Mr. Porter's about 4 o'clock. I went down to her
about 7 and found Wm. Piper, his wife and brother there, as also Tho.
Fuller, Mr. and Mrs. Hutson and Dame Durrant. We supped there
and stayed until 5 o'clock in the morning and I may I think justly say
there was not one sober man in the company. I am sure I was not, for,
finding myself in liquor I came home 1 hour or more before the rest.
This is the first time I have been in liquor since Whitsun Tuesday,
and I am sorry for this, though I know of no reason except, being in
liquor, my wife and I lost at brag between us near or quite 5s. We also
gave the servants 2s. 6d. . . .

Tues. 27 Jan. My brother Moses called on me in the morn and

---

[2] A double pistole, see below p. 150, note 26.

[3] Evidently a composite volume containing six plays: *The Revenge* by Edward Young;
A Journey to London by Sir John Vanbrugh; *The Careless Husband* by Colley Cibber; *The
Way of the World* and *Love for Love* by William Congreve; and *The Busie Body* by Mrs
Susannah Centlivre.

[4] See below p. 25, note 5.

over-persuaded me to go to Lewes with him, which I did on a horse
borrowed from Mr. French . . . Bought of Mr. Roase 18 cwt. of
Cheshire and Warwickshire cheese . . . I came home again in liquor,
but got home very well; though to do myself justice, I believe I was not
extreme good humoured when I got home, and I do think I am
prodigiously silly and apish when I am in liquor, having always, for a
great while after, a sting of conscience for the same. I will, however,
renew my former resolution and use my utmost endeavour to keep to
it; that is, not to get DRUNK again if I can avoid it . . . Tho. Davy taught
school for me.

Weds. 28 Jan. . . . After dinner my wife went up to Mr. John Vine's
Sr. I went also up in the even to talk with him about the new
schoolhouse but came to no agreement. We both came home about
7.25.

Thurs. 29 Jan. After dinnertime (for I need not say 'after dinner' on
account I ate none) I went to the funeral of Mr. Adams's daughter. We
came to church about 3.30. There was a sermon for her, the text *1
Corinthians* 15.19: 'If in this life only we have hope in Christ, we are of
all men most miserable.' After churchtime very busy. Master Piper
came and invited us up to see him in the even. We accordingly went
about 6.15 and found there John Vine Jr. and wife, Tho. Fuller, son
and daughter. We supped there on a rib seep spitted twice, a piece of
boiled (tainted) beef, a good piece of bacon, and a good butter
pudding, but all in very bad order and odd decorum, which would all
have been very excusable had it not been attended with such a mean
and stingy spirit in the old man. For that few bottles of beer etc. we got
came (as the old saying is) like drops of blood from his heart, and we
may justly attribute them more to Tho. Fuller's boldness in asking for
liquor than to the poor old man's good nature in offering it. We stayed
till 1.25. My wife and I won 3s. 6d. and gave the maid 12d., so this
brings back some part of Monday night's expenses . . .

Fri. 30 Jan. At church in the morn, but no sermon. John Jones's
wife buried in the afternoon. I was not at the funeral on account of my
being very busy . . . At home all day. My brother Moses dined here.
Borrowed of Mrs. Day near one bottle of brandy.

Sat. 31 Jan. At home all day, very busy. Master Divol brought me
from Lewes 20 cwt. goods and paid the money I sent him by his son
the 29th instant. Received from Mr. William Tooth, without a bill, as
under:

½ cwt. thick bread	9s. 0d.
Thin do.	8s. 0d.
Sweethearts	5s. 0d.
	£1 2s. 0d.

Sent my mother by Divol:

| 45 Warwickshire cheeses | 4. 2. 18 [weight] |
| 11 Cheshire | 2. 0. 26 |

In the even Charles Nebuchar came in, in his road from Burwash to Lewes; he stayed all night. Related to me his travels wherein he says he hath been in 26 countries and 15 cathedrals. Paid for bread 2d. Received of Mrs. Atkins 6s. 4½d. in full. Paid Fran. Smith in cash 8s. 0d. for 2 gallons of brandy, which he bought in London for me.

Tues. 3 Feb. At home all day . . . Had John Watford a-gardening all day for me, who I think is a mighty honest, good, sort of a man, only a little inclined to be covetous, a very necessary qualification; and then he is so harmless and innocent with his 'O's' and 'good lacks' etc. that really he is very entertaining, for he has no art to set it off. In the even my wife and I did up about 70 papers of tobacco.

Weds. 4 Feb. . . . Wrote out the form of prayer for the fast on Friday next . . .

Thurs. 5 Feb. At home all day . . . Wrote out another form of prayer for the fast . . .

Fri. 6 Feb. This being a day for public fast and humiliation for to implore the blessing of the Almighty on our fleets and armies and to beseech Him, of His infinite and unbounded goodness to spare our nation from the dreadful calamity of an earthquake which hath been lately in many places of the world whereby Lisbon[5] and many other places in Africa have been entirely ruined and brought as it were to desolation, and some slight shocks have also been felt in this our happy nation, and likewise in some of our northern colonies in America, my wife and I was both at church in the morn where we had an excellent sermon on the occasion (by the Rev. Mr. Tho. Porter, M.A.) from *Psalm* 18.3: 'And the overflowing of ungodliness made me afraid'[6]

[5] The great earthquake at Lisbon occurred on 1 November 1755. The earthquake and the subsequent deluging of the city by the waters of the Tagus was followed by fire. The city was devastated and an estimated 30,000 lives were lost.

[6] The text is actually from *Psalm* 18.4: 'And the floods of ungodly men made me afraid.'

wherein he endeavoured in a very earnest and pressing manner to show what reason we have to repent of our vicious courses and to turn to the Lord, or else, as he says, how can we expect the favour of the Almighty to be more extensive upon this our isle than upon the Portuguese nation and other places which have suffered under this dreadful calamity; and likewise endeavoured to prove that it is impossible for a king to govern a nation and keep his subjects in a steady course of virtue without the assistance of every individual. At church in the afternoon there was only prayers . . .

Sun. 8 Feb. My wife and I both at church in the morn; the text in *Romans* 4.3: 'Abraham believed God, and it was accounted unto him for righteousness.' As I by experience find how much more conducive it is to my health, as well as pleasantness and serenity to my mind, to live in a low, moderate rate of diet, and as I know I shall never be able to comply therewith in so strict a manner as I should choose (by the unstable and over easiness of my temper), I think it therefore [right] (as it's a matter of so great importance to my health etc.) to draw up rules of proper regimen, which I do in manner and form following, and which, at all times when I am in health, I hope I shall always have the strictest regard to follow, as I think they are not inconsistent with either religion or morality: First, be it either in the summer or winter, to rise as early as I possibly can; that is, always to allow myself between 7 and 8 hours' sleep, or full 8, unless prevented on any particular or emergent occasion. 2ndly, to go to breakfast between the hours of 7 and 8 from Lady Day to St. Michael, and from St. Michael to Lady Day between the hours of 8 and 9. 3rdly, my breakfast to be always tea or coffee and never to exceed 4 dishes. If neither of those, half a pint of water or water gruel; and for eatables bread and cheese, bread and butter, light biscuit, buttered toast, or dry bread, and one morn in every week, dry bread only. 4thly, nothing more before dinner, and always to dine between the hours of 12 and 1 o'clock if at home. 5thly, my dinner to be meat, pudding, or any other thing of the like nature, but always to have regard, if there is nothing but salt provision, to eat sparingly; and to eat plenty of any sort of garden stuff there is at table, together with plenty of bread and acids, if any, at table; and always to have the greatest regard to give white or fresh meats and pudding the preference before any sort of highly seasoned, salt, or very strong meat; and always one day in every respective week to eat no meat. 6thly, my drink at dinner to be always boiled water with a toast in it, or small beer, but water if I can have it, and never to drink anything stronger until after dinner. 7thly, if I drink tea at home or abroad, to be small, green tea and not more than 4 dishes; and if I eat anything, not more than two ounces. 8thly, my supper never to be meat but weak broth, water gruel, milk pottage, bread and cheese, bread and butter, apple-pie or some other sort of fruit pie, or some such light diet; my drink, water or small beer, and one night at the least in every week to go to bed without any supper. 9thly, never to drink any sort of drams or spirituous liquors of what name or kind soever. 10thly, if I am at home, in company, or abroad, if there is nothing but strong beer, never to drink more than 4 glasses, one to toast the king's health, the 2nd to the royal family, the 3rd to all friends and the 4th to the pleasure of the company; if there is either wine or punch etc., never, upon any terms or persuasions whatever, to drink more than 8 glasses, nor each glass to hold or contain more than half a quarter of a pint, nor even so much if possibly to be avoided. 11thly, if I am constrained by extreme drought to drink between meals, that to be toast and water, small beer, or very small wine and water; to wit, ¼ pint of red or white wine to one pint of water. 12thly, never to drink any small or strong beer, winter or summer, without being warmed if possible. And lastly always to go to bed at or before ten o'clock when it can be done. My wife and I both at church in the afternoon; the text in *Job* 7.20: 'I have sinned and what shall I do to thee, O thou preserver of men?' We had, both forenoon and afternoon, excellent discourses wherein that necessary and excellent duty of repentance was strongly and pathetically recommended and enjoined to be done if we hope for salvation . . .

Tues. 10 Feb. In the morn I wrote to Mr. Russell of Godstone (for Mr. James Hutson) to acquaint him that he has about 100 carps that will meet at 14 inches, and 100 that will meet at 16 inches . . . My wife went down to Mrs. Atkins's about 5 o'clock. Oh, what have I here to say—the old story again repeated—more words again between me and my wife. Sure it is a most terrible and unhappy circumstance we cannot live agreeable together. Where the fault is I cannot be a competent judge, for as I am a party concerned, prejudice in my own favour may make me partial. But this I know, that my whole desire and aim is to make my wife, self and progeny happy; and I am farther assured that I ever had, and now have, a boundless respect for her; therefore what can be the occasion of so many words I am at lost to say. I can only say this, she is a woman, but why, if she be, must we be forever unhappy? Oh! could I think of an expedient to prevent it and make us happy! For it is impossible for pen or painter to express or draw so unhappy a representation as it is for to live in a continual scene of disquietude with one that is so infinitely dear to me. But oh! let me drop a subject that is too tender for me to touch any longer. But, why? Let me recall the resolution of a man and proceed. A man, did I say? Oh! how the sound of that word makes me start. I know not scarce what I am. All that I know is I am happy in having that person, who of all the sex I ever had the greatest respect for, for my wife. But again how unhappy to have that only one in whom all my earthly facility was centred to be of such an unhappy temper as not only to make me, but herself also, miserable. How delightful and serene was it once to look forward and think, 'Such a day will all my earthly trouble be at an end and be crowned with a following scene of happiness and pleasure by being made one with the charmer of my soul.' But from that day may I date the era of my trouble.

> 'For her I've lost, alas, what have I not,
> For her my duty to my friends forgot.'[7]

But why do I accuse her? Maybe it is I am all in fault. It cannot be she; can the wife of my bosom be this person? No! She must be, she is, all charms, and I am the ungrateful man . . . My wife came home about 1 o'clock. She gave the maid 12*d*. and won at cards 2*d*.

Fri. 13 Feb.   After I came from school in the forenoon, I went over to Mrs. Virgoe's . . . I stayed there about 2 hours talking over Mrs. Virgoe's affairs concerning the sale of her house tomorrow etc. About 4 o'clock I went for Lewes on foot. Just as I was a-going, I met with Mrs. Taylor of Lewes's two apprentices who were then a-going to Lewes. We accordingly went together, and upon their account, for the sake of company, I went around by Laughton . . . I got to Lewes about 7 o'clock and called on Mr. Geo. Verral. He and I went up to Mr. John Plumer's to talk with him and found him abed with the gout . . .

Sat. 14 Feb.   . . . After dinner Mrs. Virgoe's house was put up to sale, though I believe the candle was not lighted up until near 4 o'clock.

Conditions of the sale
White Horse in Lewes, Feb. 14, 1756

Conditions of sale by the candle of a convenient dwelling-house, a good butcher's shop, slaughterhouse and stable, situate in the Parish of St. Michael near the market-house in Lewes, and now in the occupation of Mr. John Fuller, butcher; and also 2 other tenements thereto belonging and adjoining, *viz*.

---

[7] The source of this couplet has not been identified.

1st: The whole premises to be put up to the best bidder in one lot at £400 and not less than £5 to be advanced upon each bidding.

2ndly: The last bidder at the expiration of the flame of the candle is the buyer.

3rdly: If any dispute arise by 2 or more persons bidding together, such dispute to be ended by the lot being to be put up again at the price last bidden.

4thly: The buyer to deposit £30 into the hands of Mr. John Burges of Rotherfield at the time of buying as a part of the purchase, and the remainder of the purchase to be paid to the said Mr. John Burges at the time of the surrender of the premises, *viz.*, the 25th day of March next, or the deposited £30 to be forfeited.

5thly: The seller to pay all charges of the surrender and the buyer all·the charges of the fine and admittance.

The candle burnt till near, or quite, 8 o'clock. There was out and in at times during the burning of the candle the following persons: Jos. and John Fuller, Mr. Manning, Ed. Verral, Hen. Verral, W. Lee, Sam. Piggot, Mr. Feron, Mr. Saxby, Mr. Roper, Mr. Jos. Burges, Mr. John Burges, Mr. Geo. Verral (he being auctioneer), Mr. John Buckall, Mr. Charles Rand and myself and Mr. Whapham. There was no one bid anything in reality; so it was not sold. Therefore to prevent its being sold for a trifle at the expiration of the candle, I bid £420 for it. After it was all over and the people gone, Mr. Tucker, Mr. Burges and I stayed and spent the remaining part of the even together at *The White Horse.* Mr. Burges and I lay at *The White Horse.*

Mon. 16 Feb. . . . In the even Tho. Davy here; we played at cribbage; I lost 2½d. I gave him ¾ yard of shalloon for keeping school for me the 27 January last and for going to Dr. Stone's for me the 10 December last.

Tues. 17 Feb. . . . A dance in the even at Dallaway's.

Weds. 18 Feb. This morn about 6.25 it began snowing very hard and continued snowing all day, and that very hard, but as it was open weather it did not lie on the ground above 6 inches deep; but, I believe, had there been a frost, it would not have been less than 2 foot deep . . .

Thurs. 19 Feb. At home all day. A very great frost last night; a great deal of snow on the ground. Continued freezing all day . . . In the even Thos. Davy brought me a new boot to try on, being 1 of a pair I have bespoke near 12 months of Robt. Hook . . . John

Streeter, who went post from Mayfield today, says that Arundel post was lost in the snow, and, when found, almost dead.

Fri. 20 Feb. At home all day. A very sharp frost last night and a great deal of snow remains on the ground. This day Mr. Sam. Gibbs's daughter was christened. In the even John Chesham and Tho. Reeve were both at my house concerning their late brother[8] Thomas Tomsett's affairs, and Reeve sold to Chesham at my house, in the presence of myself, wife, Rob. Hook and my servant-maid, all his moiety and claim of that money due from Mr. William Rabson of Ticehurst to their late brother Tho. Tomsett at the time of his decease, which they said was in all £5, for the sum of 30s. Chesham gave to Reeve a note of hand payable to him . . . for the said 30s. and witnessed by myself and Robt. Hook. Reeve at the same time gave to Chesham a receipt for the said note, wherein he acknowledges that the said note when paid is in full for his share of the above debt. The receipt was witnessed both by myself and Robt. Hook. Reeve also gave his brother Chesham an order to Mr. Rabson for him to receive the same of Mr. Rabson, and to acquaint him that Chesham's receipt shall be his sufficient discharge, dated today and from this place; and lastly Reeve gave his brother Chesham a note to pay him, or his order, on demand, the full value of the ½ part of the watch of the late Mr. Tomsett, which Reeve has in possession, and has had ever since the 9th May last. This note Master Hook and I also attested. At home all day. In the even read 2 books of Homer's *Odyssey*, translated by Pope.

Sat. 21 Feb. . . . In the even Robt. Hook brought me home 1 pair boots, value 18s., and my wife 1 pair pumps, value 3s. 9d., both to be entered to account . . .

Sun. 22 Feb. At church in the morn . . . Halland gardener came home with me and dined with us. I gave John Cayley a susi handkerchief, value 2s. as a present for his sister for a twelve-month's use of Mrs. Browne's seat in the church for my wife and I, they refusing to receive the rent (as I may call it) that I agreed with them for. After dinner the gardener and I set out for to see Buxted Place and gardens.[9] We called at my mother's as we went and stayed there about

---

[8] i.e. brother-in-law.

[9] Buxted Park, later to be the seat of the Earls of Liverpool, was at this time under construction. It was apparently begun by Thomas Medley soon after 1713. He died in 1732 and was succeeded by his four sons Thomas (died 1735), Samuel (died 1741), Edward (died 1751) and George. It was George Medley, whose fortune had been made in the Portuguese wine trade, who was finishing the house and laying out the grounds when Turner wrote. He died in 1796. A sketch of the house in 1798 is reproduced in Christopher Hussey, 'Buxted Park, Sussex', pts. 1 and 2, *Country Life*, April 21 and 28

25 minutes. I found there Miss Fanny Smith (which, I think, seems to be the lady my brother Moses' affections are settled, or settling, upon). Whether just or imaginary I cannot, no, I will not, say, but I think I was received very coldly, not only by my mother but all the family, Miss Smith only excepted. After a stay of about 25 minutes as before mentioned, we set out for Buxted. My brother and Ed. Rowles went with us. We see the gardens and the outside of the place, neither of which is any ways near completed, but I think when they are both finished they will both be very curious in their kind. Gave Ed. Rowles 6*d.* for going with us and showing us the gardens. We came back to Framfield about 4 o'clock. Found Miss Smith there. She and we drank tea there. After tea Mr. Ed. Rowles sent for us over to his house, and the gardener and I went; we stayed and drank 1 mug of beer and smoked 1 pipe, then went again to my mother's; Miss Smith and my brother not there. My mother and I had a great many words, or at least my mother had with me. What my friends would have with me I know not; I have always done to the utmost of my power to serve them. I can with justice to myself and all mankind say I have their interest entirely at heart and never think myself more happy than in serving them; and were I assured I was to blame, I should even despise myself, and even think myself not worthy to be ranked among the rest of mankind was I to be cruel and undutiful to a mother, and one who is a widow, though doubtless I am not exempt from faults. No, I am mortal, but still how happy could I be, would my friends let but a free and sincere communication of friendship once more be opened between us, and which has of late been shut up, but upon what account I cannot tell. I am, I think, quite despairing, uneasy, drove almost to distress for want of money, and my mother has at this time £40 of mine on book debts which I never did ask for, no, nor hope I ever shall. Only do I sincerely wish I could spare them £500 was it to serve them, but still why should they estrange themselves from me? But oh, let me stop my pen and say, 'May they all be ever happy, and may the Supreme Being crown them all with His blessings in this transitory state and forever make them happy in the divine regions of eternal bliss.' We came home about 8 o'clock.

Mon. 23 Feb. . . . After dinner a man called at the door, who says his name is Matthews, with a horse loaded with horse-hair hats, corks and

1934. The house was partially destroyed by fire in 1940. For a pedigree of the Medley family, see Lord Hawkesbury, 'Catalogues of portraits at Compton Place, and at Buxted Park, in Sussex', *S.A.C.* XLVII, pp. 82–108. K. H. MacDermott, *Buxted the beautiful* (1929) provides a hodge-podge of information about the parish.

glovetops. I sold him 20½ lbs. short horse-hair at 4½d. per lb. and 1½ lbs. long *ditto* at 10d., and 17 skins for 4s. 4d., all which he paid me for in ready money except 9 pair of glovetops which I took of him at 5d. per pair. What he bought of me I am to pay the carriage of to London . . .

Weds. 25 Feb. . . . This day I bought of Dame Mott the younger of Laughton 1 gold ring, for which I gave her 8s., and 1 silver bosom buckle, for which I gave her 6d. Paid for ½ a gallon flour 3½d. In the even I posted my day book whilst my wife read that moving scene of the funeral of Miss Clarissa Harlowe to me. Oh, may the Supreme Being give me grace to lead my life in such a manner as my exit may in some respect be like that divine creature's. Oh, how happy must that life be which is spent in virtue . . .

Sat. 28 Feb. . . . In the even my wife finished reading of *Clarissa Harlowe*, which I look upon as a very well-wrote thing though it must be allowed it is too prolix. I think the author keeps up the character of every person in all places; and as to the manner of its ending, I like it better than if it had terminated in more happy consequences.

Sun. 29 Feb. . . . My wife and I both at church in the afternoon; the text in *Ecclesiastes* 9.11: 'I returned, and saw under the sun, that the race is not to the swift, nor the battle to the strong, neither yet bread to the wise, nor yet riches to men of understanding, nor yet favour to men of skill; but time and chance happeneth to them all.' From which words we had an excellent sermon to persuade us all to repent that we may avert the wrath of heaven at this time when there are now abroad in the world fires, earthquakes, and at a time when our happy isle is in hourly expectation of an invasion from a powerful enemy who wants nothing more than to reduce us to state of slavery and, what will be still worse, deprive us of our holy religion and in its stead institute popery. Then [should] we reflect on this our happy constitution and consider how unhappy we should be by such a resolution, and if we have not then the greatest reason to abandon (in the best manner our frail nature can be brought to let us) all our wickednesses and irreligion and turn to the Lord our God, who is full of goodness, long-suffering and of great kindness . . . When I drew up my rules for regimen, I mentioned to breakfast one day in every week on only dry bread for eatables, and likewise to eat no meat one day in every week, as also to go to bed at the least one night in every week without a supper. I am come to a resolution to fix the following days for a due observance of the said rules, *viz.*, to eat only dry bread on Sunday morn, to eat no meat on Friday, and to go to bed without a supper every Wednesday night.

Tues. 2 Mar.   This day was Shrove Tuesday. At home all day. Carried up to Mrs. Day's 1 cwt. raisins; drank tea there. Received of Mr. Sam. Gibbs 8s. 9d. in full except for the boy's schooling and 1 lb. gunpowder. In the even Tho. Davy at our house. We played at cribbage; I won 3d. We dined today on the remains of Sunday's dinner with the addition of 3 sausages. Paid for sweeping my chimney 6d. Heard this day that the Duke of Cumberland is a-coming into Sussex to view the seacoast.[10]

Ash Wednesday, 3 Mar.   This day I appointed to go to Lewes to meet Mr. Step. Fletcher, but did not go on account that I wanted but a small parcel, and the roads very bad; so that my trouble and expense would have been more than the profits arising from what I should have bought. At home all day. Dine on a piece of pickle pork, light pudding and greens. In the even I was sent for down to Halland. Accordingly about 7 o'clock I called on Tho. Davy, and we went down. When I came, I found that it was Ann Smith had sent for me to ask my advice in the following affair, as she tells it, to wit: About 8 years ago she kept the house of Tho. Baker in the parish of Waldron, who, being an elderly man, and at the same time in all probability might have a feeble insurrection of an unruly member which might prompt him to make his addresses to her, as she says he did; and as he found his affection slighted, and understanding she was indebted to Mr. Venner of the same parish the sum of £2 7s. 0d., and as a means, as he simply imagined, to ingratiate himself in her favour, he (as she solemnly avers) went and paid the same without her knowledge or orders; and, when he had so done, never offered to make any drawback in her wages when she left him, though she says he often told her he had paid it; and when she went away, she went and asked Mr. Venner whether she owed him anything, who answered, 'No!' So it is plain Baker had paid the money. But as Baker is now in low circumstances (though still a single person), he has lately made a demand of the same, notwithstanding it has been near 8 years since, and (as she says) never pretended to have any demand on her before, and she always looked upon it as a free gift. Now my advice was as this, to wit: if what she repeated to me was true and that he actually paid it without her knowledge and designed it as a free gift to her without any proviso to the contrary, I thought she was not obliged to pay it, only as change in circumstances and gratitude should always oblige everyone to return favours where

<hr>

[10] William Augustus, Duke of Cumberland, made an extensive tour of the South Coast viewing the fortifications at this time. On 4 March he made a review of all the troops in Sussex at Lewes. *The Gentleman's Magazine*, XXVI (1756), p. 141.

they have received any. But if it was any ways by her orders he paid it, or she was to outset it in her wages or to make him any other gratuity and did not, I thought in justice she ought to pay him. Came home about 8.10. Read part of Homer's *Odyssey*. At church in the morn.

Tues. 9 Mar. At home all day. In the afternoon I was sent for over to Mrs. Virgoe's to talk with Mr. Dungate concerning her giving Mrs. Edwards security for the money due from Mrs. Virgoe to Mrs. Edwards, and we appointed to meet Mr. Dungate at Mrs. Edwards on Thursday next at 3 o'clock. John Watford a-gardening for me. For dinner we had pork, light pudding and greens. Mrs. Virgoe drank tea with us. In the even read the writings of a farm called Chillys in Mayfield, which was entailed to Mrs. Virgoe's father and his heirs for ever, but he cut the said entailment off and entailed it again to Mrs. Virgoe and her heirs for ever after the death of her mother . . .

Thurs. 11 Mar. After dinner I carried Mrs. Virgoe to Mrs. Edwards's (on a horse of Tho. Fuller's) where she gave Mrs. Edwards a bond for £153 as a collateral security for the same sum due from Mr. Isaac Virgoe deceased, with the addition of about £6 for interest, the original debt being no more than £147 6s. 3d. The bond was filled up by Mr. Dungate and witnessed by himself and Mr. Richd. Coomber of Wadhurst . . .

Sun. 14 Mar. . . . About 2 o'clock my brother Moses and sister Sally came to see us . . . My sister Sally stayed all night . . . In the even I drew up the form of a will to send to Tho. Mepham on board His Majesty's Ship *Torbay* in order for him to write over again (I wrote it at the desire of Robt. Hook, who desired to get it wrote by Tho. Mepham), wherein he gives all the moneys that may be due to him at the time of his death, either as prize money or pay, to be equally divided between his brother Joseph and sisters Eliz., Martha and Lucy, share and share alike, and also makes them all joint executor and executrices . . .

Mon. 15 Mar. . . . After tea I went with Sally to Framfield, as did Robert Hook, who went forward to Uckfield. I went to talk with my mother concerning her leaving off trading. Robt. Hook called me at Framfield about 9.15 and came home about 10.45. As we came by Whyly, we met with Tho. Fuller the chandler who was then in his road home but, considering the coldness of the weather, thought it better to come back home by the Street on proviso I would give him a dram of gin. Oh! that baneful liquor; that mankind should be so infatuated as to give way to anything for an opportunity to drench themselves with such a slow and lingering poison! . . .

Tues. 16 Mar. . . . In the even posted my day book to this day. Trade I find to be very dull. Not that I want to get an estate; no, if it will please the Supreme Being to bless me with only enough to pay everyone their own and to maintain my family in an indifferent manner, I am satisfied. A very sharp wind these 3 or 4 days past which has dried the roads prodigiously even to make the dust steam . . .

Weds. 17 Mar. . . . In the afternoon Mr. Terry, tobacconist, called on me and drank tea with us . . . I went down to Jones's with Mr. Terry and stayed until 2 o'clock; I spent 12*d*. Mr. Terry I think to be a very pushing man in his trade; he is by birth a Yorkshireman, has been abroad in the merchant's service in Virginia 1¼ years, and rode in the tobacco trade for Mr. Albiston before he entered into trade for himself. I was a little in liquor.

Thurs. 18 Mar. . . . My brother Moses and I went down to Mr. Peckham's to advise with him about my mother's leaving off trade, but he having company, could not get to consult him in the manner I wanted to do. But he has promised me to consider of it and that I shall hear from him again . . . I came to John Jones's about 7.50, and there being a public vestry for the making a poor book,[11] I went in and found there Jer. French, Jo. Fuller, Jo. Burges, Jo. Durrant, Will. Piper, John Cayley, Robt. Hook and John Watford Jr. But some or most of them being a little in liquor, they could not agree in some of their arguments. We all broke up about 8.40.

Fri. 19 Mar. This day I found, by looking over my accounts, that when Mr. French and I balanced accounts on the 8th day of January, he charged me with only 6 load of wood in the year 1753, though I think I must inevitably have in that year 8 load, for I well remember I had both wood and faggots of Mr. French in that year, and of Mr. John Vine I am assured I bought 2 cord. Now Mr. Jam. Hutson fetched all my wood that year (for I bought it all in the place) and he charges me for the carriage of 10 load. When Mr. Vine and I balanced, he had omitted the 2 cord of wood, but that I have since told him of. As Mr. French charged me with but 6 load, I must consequently owe him for 2 load more . . .

Sat. 20 Mar. . . . After supper read the 13th book of Homer's *Odyssey*, wherein I think the soliloquy which Ulysses makes when he finds the Phaeacians have, in his sleep, left him on shore with all his treasure, and on his native shore of Ithaca (though not known to him), contains a very good lesson of morality . . .

---

[11] i.e. fixing the poor rate for the coming year.

Sun. 21 Mar. I was at home all day, but not at church. Oh fie! no just reason for not being there . . .

Tues. 23 Mar. . . . Molly French dined with us on the fillet of that leg of veal bought a-Saturday last, roasted in the oven, with a pudding under it. This day I entertained my scholars with the sight of a show which was at Jones's; the man performed in my schoolroom. I think it a very good performance of the kind. He performed several very curious balances, ate fire and red hot tobacco-pipes, brimstone etc. I gave him 12*d.* . . . After schooltime I went down to Mr. Jer. French's to ask Mr. Sam. Virgoe whether it made any difference or whether he insisted on my going out of his house at New Lady Day. He has given me leave to stay as long as I like, providing I will pay him 12*d.* per week for rent . . .

Sun. 28 Mar. Mr. Ormeroid breakfasted with us and dined, as did Charles Diggens, on a roast goose, a piece of bacon, plum batter pudding and broccoli. We smoked a pipe or two and then went down to Jones's, where we drank one bowl of punch and two mugs of bumboo; Mr. Ormeroid went away after the punch. I spent 12*d.* and came home again in liquor. Oh! with what horrors does it fill my breast to think I should be guilty of doing so—and on a Sunday, too. Let me once more endeavour never, no, never, to be guilty of the same again. I am certain it proceeds, not from the love of liquor, but from a too easy temper and want of resolution. Not at church all day. In the morn there was no churching, Mr. Porter not being well, and in the afternoon Laughton curate preached here. I this day made an offer to Charles Diggens to take him in partners with me in the Framfield shop, and which he agreed to on my proposals, to wit: to bring in half the value of the stock and, if he had not money enough, to be bound with him too anybody so far as £100, he paying the interest, and if that was not enough, for him to pay me interest at £4 per cent for enough to make up half the stock.

Thurs. 1 Apr. In the forenoon Mr. Ed. Relfe, saddler in Lewes, brought up a new pillion and dined with us on the sparerib given us by Mr. French. It was baked in the oven with a pudding under it. I paid him for the pillion etc., £1 4s. 6d.,[12] to wit:

1 quilted pillion with 2 straps of the off side 13s. 0d., lace for the cloth 3s. 6d., lining 1s. 2d., stiffening 2s. 0d., making 5s. 0d.

After dinner I set out with Mr. Relfe on horseback to go to Lewes and went with him as far as the London Gates, but not liking my horse nor

---

[12] In fact the individual sums add up to £1 4s. 8d.

the roads, I returned back again, and came home and pulled off my boots and set out on foot for Lewes about 3 o'clock, and came home again exactly at 8 o'clock. I spent nothing all the time I was gone, nor drank anything but [a] glass of orange shrub. I found it prodigious bad walking.

Fri. 2 Apr. At home all the forenoon. We dined on the remains of yesterday's dinner. About 3 o'clock I walked over to Framfield carrying with me the clothes I bought yesterday at Mr. Madgwick's. Drank tea with my mother. My brother Moses and Charles Diggens was just come home from Uckfield where they went yesterday to help the Taylers at Uckfield make their mourning for Mrs. Moon, the wife of Mr. Will. Moon of Little Horsted, who was buried this day. They were a little in liquor, so that I had no farther conversation with Charles; but I think the old man looks very odd on me, and Charles very shy so that I do imagine I shall have my offer rejected. My brother came with me as far as the middle of Eason's Green. I called in at Master Hook's as I came home and stayed there near an hour. Came home about 9.10.

Sat. 3 Apr. . . . Heard there were press warrants[13] come to Hastings and Battle.

Sat. 10 Apr. . . . My brother stayed with us about 3 hours. I also carried down to Mr. Porter's some shag for a pair of breeches for Mr. Porter; stayed there about 1 hour . . . Received of the postmaster 18*d.*, being in full for 2 magazines and 14 buttons which was lost by his late servant. Tho. Davy came in after supper and stayed with us about 2½ hours. He and I looked over Gordon's *Geographical Grammar*, and in particular the religions of all nations.

Mon. 12 Apr. At home all day . . . I borrowed of Tho. Davy 1 bottle of brandy and paid Mrs. Day the brandy I borrowed of her the 30th January last. I carried it myself and stayed at Mrs. Day's near an hour. In the even Mrs. Virgoe came over to consult me about her making of a will. Gave her the liberty to put me in as one of the trustees, and at the same time I assured her if there was ever an occasion, I would act . . .

Mon. 19 Apr. . . . Received of Joseph Fuller £3 11*s.* 1½*d.* in full on account of the parish, he being overseer. Very busy, there being a public vestry at Jones's to choose new offices. I dined there on a buttock of beef and ham, and plum pudding and greens (my family dining at home on the remains of yesterday's dinner). The company

---

[13] Royal warrants to impress men for military, or, more usually, naval service. *The Gentleman's Magazine*, XXVI (1756), p. 145 reported that 'the hottest press [was] begun for seamen that ever was known, all protection being disregarded'.

that dined at Jones's was Mr. Coates, Mr. Piper, Mr. Jer. French, Jos. Fuller, Tho. Fuller, Robt. Hook, John Browne and myself, though after dinner there came in several more of the parishioners. We paid 12*d.* a piece for dinner. After dinner I was sent for home, but after I had done my business I went down again and stayed till near 8 o'clock. The officers chosen for the year 1756 was: Jos. Fuller, churchwarden; John Vine Sr., electioner; myself, overseer; Ed. Hope, electioner.

Weds. 21 Apr. . . . Very busy all day. About 6 o'clock I went to the audit where I received of Mr. Vine £2 12*s.* 8*d.* in full for my audit bill and also 2*s.* 5½*d.* for a bill for nails. Gave the maid 12*d.* according to custom and came home drunk, but I think never to exceed the bounds of moderation more . . .

Thurs. 22 Apr. In the morn got up and went down to Mr. Porter's where I borrowed a bushel and ½ sea coal. I went forward to Framfield, but called on Mr. Peckham as I went along to consult him about going to live at Framfield, who advises me so to do . . .

Fri. 23 Apr. . . . This day I had a fire-grate set up . . . In the even my wife (I being busy) went up to Mr. John Vine's to tell him that I thought of leaving this place, so that I would have him let the school-house. After my wife came back, I went down to Mr. Porter's to tell him the same. I stayed and supped there and came home about 10 o'clock . . .

Sun. 25 Apr. . . . My wife, maid and self at church in the afternoon, and as soon as prayers was ended, Mr. Jer. French and I went out and searched the public houses:[14] to wit, John Jones's, where we found no one person but John Jones, from whence we went to Fran. Turner's where we found a man and his wife who came in overnight. They seemed to be very sober sort of people and not a-drinking so that we did not meddle with them. We came back just as the people came out of church. Mr. French went to Jones's and had a dram, and I a pint of beer. At home all the remainder of the day . . .

Weds. 28 Apr. We dined on the remains of Monday's dinner with the addition of a piece of pork, light pudding and greens. At home all day, and gardening in the afternoon. In the even read several numbers of the *Freeholder* which I think is a proper book for anyone to look into at this critical juncture of affairs.

[14] The duty of ensuring that no tippling took place on the premises of innkeepers, victuallers, alehouse-keepers or tavern-keepers outside the times laid down by statute, was placed upon constables and churchwardens. Both tipplers and publicans could be fined and the money used for the benefit of the poor. The parish officers frequently checked for tipplers during the time of divine service. See *Burn, sub* alehouses.

Sun. 2 May. . . . We had this day two very fine discourses. That in the morn was to dissuade mankind from fraud and swearing, and in the afternoon to make it our greatest concern to prepare ourselves for a future state. Tho. Davy sat with us about 3 hours in the evening. Gave notice of a public vestry to be held on Wednesday next to put out children.[15]

Weds. 5 May. At home all day. We dined on the remains of Sunday's dinner. In the even went down to Jones, there being a public vestry. Spent 2*s*. 6*d*. on the parish account, but did not pay for it. There were only Mr. French, Jos. Fuller, Peter Adams, Mr. Joseph Burges, Edmd. Elphick and myself. We agreed to take an account of Thomas Tester's goods very soon. We also put out Lucy Brazer to Edmd. Elphick at 18*d*. per week so long as the parish shall think proper; also put Ann Brazer out to Dame Trill to keep at 18*d*. per week, and to take either of them away at any time whensoever the parish shall think proper. Came home about 9.30. Found Master Hook and Tho. Davy at our house. I sold them the remains of Mr. T. Tomsett's library for 12*s*., which they carried away with them, and paid me the 12*s*. They stayed till near 12 o'clock. I paid Tho. Davy 20*d*. for the bottle of French brandy borrowed of him the 12th of April. I shall get all the books I saved for myself, together with Tillotson's works, out of the books bought of the heirs of the late Mr. T. Tomsett. I posted some London accounts.

Mon. 10 May. I was sent for down to Mr. Porter's in the morn to inform Fran. Elless, the person who is intended to take the school, when I choose to resign it, and I agreed to the 17th instant . . .

Tues. 11 May. In the morn I got up and went to Lewes; my brother having lately strained his ankle, which is become worse with stirring upon it a little yesterday, so that he could not go with me. I got to Lewes about 8.30. I breakfasted at *The White Horse* in company with Mr. Fletcher, Mr. Geo. Beard and Mr. Relfe of Ripe. After breakfast I gave Mr. Fletcher 1 bill . . . which I drew on Messrs. Margesson and Collison, value £14 19*s*. 0*d*. and in full on account of Mr. Samuel Ridings. I bought a parcel of goods of him, which I left at *The White Horse*. Paid Mr. John Madgwick 5*s*. 1½*d*. for goods bought today as under:

⅝ yd. cloth [at] 7*s*.   4*s*. 4½*d*.
½ yd. green napped   9*d*.

[15] To board out pauper children at the expense of the parish.

I dined at Mr. Isaac Hook's on a leg of mutton and caper sauce (my family at home dining on the remains of yesterday's dinner). After dinner I went to *The White Horse* where I smoked one pipe with Mr. Fletcher and drank part of one pint of wine, which was all I drank the whole day. I looked out some earthenware at Mr. Will. Roase's. Spent 7*d.* Got home about 4.30. After I came home, I agreed with Robt. Hook to take all his brother Jenner's butter which he makes this year at 6*d.* per pound, delivered here or at Lewes. I am to find crocks.

Weds. 12 May. ... About 6 o'clock Jos. Fuller and I went and took an account of Tho. Tester's goods, which we valued at about £6 13*s.* 0*d.* though I believe it is justly worth £8, and then not over-valued ...

Thurs. 13 May. At home all day ... In reading Homer's *Odyssey*, I think the character which Menelaus gives Telemachus of Ulysses, when he is a-speaking of his war-like virtues in the 4th book, is very good ...

Fri. 14 May. At home all day. My brother William came to see us. He dined with us on the remains of Wednesday's dinner with a few ratios of pork. I appointed to go to Uckfield Fair, but my brother's coming prevented my going. In the afternoon Tho. Cornwell, Tho. Durrant, and myself sued my pond in the orchard and caught two small perch only. My brother went away about 10.15. In the afternoon I was sent for over to Mrs. Virgoe's to talk with Mr. Tho. Tourle Jr. who has a mortgage on Mrs. Virgoe's house of £100 at £3½ per cent per annum. There is now about 2½ years' interest due, and he seems to be very impatient for his interest, so that is to have the whole 3 years' interest paid when it becomes due, or otherwise (he was so kind as to tell Mrs. Virgoe) he should enter upon the house.

Sat. 15 May. At home all day. This day I resigned up my school to Francis Elless ...

Mon. 17 May. At home all the morn and busy ... In the afternoon went down in the park to a cricket match, to wit, the Street against the Nursery quarter.[16] I played and we beat them, having 6 wickets to go up the last hand. Came home about 10 o'clock. Spent no more than my shilling for the game[17] ...

Thurs. 20 May. ... Geo. Richardson called and breakfasted with us, who informed me that war was declared against France a-Monday last.[18]

[16] Nursery quarter was otherwise known as the hamlet of Halland.
[17] See above, p. 9, note 9.
[18] War was formally declared against France on 15 May following the news of the French attack on Minorca. This was the start of the Seven Years' War.

Fri. 21 May. In the morn my sister Sally and Molly Dine came over. Molly Dine altered 2 gowns for my wife and took another home to make. About 12 o'clock Charles Nebuchar came in, who also dined with us, as did my sister and Molly Dine, on the remains of yesterday's dinner. Miss Day and Molly Fuller, with the above that dined with us, drank tea with us, and after tea my wife, Miss Day, Mrs. Fuller, Molly and Elizabeth Fuller and Fran. Cole went for Whitesmith Fair. But as soon as they were gone, I received a letter from my brother Moses to acquaint me that my cousin Char. Hill was at their house and that I would come over immediately. I then sent for my wife, who came back directly, they not being got more than ¼ of a mile, and proposed to ride over to Framfield. But I was no sooner astride on my horse than dismounted, he throwing me off in the court, on which I made a resolution never to get on his back again. So my sister and I walked over to Framfield, and Molly Dine walked with us as far as Eason's Green. Charles Nebuchar set out for Lewes when we did for Framfield. I got to my mother's about 7 o'clock where I found Dr. Hill who was to set out tomorrow for Portsmouth in his way directly for Plymouth where he is to go on board of His Majesty's Ship *The Monarch*, a third-rate, 74 guns, in the capacity of Surgeon's First Mate.[19] He stayed about 40 minutes after I came in, and my brother and I went to bring him on his way to Little Horsted so far a half across Tealing's Common. We came back to my mother's who lent me her horse to ride home upon, I hurting my hip a little in my fall. Came home about 11 o'clock. This day received of Mrs. Atkins 1 hind quarter pork, weight 17¾ lbs.

Sun. 23 May. . . . In the afternoon we called a vestry about Tester's affair, but came to no resolution.

Mon. 24 May. In the morn I went up to Mr. John Vine's Sr., on an invitation from himself, where I saw two ponds emptied and the fish taken out. Mr. Vine, according to his promise on his invitation, made me a present of a brace of fine carp and 9 eels. Came home about 10.45. In my absence Mr. and Mrs. French was at our house and took up a coat etc. for John. My wife also delivered my horse to Mr. Clinch, who paid her 50*s.* for him as agreed for a-Saturday. Mr. French and I talked of Tester's affair, and he desired I would come to the cricketing in the afternoon, where, if any more of the parish were there, he would come to a resolution concerning the affair and agree to something for

---

[19] For Dr Charles Hill see Appendix B. *The Monarch* or *Monarque* was to achieve notoriety as the ship on which Admiral Byng was executed on 14 March 1757. See below, pp. 84–5, 92–3.

the support of Tester. We dined on a loin of pork roasted in the oven. Just as we had dined, the post came along who ate a piece with us. About 3.25 I went down into the park where our boys played with Framfield and were beat near 100 runs.[20] There being none of the parishioners there except Mr. French, Tho. Fuller and myself, so that nothing more was done in Tester's affair . . .

Tues. 25 May. At home all day. Will. Burrage and his boy at work for me today about ¾ of a day a-mowing of my orchard and gardening. We dined on 1 carp boiled, a piece of bacon and greens, with the remains of yesterday's dinner . . . In the afternoon my brother was at our house. This day put the water to the raisins in order to make raisin wine.

Weds. 26 May. This morn relieved two women with a pass with 12*d*. They were passed from Dorchester to Canterbury, having one child each, and one big with child[21] . . .

Holy Thursday, 27 May. In the morn got up and went over to Framfield in order to get my brother Moses to come over to stand the shop whilst my wife and I went to Dicker Fair. I rode to Framfield on a horse borrowed of Mr. French and brought Philip home in my lap. My brother coming out with us but, forgetting something, he went back again so that he did not come till about 30 minutes after I came home. We dined on the remains of Monday's and Tuesday's dinner with the addition of a green salad. About 12.30 my wife and I set out for Dicker Fair on Mr. French's horse and went to Mr. Crowhurst the potter's at Bayley's Lane in order to buy some earthenware, but he had no quantity of ware burnt, only what was in the kiln so that I did not look out any. We came back to Dicker Fair about 3.30 where, meeting with Miss Day and several more, I had them into a booth to treat them. As we sat drinking, Mr. John Kenward of Westham came in, as did Mr. French and Richd. Hope. Spent on the whole at the fair 3*s*. Came home about 8.30 in company with Mr. French, John and Molly French, Molly and Elizabeth Fuller, James and Ann Fuller, Miss Day and Richd. Hope. After I came home, my brother went home and left Philip at our house. I also went up to Mrs. Day's to carry up 5 yds.

[20] See above, p. 9, note 9.

[21] Turner here acted in his capacity of overseer of the poor for the parish. The two women were evidently vagrants who had been apprehended in Dorchester; the justices there, having examined them and discovered that their legal place of settlement was Canterbury, equipped them with a pass. The pass required constables and overseers on their direct route back to Canterbury to relieve them and to convey them to the next parish. Such a pass is illustrated in *Tate*, fig. XIV.

ferret which I got at Lewes for her. I stayed at Mrs. Day's about half an hour.

Mon. 31 May. . . . Saw in the Lewes newspaper of this day that on Saturday last there was several explosions heard in the bowels of the earth like an earthquake in the parishes of Waldron and Hellingly, as also by one person in this parish . . .

Thurs. 3 June. . . . In the even Fran. Elless the schoolmaster at our house a-learning the contracted division of decimals.

Sat. 5 June. At home all the forenoon. Dined on the remains of Tuesday's dinner. After dinner I was sent for over to Mrs. Virgoe's, who desired I would go to Lewes that afternoon in order to get her will made, which I did accordingly on Mr. Burges's mare. But when I came there, I went to Mr. Geo. Verral's to consult him in the affair, who recommended me to Mr. Wheeler the attorney and sent for him up to his house. But on Mr. Wheeler's seeing the will of Mrs. Virgoe's husband, he said it was his opinion that Mrs. Virgoe could have no right to make a will; so declined making it. We both stayed and smoked two pipes with Mr. Verral. I came home about 12 o'clock. It was about 5 when I went for Lewes. Spent 10*d.* which I can think but just for Mrs. Virgoe to pay me again. This day wrote a letter to my brother Will. after I had asked Mr. Porter's advice whether he might receive the communion here tomorrow whose answer was in the affirmative if he had been confirmed. Left Mr. Virgoe's will to have counsel on.

Mon. 7 June. This day my brother having affixed to go to Brighthelmstone[22] in order to bathe in the sea and from which I dissuaded him from doing, thinking he would get into bad company and get in liquor, but upon these conditions: that I would accompany him to some other part of the sea coast; so accordingly about 8 o'clock he and I set out from Hoathly on foot for Seaford where we arrived about 11.30. My brother and I went down to the sea, wherein he bathed and came back and dined at *The Tree* at Seaford on veal steaks (not deserving the name of cutlets); and for which dinner we paid 9*d.* apiece, though I think as badly dressed as I ever saw a dinner, and nothing set at table but salt . . . Seaford is a small town with many good buildings in and near it, but it doth not stand compact, for the houses are very much separated. It lies about ½ mile from the sea, and upon the cliff near the sea is built a sort of fort, but no guns in it, nor in reality is it of any service, for it lies so much higher than the sea that I think they could not point the guns to do any great execution; and the

[22] Brighton.

walls, being built of flint, and so very thin, that if a cannon of any large weight were fired against it, I think the flints must of consequence destroy all the men in the fort. Between the town and fort there are 4 18-pounders laid, which, I think, if rightly ordered, might be of signal service in war-time to protect their fishery from the insults of privateers . . .

Tues. 8 June. This day at the request of Mrs. Virgoe I went with her brother to Lewes on foot to know the result of counsellor Humphrey's opinion on her late husband's will, which was that by that will's being badly made she had no power to make one; and he also said Mr. Tourle's mortgage, or, as he expresses it, 'He (Tourle) was damned bad security.' My brother and I dined at Mr. T. Scrase's on a cold quarter of lamb and green salad. Paid Mr. George Verral 8s. 8d. in full for 2 doz. soap. Now what I am a-going to mention makes me shudder with horror at the thought of it. It is I got very much in liquor. But let me not give it so easy a name, but say I was very drunk, and then I must of consequence be no better than a beast. And what is still more terrifying, by committing this enormous crime I plunged myself into still greater; that is, that of quarrelling, which was this: my walking yesterday and again today, my feet were very sore, so that, meeting with Peter Adams, I asked him to carry me home, which he agreed to; and I accordingly got on horseback at *The Cats* after first having some words with a person and I can think for no other reason but because he was sober, or at least I know it was because I were drunk. We then proceeded on the road home and, as I am since informed, oftentimes finding an opportunity to have words with somebody, and, doubtless as often, giving somebody the opportunity to sneer and ridicule myself, as well in justice they might. And, I suppose I, to gratify Mr. Adams for his trouble, told him [if] he would go around by Will. Dicker's, I would treat him with a mug of 6d., which he readily accepted of (though he, I understand, was very sober). There we met Mr. Laugham and several more, but who I cannot remember, and I suppose also in liquor. Now there was formerly a dispute between Mr. Laugham and I about a bill wherein I was used ill, and I imagine I must tell him of that. Or whether they, seeing me more in liquor than themselves, put upon me, I do not remember, but Mr. Laugham pulled me by the nose and struck at me with his horse-whip and used me very ill, upon which Mr. Adams told 'em he thought there was enough for a joke, upon which they used him very ill and have abused him very much. And whilst they were a-fighting, I, free from any hurt and like a true friend and bold hearty fellow, rode away upon poor Peter's horse leaving him to shift for himself and glad enough I got away with a whole skin. I got home about 10 o'clock. But what can I say in my own behalf for getting drunk? Sure I am a direct fool, so many resolutions as I have made to the contrary, and so much as I am desirous of living a sober life, that I should suffer myself to be so easy deluded away when I know almost the sight of a bottle of wine will make me drunk. But, Oh! may the Supreme Director of all events give me grace to be wiser for the future; and as I have in so miraculous a manner several times been preserved from danger, I hope I shall never more be so weak, but have resolution enough to make this the last time . . .

Sat. 12 June. . . . After dinner Robt. Hook and I went down to see Peter Adams. After we had stayed there a while, he went to Waldron and I went to see one John Baker (at Mr. Adams's request), who the same people that abused Mr. Adams have in a manner abused; to wit, as he was a-coming home from the fair, they overtook him and whipped his horse and made him throw him off, and many more such-like actions. Came home about 8 o'clock not very well. With being so continually haunted with my conscience, it hath thrown me into a slight fever . . .

Thurs. 17 June. In the forenoon I went down to see Peter Adams. Found him in bed and determined to prosecute the men that abused him. Stayed and smoked one pipe and came home about 1 o'clock . . . Our parish played at cricket with Chiddingly Parish on Broad Oak and had not time to play it out, but to all appearance it was in the favour of this parish.[23] At home all day except for going to Peter Adams's . . .

Sat. 19 June. A very great tempest of thunder, lightning and rain, but in particular of lightning. It began about 12 o'clock in the morn and continued until near 2. This morn about 7.30 Mrs. Porter was safely delivered of a girl . . . We having fixed on this day for my wife and I to go to Hartfield on, my wife endeavoured to borrow a horse (being both disappointed of having one as we expected from Hartfield and Francis Smith) of Jos. Fuller, Tho. Fuller, Will. Piper, Jos. Burges and Richard Bridgman; though I believe they had no other reason for not doing it but want of good nature and a little gratitude, but I make no doubt they will, some one or other of them, be so good-natured as in a little time to come: 'Do write this land tax, or window tax, book for us.' And then I always find good nature enough to write them and the same time to find them in beer, gin, pipes and tobacco whilst I are a-writing them. And then, poor ignorant wretches, they sneak away and

[23]   See above, p. 9, note 9.

omit to pay for paper. But God bless them; and I'll think it proceeds
more from ignorance than ill nature. After Fran. Smith and I had
balanced accounts today, I gave him £2, which he is to pay in London
for me. I sent our maid to hire a horse of Will. Sinden, who replied he
had never a horse able to carry us (though I should not have sent, had
not his man told me he had a very good double horse I could have)
though he daily carries a man almost as heavy as us both, beside a sack
of wheat. My wife then applied to John Watford, who hired us his
horse. My brother came over to stand the shop for us in my absence
(about 3 o'clock). And about 4 o'clock we set out on our journey for
Hartfield where we arrived about 8.10. As we were a-riding along near
to Hastingford, no more than a foot-pace, the horse stood still and
continued kicking up until we was both off. Our fall happened to be in
a very dirty hole, but, thanks be given to God, we received no hurt. My
wife was obliged to go into Hastingford House to clean herself. Gave
the woman of the house 6d. Called at Mr. Browne's at Withyham to
know about hop-bagging, but he was not at home. But his man told me
he had not as yet bought any. This day received of Jos. Fuller 1 neck of
veal, weight 6½ lb. at 3d. per lb. My wife and I spent the even at my
father Slater's. We dined off some rashers of pork and green salad.

Thurs. 24 June. . . . Paid John Streeter 6d. which was a wager I laid
with him concerning a cricket and lost[24] . . .

Sun. 27 June. In the morn after breakfast I got up on one of Mr.
Batchelor's horses and rode with him over to Summerford in order to
try him. Stayed and drank one bottle of cider with Mr. Mills in
company with Mr. Jackson, Abra. Spencer, Mr. Heywood and Mr.
Batchelor. Came home again about 11 o'clock. Dined at my father
Slater's on a roasted loin of veal and a plum pudding. After dinner my
father Slater, Sam. Slater, Master Pain and myself, with Mr.
Batchelor, went beyond Buckhurst to see another mare of Mr.
Batchelor's, which we caught and brought to Hartfield, calling, as we
came along, at Buckhurst and drank 2 or 3 mugs of beer. This mare I
like very well and am to give 8 guineas for her and to send him word by
the carrier whether I will have her or no. We drank tea at my father
Slater's, and about 5.45 my wife and I came away. She gave the maid
12d., and I the man 12d. Sam. Slater brought us on our road about
6 miles. We came home about 9.15, just before there came on a very
great tempest, for we had not been at home above ½ an hour before it
began to thunder and lighten at a prodigious rate. There also was a

---

[24] This reference to betting on cricket is not included in the article in *The Cricket
Quarterly* referred to above, p. 9, note 9.

very great quantity of hail and rain fell, and it was also very windy. Our maid at home dined on the remains of yesterday's dinner. Received of my father Slater £3 16s. 4d. in full. He gave me, in the way of balancing accounts, about 20s., but unknown to his wife, who I think is the very picture of ill nature, and to whom in a manner I may partly charge some of my misfortunes upon. For her temper is not unlike that of Xantippe, nor my father's far distant from that of Socrates.

Mon. 28 June. At home all day. In the morn Master Piper came along with: 'I have got a little job I want you to do.' And which I did, but the poor old wretch sneaked away without ever offering to pay for paper etc., though all of it together, to wit, paper, sealing-wax and thread, amounting to 1d. His business was for me to write a letter to send with his rent and for me to send it by the carrier, which I did, as also wrote a bill and a receipt for land tax to go with it. But for all this, were I only to ask the poor creature to lend me a horse, it would be: 'Pho! I've ne'er horse. There is Jack Vine has 2 or 3 that do nothing but run the streets', not thinking all the time that he had not worked his above a month. Well, God bless him, and so let him go with only this: that he must be very ungrateful or very ignorant . . . I also sent word to Mr. Batchelor that I intended to have his mare at £8 8s. 0d. and he should accordingly send it by Smith a-Friday, and which mare is to be kept in common between my mother and myself, both to be joint purchasers of the mare. My wife very bad.

Fri. 2 July. At home all day. Dr. Stone here in the morning. We dined on the remains of Monday's dinner with a butter pudding cake. After dinner our maid went over to Framfield. This day I received the horse from Mr. Batchelor of Hartfield and lent Master Darby in cash 10s. My brother here in the even but did not stay. Heard also of Elizabeth Elless's being with child.

Sat. 3 July. In the morn Mr. French and I went up to talk with Eliz. Elless, who acknowledged she was with child and not above 2 or 3 weeks more to go of her time. We asked her to inform us of the father, which she seemed very unwilling to do, but she agreed with us to go and swear her parish,[25] though we were almost confident she did belong to our parish, but thinking she might be persuaded to swear the father,[26] we concluded to carry her to Lewes and accordingly came

[25] To go before a magistrate and to declare on oath the parish in which she had last gained a legal settlement within the terms of the Poor Law. The conditions which it was necessary to fulfil before settlement could be gained in a parish other than that of one's birth are fully explained in *Tate*.

[26] To name the father on oath before a magistrate. See below, p. 50, note 31.

back to prepare for our journey. My brother being come over, I got him to stand the shop for me in my absence, which he agreed to. Then Mr. French and I went down to Mr. Porter's to get him to meet us as we were a-carrying my lady along and to talk to her to inform her of the reasonableness and justice of her either informing us who the father was or swearing it, which he promised to do. I then went home with Mr. French to get his horses and dined there on a piece of pork and greens and a beef pudding (my family at home dining on some fried veal and bacon). After dinner we set out for Lewes, Mr. French's servant, J. Shoesmith carrying her, and he and I rode single. Mr. Porter, according to his promise, met with her and talked to her very much but all to no purpose; so we proceeded on our journey. When we came to Lewes, had her examined, and being informed by Mr. Verral, the justice's clerk, she would belong to us, we did not have her sworn; and with all the persuasions Mr. Verral, Mr. French and myself were masters of could not prevail on her to confess the father, though I think we tried all ways to come to the knowledge of him. We came back again about 9 o'clock and spent 4s. 9½d. . . . to wit: At Turner's 3½d., at Martin's 1s. 1½d., at *The White Hart* 1s. 2d., examination etc. 1s. 0d., *do.* at Martin's 5d., *do.* at Turner's 3½d., Turnpike 6d.

Sun. 4 July. . . . About 3 o'clock my brother went to my uncle Hill's in order to get a horse to go with me to Cuckfield upon tomorrow . . .

Mon. 5 July. In the morn I was called up about 5.25 by Master Mugridge to inform me [he] had some wheat to dispose of, and I agreed for Sinden to look on it. About 5.50 I set out for Framfield on foot, but met my brother about the London Gate with a horse where I got on horseback and went and breakfasted at Framfield. After breakfast we set out for Cuckfield in order to treat with Mr. Hesman concerning hiring of his shop. We arrived at Cuckfield about 11 o'clock where we treated with Mr. Hesman and saw his stock, and he agreed to leave it immediately by appraisement or to cry a sale and keep on till St. Michael and then leave it by appraisement. But we agreed to consult within ourselves and let him know our minds in about 14 days' time. We dined at *The King's Head* on veal cutlets. We just called on Mr. Jos. Hills, and from Cuckfield we went to a fair at St. Joan's Common. From thence we came to Chailey and drank tea (meeting with Mr. Beard a-coming from the fair by accident). We came together as far as Uckfield where we parted, my brother going to Framfield, and I came home about 10.30. I spent about 4s. 6d. This day I received of my mother £4 4s. 0d. for her half part of the mare bought between us . . .

Weds. 7 July. At home all the morn . . . In the even Master Darby, myself and one of Master Rice's men went down and looked at Master Trill's house, and it seemed very much out of repair. I gave orders to have it repaired. Read some of Tillotson's work. Master Piper and Mr. French again come with: 'I wish you would write the land and window tax books for us.'

Thurs. 8 July. At home all day. Busy a-writing the land tax book. Dined on the remains of Sunday's dinner with the addition of an eel pie. Read some of *The History of England*. Finished writing the land tax book.

Fri. 9 July. In the morn wrote out one window tax book. About 9 o'clock Mr. French called me to go to Laughton with him in order to see a funeral there, to wit, the Hon. Lady Frances, dowager of Castlecomer,[27] sister to his Grace the Duke of Newcastle, and accordingly we went. She was brought to Halland about 11 o'clock, but not taken out of the hearse, and was interred in their family vault at Laughton about 1.30, and in the 69th year of her age. The pall was supported by the Hon. Col. Pelham, Sir Francis Poole, ---- Campion, Esq., T. Pelham, Esq., John Pelham, Esq., and Henry Pelham, Esq. The funeral service was read by the Bishop of Chichester. There were two mourning coaches, Mr. Pelham's and Col. Pelham's and the bishop's, but not so great a number of people as might be imagined. We called at the keeper's as we went as also went into Markwick's at the Pound both before and after the funeral. For Mr. French is that man who would willingly never be without a dram of gin in his hand. Oh, that mankind should be so abigated[28] to that baneful liquor, a liquor more surer to kill than even a cannon ball. We called at T. Jones's as we came home, but, however, I got home about 6 o'clock, and sober. After I came home, I dined on what my family left, to wit, a cold eel pie and batter pudding and cold bacon (but that I ate none of). Afterwards I finished writing out the land and window tax books, and in the even Mr. Elless and I walked up to John Vine the younger's to borrow his little horse to ride to the Wells[29] upon, but he was not at home, which was a good excuse for not lending me it . . .

Sat. 10 July. At home all the morn. Mr. French and I went with Gilbert the bell-hanger up in the steeple to see what was the matter

---

[27] The sister of Thomas, first Duke of Newcastle, and the widow of Christopher Wandesford, Viscount Castlecomer. She had died on 27 June.
[28] Turner seems here to have committed a malapropism. To abigate is to drive away. What he appears to have meant is 'addicted'.
[29] Tunbridge Wells.

with the great bell. We found the gudgeon broken. We accordingly
agreed to give him for mending it 7*s.* 6*d.* and to pay Joseph Durrant for
his work besides, and it must ensue as a natural consequence for Mr.
French to go into Jones's, which we did where he spent 5*d.* and had
most of the liquor to himself . . . After dinner my brother came to call
me to go to Tunbridge Wells, and according about 4.30 we set out, he
upon Mr. French's little horse and I upon my own. We called at my
aunt Ovendean's at Boarshead and stayed about 30 minutes. We got to
my brother's at the Wells about 7.30 and stayed and chatted with my
brother until past ten, when we adjourned to *The Angel* and stayed
there before we broke up till past 2, but all very sober. We should not
have stayed so late had it not been for a little diversion we were
partakers of, occasioned by some words happening between the
gentleman of the house and his barkeeper, or whore,[30] or, I am pretty
well assured, both. But what a surprising thing it is to think a man
should suffer himself to be used in such a manner by an almost
common jilt. My brother paid the whole reckoning. Moses and I lay at
The Angel and *Crown.* I bought of Mr. Ed. Baker the 1st volume of *The
Tatler* for T. Davy, which cost me 18*d.* and also Ainsworth's *Dictionary*
for Mr. Fran. Elless, which cost me 12*s.*

Mon. 12 July.   At home all the morn and very busy. My brother went
home as soon as we had breakfasted. In the morn Mr. French and Mr.
Piper came for the land and window tax books. As Mr. French has so
often favoured me with a horse, I could not expect any gratuity from
him. But as to poor old Piper, who never conferred a favour on me, he
sneaked away with, 'Sir, I thank you', but forgot either to pay for the
paper, which was 2*d.*, or so much as to say he should be glad to have an
opportunity that he might have it in his power to serve me as far. Oh!
thou black fiend, ingratitude. But what can one expect from a wretch
who sets his whole delight in nothing but money and knows not the
real use of it farther than it is conducive to the adding of store to store? . . .

Tues. 13 July.   This day died Eliz. Elless, and immediately after she
was dead, Mr. Adams told me Mr. French and I would be fined on
account of her death. The reason was because we carried her before a
justice and asked her to swear the father.[31]

[30] Turner, even in his private diary, found it necessary to write this word 'w---e'.
[31] Had Turner and French taken Elizabeth Elless before a magistrate on 3 July, as
they had originally intended, and had her questioned as to her pregnancy, they would
have been breaking the law. The statute 6 George II, c. 31 forbade such action until one
month after an illegitimate child was born. However, Adams's threat was an empty one,
since Turner and French had eventually decided that there was little point in getting her
to swear the father. See above, p. 48.

Weds. 14 July. At home all the forenoon. Paid Jos. Fuller for a bullock's heart 1½*d.* We dined on the heart baked in the oven and stuffed and a pudding under it. About 4 o'clock Mr. Porter came to me and told me he thought it was the parish's duty to examine into the death of this poor creature who died yesterday, and have her opened. For there was, according to all circumstances, room to suspect she or some other person had administered something to deprive herself or child of life. For they had agreed with a nurse to come a-Monday, which she accordingly did, and was agreed with for only a week, and a person an entire stranger. Now this creature was very well all the day a-Monday and baked. And after she had taken the bread out of the oven, she took a walk and returned about 8 o'clock. And about 10 o'clock, or between 9 and 10, she was taken with a violent vomiting and purging and continued so all night until Tuesday, 5 o'clock, at which time she expired. And the latter part of her time she was convulsed, and if asked where in pain, she would answer 'All over.' Now what was very remarkable, she had not above 2 or 3 days more but her time as to child-bearing was expired. And during all the time of her sickness she never had any pangs or throes like labour, nor no external symptoms whatever, and complained of great heat, and was afflicted with an uncommon drought. And what more increased our suspicion was as Mr. John Vine's two men and apprentice was a-coming home from work a-Monday night, they saw Peter Adams's horse stand tied up at a pair of bars which lead into a very remote and obscure place in a wood, and upon which they immediately concluded to see whether he was alone and accordingly placed the boy at or near the bars whilst they went into the wood. Before they had went far, they saw Mr. Adams, who made directly for the bars where the boy see him get on his horse and ride off. And as the men also knew him, they went forward, but not far, before they found where two people had stood and also two places where people had lain down. They then agreed to separate and endeavour to find out his partner, and one of them had walked but a little way before he see this unhappy creature, with whom he shook hands and talked to. And afterwards they all three see her together. This the men offer to swear before any magistrate. And as the affair has occasioned much talk, it led Mr. Vine the elder to see if there was anything in what they said as to there being a place as if people had lain down, where he found two as they described and also found a horse had been tied up at the bars. They were also seen on Saturday night by another person, conversing over a pair of bars, he on horseback, leaning over his horse's neck, and she a-leaning over the bars. And during the whole time of her illness they never sent for any midwife or apothecary, nor did not call in any neighbours till near noon on Tuesday, and then only 2 or 3 simple creatures. And he, Peter Adams, were with her a great part of the day on Tuesday until she became speechless, and then shook hands with her and parted. And for a great while past they have been as conversant and familiar as if they were lovers though he was a married man. To do him justice he has had one child before by another woman, and his wife, poor woman, is now big with child. Upon this suspicion we went down to Mr. Jer. French's to consult him, whom we found of the same opinion. From there we went to Mr. Coates to consult him where we found him already very strong in the same opinion. And we all agreed to have her opened in order if possible to discover whether she or any one else had administered anything to deprive her or the child of life. We stayed and drank a mug of beer and all came away together, Mr. French going home and Mr. Porter and I came up the street, it being then about 6 o'clock. Mr. Porter lent me a horse upon which I immediately set out for Luke Spence's Esq., to ask his advice and which way to proceed, but he not being at home, I went forward to John Bridger Esq. and very luckily met him a-walking in his garden near Offham. He told me he thought it was our duty and also very proper to have her opened, and as she was an inhabitant of the parish, her friends nor no other person could prevent our doing it. I then went to Lewes to get Dr. Snelling to perform the operation, and whom I found at T. Scrase's. But he told me if there should be anything found in the midwifery, he could not report it; so it would be proper to have a man midwife to assist him, and on that account I did not agree with him to come until such time as I had again consulted the parish. I saw Mr. Tucker at T. Scrase's, who informed me that at Windsor Fair wool sold for no more than 6*d*. per lb, which he said was about of equal goodness with our common wool, but not so clear from filth; and lamb wool was from 5½*d*. to 6*d*.; and farther added that Mr. Tho. Friend's orders out of Yorkshire were all stopped. I stayed at Mr. Scrase's while my horse was a-baiting and drank one mug of mild ale between Mr. Tucker and Scrase and myself. I came home just at 10 o'clock. It lightened very much all the way I came home at times. I went directly to Mr. Porter's to consult him again in the affair. He seemed to blame me a little for not getting Dr. Davy, or some other man midwife; but however, we agreed that I should set out early tomorrow morning in order to get Snelling and Davy both to come along with me as early as possible. I then went to Joseph Fuller's and borrowed a horse to go upon tomorrow (who I found all in bed, and who I called, and they accordingly promised me I should have one). I came to my own house about 11 o'clock . . .

Thurs. 15 July.   About 4 o'clock in the morn I rose and went down to Jos. Fuller's and called up Jo., T., and R. Fuller and got their horse and set out about 5 o'clock and called at Whyly to inform Mr. French of our intentions (whom I found abed) but called him up; and as Mrs. French was just going to breakfast, I stayed and breakfasted with them. I got to Lewes about 6.20 where I called up Mr. Davy as also Mr. Snelling. I borrowed of Mr. Snelling in cash £7 4s. 0d. I also left with Mr. Tho. Scrase, who I called up, £6 15s. 0d. which he was to pay to Mr. Geo. Kemp, and take up a bill which he had of the same value, drawn on me by Mr. Richd. Waite, which bill he was to send me by the post. Mr. Snelling, Mr. Davy and myself came to Mr. Porter's about 10 o'clock, where we went in and stayed just the time of eating a bit of bread and drinking a glass of wine. We came up to my house where we provided ourselves with all things necessary for the operation, to wit, a bottle of wine and another of brandy and aprons and napkins, together with a quantity of fragrant herbs such as mint, savory, marjoram, balm, pennyroyal, roses etc., and threaded all the needles. We then proceeded to the house when we duly examined the nurse, who confirmed all we had heard before, with the addition that it was such a case as she never saw before and that she was fearful all was not right. The doctors then proceeded to the operation after they had dressed themselves and opened their instruments. They first made a cut from the bottom of the thorax to the os pubis and then two more across at the top of the abdomen as:

The operation was performed in mine and the nurse's presence. They also opened the uterus where they found a perfect fine female child, which lay in the right position and would, as they imagined, have been born in about 48 hours. And as the membranes were all entirely whole, and the womb full of the water common on such occasions, there was convincing proof she never were in travail. The ilea were all very much inflamed, as was also the duodenum, but they both declared they could see no room to suspect poison. But if anything else had been administered, it had been carried off by her violent vomiting and purging (though they said circumstances looked very dark and all corroborated together to give room for suspicion). We came back to my house about 1 o'clock, and Mr. Snelling and Mr. Davy went to

Mr. Porter's. The doctors both allowed this poor unhappy creature's death to proceed from a bilious colic (so far as they could judge). After dinner they both came up to our house when I paid Dr. Snelling the £7 4s. 0d. I borrowed of him in the morn and also gave each of the gentlemen one guinea for their trouble . . .

Sat. 17 July. In the morn after breakfast went down to Mr. French's to get him to bring me from Lewes ½ oz. cauliflower seed, and when I came there, I found Mr. French, his servants, and Tho. Fuller a-catching of rats; so I stayed and assisted them about 3 hours, and we caught near 20. The method of catching them was by pouring of water into their burrows, which occasioned them immediately to come out, when either the dogs took them or [we] killed them with our sticks. Just as we had done, Mr. John Vine came in. We stayed about ½ an hour and came all away together, Mr. Vine and T. Fuller coming round by our house and only for the sake of a dram. What a surprising thing it is that a man of Mr. Vine's sense and capacity should so much give way to the unruly dictates of a sensual appetite. Mr. Vine, as we came along from Mr. French's was making several observations with regard to good economy in husbandry. To wit, he said that man who went the road a-timber-carrying etc. never hurt his horses if he did not overload them, and he very plainly demonstrated that going with a light load turned out most to the master's advantage in time, and therefore he must of consequence get more money by light loading than heavy. He also observed that the only way to eradicate the weed vulgarly called 'kilk'[32] out of the ground was by pulling it up, for was it once permitted to stand to seed, it would be difficult to get it out of the ground because every time it was ploughed, and the seed turned up to the surface of the ground, it would then grow. He said he had been credibly informed by gardeners that the seed would lie in the ground 50 years, which he in some respect confirmed by the following instance of his own observation, to wit, he sowed a border in one part of his garden with lettuce, some of which he let stand and seed, but never tilled the border nor did anything to it for 3 years, nor all the time had any appearance of any young lettuces, as might justly have been expected to spring from the seed that must consequently have shed itself. But at the expiration of 3 years when they came to till the border again, they had as fine a crop of lettuces as if the border had that year been sown with new seed. He also observed that ground designed to sow wheat on, if it be laid up fallow and exposed to the summer sun and well stirred, it will mend more than a coat of lime and the best

[32] Charlock.

method that can be taken to destroy weeds . . . After dinner I went
down to Messrs. Merrick's and Rothfield's for some scythes which
were brought from Lewes and left there for me.

Sun. 18 July. . . . After churchtime my wife and I went up to Mr.
Jos. Burges's and drank tea there and stayed till between 9 and 10. But
I think I never saw children humoured more to their ruin than theirs,
Mrs. Virgoe's excepted. I this day heard of the loss of Fort St. Philip
and the whole Island of Minorca after being possessed by the English
nation 47 years and after being defended 10 weeks and 1 day (to wit,
from the 18th of April to the 28th of June) by that truly brave and
heroic man General Blakeney, and at last was obliged to surrender for
want of provision and ammunition. No man, I think, can deserve a
brighter character in the annals of fame than this. But, oh, he were, as
one may justly say, abandoned by his country, who never sent him any
succours. Never did the English nation suffer a greater blot than in this
affair, nor, I doubt, a greater blow. Oh, my country, my country! Oh,
Albion, Albion! I doubt thou art tottering on the brink of ruin and
desolation.[33] This day Tho. Roase and Catherine Clarke were out asked.

Thurs. 22 July. . . . Mr. F. Elless and I walked down to the Nursery.
My business was to get some money of Edmund Elphick, but could
not. Never, never was money so scarce as now. We came home about
8 o'clock. As we went along, we laid 1 lb. of gingerbread concerning the
length of the church field footway. I laid it were 38 rods, and he that it
was not. So accordingly as we came home, we measured it and found it
to be 39 rod and 3 feet, so that I won this great but innocent wager, and
a far more prudent one than had it been £100. This day Mr. Porter's
daughter was baptized by the name of Elizabeth.

Fri. 23 July. In the morn Mr. French and the keeper drawed the
pond before our door and made us a present of a brace of carp.
Afterwards we went down to the church to take Peter Adams's bond
out of the chest in the church to ask Mr. Poole's advice on it, who is at
Mr. Porter's. We went into Jones's and spent 5*d.* apiece. Dined on a
piece of pork and peas with a baked beggar's pudding.[34] After dinner
went down to Mr. Porter's to ask Mr. Poole's opinion on the bond,
who says without the order he could do nothing, but we must execute
the bond against him. Afterward searched the church chest to find the

[33] The loss of Minorca to the French early in the Seven Years' War was greeted as a
national disaster. Turner's diary reflects both the nation's mood and also its admiration
for General William Blakeney, the eighty-four-year-old defender of the island, who was
said not to have gone to bed throughout the entire 70-day siege.

[34] Perhaps the pudding made with bacon which elsewhere in the country is called
Badger pudding. See Lizzie Boyd ed., *British cookery* (1976), p. 309.

order but could not. My brother came over in the afternoon and stayed and drank tea with us. In the even I went up to Mr. Piper's to ask him after the order where I found about half of it. Came home about 8.25. I gave my brother the best of the 2 carp.

Sat. 24 July. After breakfast went down to Mr. Porter's to have Mr. Poole's advice again upon the bond. What I wanted to know was how to proceed against Peter Adams in [this] affair. He, the said Peter Adams, had a female bastard child by Ann Caine, now the wife of Thomas Ling, and as security to the parish for the said child that it should not become chargeable, he gave the said parish a bond, dated 4 April 1752, wherein he binds himself, his heirs, executors and administrators to pay the churchwarden and overseer of the parish for the time being the sum of 18*d.* a week and every week from the birth of the child for so long time as the said child shall continue to be chargeable to the parish. Now during the 14 last months of Jos. Fuller's being overseer, he (Fuller) paid the said woman the 18*d.* per week for keeping the child, but as yet hath not been paid it again, and when I first came into office, I asked Mr. Adams about it, and he seemed to make use of a great many quibbling expressions concerning paying it. But after 2 months had been elapsed and he often asked about it, he at last told me I should pay it and he would soon pay me again. But now there is almost 2 months more past and mine is not paid; and Ling and his wife continually harassing me for the money, I have again mentioned it to Mr. Adams and told him I would pay no more and that he must pay it soon, upon which he told me it was my business to pay it. Mr. Poole gave me a summons to oblige Adams to appear before him at *The White Hart* at Lewes a-Saturday next to give his reasons for not paying; and then if he could give no reason for not paying and could not be brought to do it by the justices, we must then immediately execute the bond against him and sue him for the same. I gave Mr. Poole's servant 12*d.* for the summons . . . In the even talked to Peter Adams again who still quibbles on but will not absolutely deny paying it.[35]

[35] The East Hoathly parish officials had got themselves into a muddle. Evidently when Adams had admitted being the father of Ann Caine's bastard child in 1752, they had obtained an order from the justices obliging Adams to pay for the child's upkeep; they had then, as they thought, insured this order by making Adams enter into a bond to pay 18*d.* a week for as long as the child remained a charge on the parish. The taking of this bond had, by 6 George II, c. 31, rendered the justices' order unenforceable, and in order to extract payment the parish officers would have had to sue Adams on the bond at common law. Richard Burn in *The Justice of the Peace and Parish Officer* (*sub* bastards) advised churchwardens on just this point: 'the taking of a bond in any kind seemeth not

Sun. 25 July. . . . This day I had a sailor at the door who asked charity (and whom I relieved), who could speak 7 tongues . . .

Mon. 26 July. Paid Joseph Fuller for 1 loin of lamb, 2½ lb. at 3½*d.*, 8½*d.* My brother came over in the forenoon. I also went down to Mr. French's to take the measure of Sam for a hat. Paid Dame Trill 3*s.* in full for keeping Ann Brazer, due this day. We dined on a roasted loin of lamb and cauliflower with remains of yesterday's dinner. My Dame Trill complained that the 4*s.* a month allowed them was not enough; so I made her an offer to give her 5*s.* this month, and for it to be continued. But she would not accept of it and said she would have more or none. In the afternoon, wrote out Mr. John Vine's bills. This day received of Mr. Fran. Elless 12*s.* for the dictionary I bought for him at the Wells[36] the 11th instant. Also received of him 15*s.* 9½*d.* in full on account except the forms, tables etc. This night our servant sat up with Joseph Mepham. Read part of Hervey's *Theron and Aspasio*. This day gave Dame Dan a pair of stock and hand cards on the parish account.

Weds. 28 July. In the morning my brother came over . . . After breakfast I went over to Framfield with my brother to borrow some 6*d.* sugar. Came back again about 11.10 and brought my brother John's little boy home with me [37] . . .

Thurs. 29 July. In the morn paid the gardener at Halland 8*d.* for 1 gallon of currants for myself and 1 hundred of walnuts for my mother. We dined on a piece of pork and beans. Yesterday I see Peter Adams and asked him again for the money which is due from him to me and Ann Caine for her keeping the child which she had by him, but could not get it. In the even I went down to Mr. Porter's to consult him about it, and who is altogether for my serving the summons upon him, but I must think that is a wrong way of proceeding. For as he gave the parish a bond, we can have nothing to do with a justice in the affair because the bond must I think make their order (if we really had a whole one) invalid. So proceeding in this manner will subject us to his ridicule and

so convenient for the parish, as an order made by the justices; because the suing upon a bond is both tedious and expensive, whereas the course of carrying of an order into execution is very short and easy. But then, on the other hand, a bond will bind a man's executors; but the order of the justices being obligatory only upon the man himself, when he dies, the order dieth with him'. Turner had a copy of Burn's book and appears to have consulted it by 29 July, but it had only been published in the previous year: the mistake had been made three years before that.

[36] Tunbridge Wells.

[37] The boy stayed with Turner until 11 January 1757, and returned to stay intermittently thereafter.

be expending the parish's money to no purpose. Now I think the only way to act would be to empower an attorney to give him a letter, and if he did not pay it on receiving that, for him then to execute the bond against him and to sue him until such time as he should either pay it, or we could carry him to gaol, unless he could give bail . . . In the evening about 7 o'clock went down to Mr. French's as agreed, from whence we went to Halland in order to have Mr. Coates's opinion on this affair, who says our best way will be to serve the summons, and then the justices will oblige him either to pay it or, he thinks, commit him to the house of correction. But, oh, those are all vain and chimerical notions formed in the brain by ignorance! For by an Act of the 6th of George II it says if a bond is made subsequent to an order, it invalidates the same, and we must sue him at common law for the same.[38] But however, Mr. French and I agreed at their request to serve the summons upon him tomorrow. Came home about 9 o'clock. Mr. Coates informed us that Admiral Byng was arrived at Spithead where he was under an arrest for cowardice and misbehaviour in the Mediterranean relating to the losing of the Isle of Minorca.[39]

Fri. 30 July. In the morning Mr. Jer. French came to call me in order to go and execute our summons and went to call Joseph Fuller in order to get him to go as a witness for us that we actually did serve him with the summons. As we were a-going to Peter Adams, we were so fortunate as to meet him not far from Master Potter's, where after some talk and persuasions for him to consider and not give us trouble and put himself to an unnecessary expense, I served the summons on him. He seemed a little surprised at the sight of it and actually did behave better than I expected and promised to see us again tonight or tomorrow morn. I came home and found Mr. Roase at our house, who stayed and breakfasted with us, but went away directly afterward . . . I then rode over to Framfield to borrow some more sugar of my mother and also carried her walnuts. I received for them 2*d.* and ½ gallon of black currants. Came home and dined on a piece of bacon and beans . . .

Sat. 31 July. Mr. French and I in the morn hearing that Richd. Hope's servant was with child went up and talked to her about it, as

[38] See above, p. 56, note 35.

[39] Admiral John Byng was made the scapegoat for the loss of Minorca. He arrived at Spithead on 26 July and remained there under arrest until court-martialled. The court sat from 28 Dec. to 27 Jan. 1757 and found him guilty of neglecting his duty in not doing his utmost to destroy the French fleet and relieve Minorca. He was sentenced to death and was shot on the quarter-deck of the *Monarque* in Portsmouth harbour on 14 March 1757. A fierce pamphlet warfare raged on the question of his culpability.

also to her master and mistress concerning her being a parishioner by living in their service. To the first of which the poor ignorant creature answered she hoped she was not with child, though 'twas true, she said, she had deserved for it more than once or twice. But however I am very doubtful there is reason to suspect she is with child. We then talked to her mistress (her master not being at home) about her settlement, who says they agreed with her from about 1 July 1755 to new Lady Day 1756, after the rate of 35*s*. per year, and at the expiration of which time they agreed with her from Lady Day 1756 to Lady Day 1757 after the rate of 30*s*. per year. As we came home, we met with Richd. Hope who confirmed all his wife had told us before. About 12 o'clock Mr. French and I went to Lewes in order to meet Peter Adams according to his summons, but he never came, and the justices, to wit, John Bridger and William Poole told us the order was made invalid by accepting of his bond and we must sue him on the bond at common law. Mr. Verral, Mr. Plumer and Mr. Aldersey told us the same, the last of which wrote us a letter to Peter Adams. We also asked of their worships (as Mr. French called them) concerning where the poor girl belonged to, who all assured us to our parish. Mr. French and I dined at *The White Hart* on a piece of boiled beef and greens, a breast of veal roasted and a butter pudding cake . . .

Weds. 4 Aug. . . . About 12 o'clock Peter Adams called on me to go with him to Ringmer in order to see the people who were eye-witnesses of his abuse a-Whitsun Tuesday, which I accordingly did. We found James Carter, John Mitchel, David Tippings and Dorothy Presnal, all eyewitnesses of his and my abuse, which they say was very great and, in their opinion, insufferable. We called at John Martin's, Wm. Dicker's, T. Peckham's and Fran. Turner's. I spent 12*d*. Mr. Adams (though I went in a manner purely to oblige him) could not agree very well for going out in a hurry. I went without any money in my pockets and upon that account got him to pay 3*d*. for me, upon which he very tauntingly asked me whether my wife would not let me carry any money in my pockets for fear I should either lend him some or pay some for him. So one word rose to another until I fairly brought him under . . .

Sat. 7 Aug. At home all the morn. My brother dined at our house on beans and pork. After dinner he and I set out for Lewes, both on one horse. My business was to borrow £40 of Mr. Rideout on my mother's account, but could not. Drank tea at Mr. Hook's. I was in company with my uncle Hill at *The White Horse* where he informed me that Mr. Geo. Beard and his wife intended to come and dine with me tomorrow. As Mr. Hill, Mr. John Fuller, Mr. Atkins and myself sat a-drinking, Mr. Hill and John Fuller laid the following bet, to wit, provided John Fuller is not married before Xmas next, he is to give Mr. Hill a fore-quarter of beef weighing 20 stone. And if he is married before Xmas next, then Mr. Hill is to give Mr. Fuller £4 for a fore-quarter of beef of 20 stone[40] . . .

Sun. 8 Aug. . . . Just before churchtime my mother and brother and Mr. Beard and his wife came in, and they with myself and nephew went to church . . . They all dined at our house on 2 roasted ducks (of our own breed), a piece of bacon, a leg of mutton, cauliflowers and carrots, with a currant pond pudding boiled . . .

Tues. 10 Aug. . . . I yesterday and today endeavoured to borrow some bottles of Mr. Atkins and Mrs. Fuller, as also Mr. Porter, but they could not lend me any though I have lent them some several times. Oh, ingratitude, thou blackest of fiends!

Thurs. 12 Aug. In the morn Mr. Jeremiah French and myself signed a certificate for Edward and Mary Gibbs to the Parish of Wittersham in the Isle of Oxney in the County of Kent, and it was attested by Thomas Prall and John Streeter. It was dated the 10th of August, and Streeter carried it with him to Lewes to have it signed[41] . . .

Sat. 14 Aug. . . . Paid James Fieldcox of Battle £1 2s. 6d. in full for the goods this day received from him: 2 doz women's pattens 18s. 0d.; 1 doz. *do.* clogs 4s. 6d. . . . About 3 o'clock my brother Moses came over to stand the shop for me whilst my wife and I went to the Wells. Accordingly about 3.05 my wife and I set out for Tunbridge Wells in company with Tho. Davy. We called at my Aunt Ovendean's at Boarshead and stayed about 30 minutes. We got to the Wells about 6.50. We drank tea at my brother's and then took a walk about the Wells. Saw Edward Smith. My wife lodged at my brother's and I and T. Davy at *The Angel*. We went to bed about 11 o'clock. My brother stayed all night. Received by Streeter the certificate sent by him a-Thursday, which is signed by J. Bridger and T. Fuller.

Sun. 15 Aug. In the morn we got up about 5 o'clock and

[40] £4 for a 20-stone fore-quarter would have been about twice its value.

[41] French and Turner were here issuing a certificate, which had been asked for by the parish in Kent to which Edward and Mary Gibbs had moved, acknowledging that their last legal place of settlement was East Hoathly. Thus if the Gibbses subsequently became a charge on the rates of the parish into which they were moving, and had not fulfilled any of the conditions necessary to gain a settlement there, they could be sent back to East Hoathly. The certificate, called a Settlement Certificate, was issued on a printed form and needed the counter-signature of two justices. One of the witnesses (in this case Streeter) was required to swear before the justices that he had seen the parish officers sign the document. A settlement certificate is shown as Plate 6.

breakfasted at my brother's. And then my wife, Sally, T. Davy and myself set out for the camp on Cox Heath where we arrived about 11 o'clock, just as they were all got to their devotions, to wit, 12 congregations and 1000 in each congregation. They seemed to be very attentive at their devotion, and minister seemed to have a fine delivery. I think the camp as fine a sight as I ever see.[42] We went from the camp to Maidstone where we dined at *The Bird-in-Hand*, together with many more, on a piece of fine boiled beef, carrots, and cabbage and a fore-quarter of lamb roasted and French beans. We stayed till about 3 o'clock, and came home by the camp and got to the Wells about 9 o'clock . . .

Mon. 16 Aug. . . . We came away about 3 o'clock, and called at my Aunt Ovendean's, drank tea and came home about 8.35 . . . Bought at the Wells Locke *On Human Understanding* which cost me 4s. 6d., and Mead *On Poisons* which cost me 4s., but did not bring them away because he was to send them me by the news man.

Thurs. 19 Aug. . . . After breakfast went to Mr. Dicker's for to receive a small bill (along with Robt. Hook), but did not get it. From thence we went to a cricket match played on Cliffe Hill between the Parish of Mayfield and an eleven pretended to be chosen out of the whole county—but it were only to draw people together. Mayfield went in first and got 78 runs. The pretended county eleven got 55. Then Mayfield went in and got 73, and the county men got about 10 and 3 wickets down, when their time expired.[43] See Mr. Beard, who told me I could not have above 1 or 2 at the most of the 4 pieces of hop-bagging I bought of him the 20th of June. Oh, scandalous! . . .

Fri. 20 Aug. . . . Oh, how dull is trade, and how very scarce is money! Never did I know so bad a time before. To think how much I have due to me and cannot get in! What shall I do? Work I cannot, and honest I always will be if the Almighty will give me grace. I that used at this time of year to take £15 or £20 a week, and sometimes £25 or £30, now seldom take above £5 or £10. To what can I attribute this loss in trade? I sell my goods as cheap as ever I did and buy them as well, so far as I can judge; and my design is to use my customers with as good manners as I ever did. And I do my utmost endeavour, so far as I know

[42] Previous editors have regarded this gathering as a church parade of 12,000 troops; and, indeed, Coxheath near Maidstone in Kent was the principal training ground for the military in the second half of the eighteenth century. See C. Herbert, 'Coxheath camp 1778–1779', *Journal of the Society for Army Historical Research*, 45 (1967), pp. 129–48. Turner's somewhat cryptic entry, however, makes this assembly sound more like a huge Methodist field-meeting.

[43] See above, p. 9, note 9.

to do it, but trade in all places, and more particularly in a country place, is very precarious. For I observe trade has declined ever since I have been married, but why I know not . . . This day the King's Plate was run for on Lewes Downs when only Mr. Martindale's horse *Adolphus* started for the same. Afterwards there were several hacks started for a saddle.

Sat. 21 Aug. In the morn about 8 o'clock I was sent for down to Mr. Porter's to be a witness to something, but what I do not know. (But I conjecture it to be articles of agreement between Mr. Porter and some other for the sale of a farm of Mr. Porter's lying in Essex, and now in the occupation of Tho. Wright). Mr. Porter signed one paper, and the gentleman's name that signed the other was, I think, John Benison. He was dressed in a band, and had a gown or cassock on and seemed as if he belonged to some college or hospital. Miss Dinah Binge and myself witnessed both the papers . . . This day the subscription purse of £50 was run for on Lewes Downs when there was only 2 started for the same; *viz.*, Lord Craven's bay mare *Princess Mary*, and Mr. Howe's chestnut horse, which was won by the mare with ease, there being only two heats, though the odds at starting were 12 to 10 on the horse.

Sun. 22 Aug. . . . After churchtime went down to Mr. Sam. Gibbs's (I being sent for), they having this day lost their youngest child. They wanted to know if I had any shrouds small enough. I drank tea there in company with Mrs. Fuller, Mrs. Gibbs, and Mrs. Thunder . . .

Mon. 23 Aug. . . . In the afternoon very bad with tooth-ache. Read the newspaper wherein I find the nation is all in a ferment upon the account of losing dear Minorca. A remarkable wet day. In the even put up about 60 papers of tobacco.

Thurs. 26 Aug. In the morn went to Mr. Peckham's to weigh his wool which I bought of him the 2nd of August for my mother. Accordingly I weighed and packed it, my brother coming there to me. There was 165 fleeces which weighed 13 tod 24 lbs,[44] which at the price agreed for (to wit, 20s. per tod) amounted to £13 15s. 0d., and which sum of money my brother went and brought from my mother's, and I accordingly paid it to Mr. Peckham and took his receipt in full of all demands from my mother. Mr. Peckham gave me 12d. for a packing fleece and which my brother and I divided between us . . .

Mon. 30 Aug. . . . About 11.20 Tho. Fuller and I set out in order to see a cricket match at Mayfield, *viz.*, Mayfield against the county (or the pretended county), but there was not time to play it out. The

[44] The tod was equal to 32 lbs. See above p. 15, note 22.

county got the first innings 52, and Mayfield headed them 18. Then with 4 of the county out in the 2nd innings, and that 18 got, and 14 ahead of Mayfield when they left off.[45] I won 6*d.* and spent 12*d.* We came home to T. Fuller's about 10 o'clock, sober. But before I could get home I was a little high, which proceeded from what I drank there . . .

Sat. 11 Sept. . . . In the afternoon my brother came over to bring the mare for my wife and I to go to Chailey upon tomorrow, but did not stop . . .

Sun. 12 Sept. . . . In the morn about 5.20 my wife and I set out on our intended journey and went to Framfield . . . We got to Chailey about 9.05. We breakfasted at Mr. Beard's and went to church there . . . My wife, my brother, and myself dined at Mr. Beard's in company with a man and woman (whom, I understood, came from Lindfield) on a piece of bacon boiled and turned, cabbage, a piece of beef roasted, plum suet pudding, horse radish and cucumbers . . . We got home about 9 o'clock, and very sober. I paid Mr. George Beard 20*s.* in full for a piece of poking (hop-bagging) received from him the 4th instant . . . I think Mr. Beard is very happily situated with regard to trade, but I doubt not very much so in his wife. But, as he got to bed to her beforehand, he must be said to have done very justly in marrying of her.

Mon. 13 Sept. At home all the morn. About 11.30 I went to Mayfield in order to see a cricket match, viz., Lindfield against Mayfield. When I came to Mayfield, there were four of Lindfield out, but Lindfield got 52 runs the first innings and Mayfield 48, which made Lindfield 4 ahead. Lindfield got 31 the second innings, which made them 35 ahead of Mayfield, who went in and got them with only one wicket up. I came home about 6.50. I paid for the standing of my horse 2*d.*, which was all I spent, for I neither ate nor drunk while I was gone, nor nothing before I went but dry bread and cocoa, tea and some coffee mixed . . . My opinion concerning the game was that Lindfield kept the field best and batted best in general, but could not bowl. And what is remarkable, they was all tradesmen, and but one above 25 years of age, and I think eleven of very civil men.[46]

Tues. 14 Sept. . . . Sent to Messrs. Barlow and Wigginton to buy me two sixteenths of two lottery tickets.

Weds. 15 Sept. . . . Bought of a man who called at the door, and

---

[45] The grammar of this entry is not obvious, but the meaning is clear. See above, p. 9, note 9.

[46] See above p. 9, note 9.

whom I took to be a Jew, 3 doz. lead pencils, for which I gave him 2*s.* 7½*d.* . . . About 5 o'clock Mr. Elless and I walked down to Laughton to inquire after a dozen of scythes I had left at Gurr's which I found there, and also to see the church which is now an-ornamenting . . .

Sat. 18 Sept. . . . Received from Barlow and Wigginton 1/16th part of 2 lottery tickets which they bought for me, and for which I am to send them 32*s.*, viz. 16*s.* for each 16th share[47] . . .

Mon. 20 Sept. In the morn my brother Moses and Mr. Will. Bennett called on me in their way to Lewes, but did not stay . . . We dined on the remains of yesterday's dinner. Wrote a letter for Mr. Will. Piper to Mr. Geo. Tomlin with 2 pockets of hops marked 'W.P., No. 1, 2.' Also wrote another for Mr. James Hutson to Mr. Geo. Tomlin with 1 pocket of hops marked 'J.H., No. 3;' and also a letter for Mr. Jer. French to Messrs. Swanson and Hedges with 1 pocket of hops, marked 'J.F., No. 3.' In the even Mr. Porter's hoppers bought their pole-pullers' neckcloth.[48]

Thurs. 23 Sept. . . . This day there was a cricket match played on the common, *viz.*, between the tradesmen and farmers of our parish which was won with great ease by the latter.[49] In the even Halland hop-pickers bought their pole-pullers' neckcloth and, poor wretches, many of 'em very unsensible . . .

Sun. 26 Sept. In the morn my wife and self went to church, and soon after we was at church, my brother Moses and sister Sally came to us . . . Surely, as we expected my brother and sister, nothing could be more rude than for not one of us to stay at home besides a servant. But, however, their good nature overlooked it and they came home and dined with us on 2 roast ducks and gravy sauce, a piece of beef boiled with turnips and carrots and a suet plum pudding. My whole family at church in the afternoon, and also my brother and sister . . . My brother and sister stayed and drank tea with us and went away about 6 o'clock. About 9 in the evening we were all alarmed by a drunken travelling woman, swearing and rolling about the street. In the evening read 3 of Tillotson's sermons.

Mon. 27 Sept. . . . Just before dinner my father and mother Slater,

[47] See below, p. 115, note 43.

[48] Hop-picking or hopping attracted a large number of migrant workers in September. The pole-puller held a key position in the work. He cut the hop-bines and lifted the supporting poles out of the ground in order to bring them low enough to be stripped by the pickers. It was necessary that he should stand out in the hop-gardens and he therefore wore a brightly-coloured neck-cloth. For hop-picking see George Sturt, *A small boy in the sixties* (Cambridge, 1927), Ch. VII.

[49] See above, p. 9, note 9.

Mrs. Elliot and son, came in and dined with us on the remains of yesterday's dinner. Soon after dinner Mrs. Elliot and her son went for Hailsham, and I, out of complaisance to my father Slater walked with him to Eason's Green to see a cricket match, *viz.*, Lindfield against Framfield and Maresfield which was not played out but was like to be won with great ease by Lindfield, they most often running out designedly or striking up their wickets the second innings[50] . . .

Mon. 4 Oct. This day Thomas Roase and Catherine Clarke were married at our church and are, I believe, an old maid and an old bachelor . . .

Tues. 5 Oct. . . . My brother came over in the forenoon for the mare and for a basket [of] bullace. Received of him 16*s.* for his share of the 2 shares of the 2 lottery tickets I bought. Will. Burrage at work for me in the forenoon a-picking up of apples etc. . . . In the morn wrote out about ¾*d.* worth of paper in land tax receipts for Mr. Piper and Mr. French, but they neither paid for the paper nor thanked me for my trouble. But there, I'll suppose it was want of knowing better. Paid my brother 2*s.* for a Bible I had of my mother a-Sunday.

Fri. 8 Oct. . . . In the even read one of Tillotson's sermons and which I think a very good one. Oh! may the God of all goodness give me the grace to mind what I read, that the same may sink deep into my heart and mind, and that I may every day become a better Christian. Oh! how weak and feeble are my best resolutions . . . Daily and hourly do I sin, my own righteousness being but filthy rags . . . and may the God of all goodness pour into my heart his holy spirit that I may live by faith and not rely on my own works, which are vain, and that I may work out my salvation with fear and trembling.

Sun. 10 Oct. In the morn Dr. Snelling came and ate some breakfast with us and afterwards opened one of the capillary arteries of my temple for the benefit of my eyes. I asked several people to assist Mr. Snelling in doing it, but could get none till I asked Dame Durrant, who assisted in doing it. The artery lying deep, the operation was obliged to be performed with a dissecting knife. The first cut did not hurt greatly, but the incision not being big enough at the 1st cut, he was obliged to cut a second time, which hurt me very much. Mr. Snelling did not stay, but went away very soon. Only our maid at church in the forenoon, who stayed the communion. My wife very ill . . .

Mon. 11 Oct. . . . After breakfast borrowed a horse of Joseph Fuller to ride to Lewes upon to have my temple dressed. I got there about

---

[50] See above, p. 9, note 9.

12 o'clock, where I found Mr. John Snelling in bed, who arose and dressed my temple with a pea in the same nature as an issue, and did, in a manner, ask me to dine with him, but his behaviour was such as gave my imperious temper disgust, so I did not dine with him, but went and spent an hour or two at *The White Horse* with Mr. Tucker. I came home about 6 o'clock, and, with eating nothing all day and drinking but little, yet I was somewhat in liquor . . .

Thurs. 14 Oct. . . . The 2 Master Watfords a-gathering of apples for me all the afternoon, and stayed to beat out some apples for me in the even to make some cider with, as did also Tho. Durrant, who came in to cut my wife's and nephew's hair . . .

Fri. 15 Oct. . . . Tho. Davy spent the even with me, and also supped with us. We played at cribbage until past 11 o'clock; I lost 2*d.* Mr. Elless sat with us a little time in the first of the even. This is the day on which I was married, and it is now 3 years since. Doubtless many have been the disputes which have happened between my wife and myself during the time, and many have been the afflictions which it has pleased God to lay upon us, and which we have justly deserved by the many animosities and dissensions which have been almost incessantly continued and fermented between us and our friends from almost the very day of our marriage, but I hope I may now say, with the holy Psalmist, 'It is good for us that we have been afflicted,' for, thanks be to God, we now begin to live happy, and I am thoroughly persuaded if I know my own mind that if I was single again and at liberty to make another choice, I should do the same, I mean, make her my wife who is so now.

Sun. 17 Oct. About 9 o'clock my wife, self, and nephew set out upon a horse borrowed of Fran. Smith to see Mr. Hill of Little Horsted where we arrived about 10.30 and found ourselves disappointed of going to church as we had proposed, my uncle being already gone to church. We dined at my uncle's in company with my mother and brother (who came to Horsted Church and so came home with my uncle) on a leg of very ordinary ewe mutton half boiled, very good turnips, but spoiled by almost swimming in butter, a fine large pig roasted, and the rind as tough as any cowhide (and it seemed as if it had been basted with a mixture of flour, butter and ashes), and sauce which looked like what is vomited up by sucking children, a butter pond pudding, and that justly called, for there was almost but enough in it to have drowned the pig, had it been alive . . . We came home just at 7 o'clock and, considering the house I had been at, sober, though it may with justice be said I was the worse for drinking . . .

Tues. 19 Oct. . . . About 4.30 I went down to Jones's according to notice given a-Sunday of a vestry. Our company were as under: Mr. Coates, who did not stay long; Mr. Piper, Jos. Durrant, Ed. Hope, Rd. Hope and John Watford went away about 7 o'clock, but the reason was because they found if they stayed they must spend their own money and not the parish's; Mr. Jer. French, Mr. John Vine, Mr. Adams, Jos. Fuller, Mr. Hutson and myself stayed till 11.40. We spent upon the parish account 3s. 6d. which I paid, and we afterwards spent our 7d. apiece. We agreed upon the following particulars; *viz.*, to allow Dame Burrage ½ bushel flour per week; to allow Edward Babcock ½ cord wood and ¼ hundred faggots; and also to allow John Dan the same quantity. Also it was agreed to pay Tester's debt by a majority of 5 or 6 voices. I borrowed of Mr. French 18d. Mr. French and Peter Adams laid 5s. each; Mr. Adams laid Mr. French should go to the parish church of Hellingly and there hear a sermon (to be preached by the Rev. Mr. John Stone) and that when he comes from church he cannot repeat the text, which Mr. French laid he did name the chapter and verse of the text, and also each particular word contained in the text. At home all the forenoon a-writing.

Weds. 20 Oct. . . . As I were this day a-considering of the particulars that passed at the vestry yesterday, I think nothing sinks so deep in my heart as Dame Burrage's affair, to see a poor woman supplicating our charity with 6 poor helpless children (all small) deserted by a husband (who was well-known to be a more than common industrious man and also one who did not spend his money, but readily and with cheerfulness shared it in his family) and who has been eloped from his family about 10 days. Sure the thought of it must pierce any heart that is not entirely shut up and has bid adieu to all humanity, and so must it more, if we only reflect and consider something great must be the reason to force a man from his beloved family and which undoubtedly has been this poor man's cause. Many of the richest and leading men of our parish (though I think not the wisest) have long since been endeavouring to pull down the price of this and some more poor men's wages (though not a man of 'em can say he ever asked more for a day's work than he earned) by bringing in many poor into the parish from other parishes, some with certificates and some without, until the parish is full of poor, and those wise gentlemen's scheme almost come to take effect. Here was at this time everything remarkably dear; *viz.*, malt 3s. 9d. and 4s. a bushel; Warwickshire cheese 4¼d. a pound; beef and mutton 3d. and 3½d. a pound; and wheat 5s. a bushel, and nothing but daily expectation of working for small wages (nay, even for less than his due). Now let any of those cunning men, or even any other person, only lay his hand upon his breast and put it to his conscience, and at the same time to let his tongue utter the dictates of his heart; then must the following sentence be pronounced: Oh, cruel and inhuman usage, oppression, fraud and grinding the face of the poor are our guilt! Oh, may the annals of future times never record so much barbarity![51]

NB: I do not any ways commend Burrage for leaving his family, for I think it a very unjust and imprudent thing in him, and more particularly so, as he the night before he went away received of Mr. Jer. French £3 and which he carried away with him. The only thing I endeavour to point out is the motive which occasioned him to abscond.

Thurs. 21 Oct. . . . I went down to Jones's with Master Hook to consult with him about distraining Dallaway. Spent there 2*d.* Spent at Uckfield 16*d.* on the parish account and 2*d.* on my own. Received of my mother 8*s.* for 2 gallons of brandy. I must, I believe, drink nothing but water, for I find a glass or 2 of liquor makes me drunk, for today I could not get home sober; but I will once more try if I cannot live without being guilty of this vice.

Fri. 22 Oct. In the morn went and asked Dallaway for the money due to me; who seemed to make light of it . . . In the even Master Dallaway and his wife came and talked with me again about their debt and seem to talk to me in a very imperious and insulting manner, which I find is all I get by my good nature in trusting people so long, and then if I but ask them for it, I am insulted in this manner . . .

Sat. 23 Oct. Paid Jos. Durrant in cash £2 5*s.* 9*d.* being for the same sum paid for my use by Mr. Geo. Tomlin, hop factor (who sold Durrant's hops), to Messrs. Margesson and Collison. Notwithstanding Durrant owes me money and has done these 18 months past, he was so unreasonable and I think I may say unjust as to ask me for the above; but Oh, such meanness of spirit! . . . The gardener at Halland made my wife a present of some grapes in gratuity for my trusting him sometimes (only let me observe the difference in people's tempers) . . .

Tues. 26 Oct. . . . My wife paid Mrs. Atkins a visit in the afternoon and drank tea with her, but I was obliged to send for her home about 5.45 on account of my brother's and Mr. Soundy's (servant to Mr. Humphrey of Horsham) coming to our house to treat with me about letting the shop at Framfield to the said Mr. Soundy, who stayed at our

----

[51] See below, p. 239, note 19.

house until near 7 o'clock when I went with them to Framfield and spent the even there (my mother not being at home, having been at Mr. Beard's at Chailey ever since Sunday last). The gentleman, I think, seems to be about 26 years of age, born in the suburbs of London, also served his time in Southwark, and has lived journeyman in several places in London, and made 1 voyage to our American plantations, the agreeableness of which he greatly extols; and he seems to be a very sober person, and one who, I think, appears qualified to move in a higher sphere of action than keeping a little shop at Framfield. We did not go to bed until near 1 o'clock. Mr. Soundy and I proposed going to Chailey tomorrow to talk with my mother. Laid at Framfield all night and went to bed sober.

Weds. 27 Oct.   We breakfasted at Framfield, and about 10.10 Mr. Soundy and I set out on our intended journey. We arrived at Chailey about 12.05 where, after talking to my mother about 40 minutes, Mr. Soundy proceeded on his journey to Horsham. The proposals I made to him are as under, *viz*:

To let him the shop with all the appurtenances thereunto belonging yearly, or for any term of years, for £8 a year; and for him to take the whole stock at the appraisement of 2 persons indifferently to be chosen betwixt him and my mother; and also for him to take all the fixtures, and to pay one half of the money down and the remaining at 6 months' end, and to enter upon it at Lady [Day] next; and [he] has agreed to consider of it and let me know his further resolutions soon.

I dined at Mr. Beard's on part of a leg of mutton hashed (but spoiled in doing), a pigeon pudding and cold plum pudding, but all in very bad order . . . Were I to speak my sentiments freely, I think Mr. Beard seems not happy in his choice. And really Mrs. Beard appears to me to be a very indifferent partner . . . I think I never came home with more pleasure in my life than tonight. Not that it proceeded from any material business I had done in my journey, but to find my wife well and also myself sober and not to have spent any money in my journey. Oh, the reflection is pleasure immense! . . .

Thurs. 28 Oct.   . . . Mr. Porter, for himself and also for Mr. Coates, Tho. Fuller, Mr. French, Mr. Gibbs and myself—in all 6—agreed with John Fielder to bring us from Hastings a-Saturday next 1 thousand herrings, Hastings tale,[52] and one thousand of pandles,[53] for the doing of which we are to give him 15*s.* 6*d.* and also to pay what the herrings and pandles cost on the beach; and then we are to share the

---

[52] i.e. according to the Hastings measure.   [53] Shrimps.

herrings and pandles between us in equal shares. Mr. Millward has promised us to take care we shall have good fish and as cheap as if we bought a last.[54] Very busy all day.

Sat. 30 Oct.   After breakfast Mr. Sewell of Littlehampton came to talk with me concerning hiring my shop. I agreed to let him know my further resolutions in about 3 weeks . . .

Sun. 31 Oct.   In the morn Fielder brought our herrings, and one hundred over for Mr. Vine but could get no pandles. I paid him for the 1100 herrings 33s., *viz.*, 16s. 6d. the herring cost on the sea beach and 16s. 6d. his charge of bringing, with Mr. Vine's. They were parted in the manner:

Mr. Porter	213		5s.	0d.
Mr. Coates	213		5s.	0d.
Mr. Gibbs	213		5s.	0d.
Mr. French	213		5s.	0d.
Mr. John Vine	64		1s.	6d.
Do. Jr.	64		1s.	6d.
Mr. Peckham	96		2s.	3d.
R. Hook	32			9d.
Jos. Fuller	32			9d.
Jos. Durrant	32			9d.
Myself	24			5d.
	1196	£1	8s.	0d.
Tho. Fuller	213		5s.	0d.
	1409	£1	13s.	0d. . . .

Mon. 1 Nov.   . . . Wrote to Mr. Soundy to beg him to send me his answer as soon as possible, whether he likes to take my mother's shop or not . . .

Thurs. 4 Nov.   . . . About 2 o'clock my wife went down to Whyly to pay Mrs. French a visit, and I drank tea at Mrs. Weller's and talked with her about my leaving the shop. About 6.30 I went down to walk home with my wife. I stayed until near 9 o'clock and had the good fortune to get away indifferent sober. While we were at Mr. French's, Mr. Piper came to call Mrs. French to his wife, who, it seems, had sent for the midwife and many more good women—now, the poor old creature's purse!

[54] A last of herrings is ten thousand. It is not clear what this consortium did with the herrings. The members presumably either pickled or smoked them.

Fri. 5 Nov. In the morn went to church, and about 2 o'clock I went to Mr. Porter's tithe feast where I dined (in company with Mr. Porter, Mr. Coates, Mr. Piper, Mr. Burges, Jos. Fuller, Jos. Durrant, Fran. Turner, Richd. Bridgman, Edmd. Elphick, Will. Jenner and John Vine Jr.) on a buttock and brisket of beef boiled, 5 or 6 ribs of beef roasted, 3 boiled plum puddings, carrots and turnips. I paid Mr. Porter 8s. for 1 year's tithe due at St. Michael last . . . I came away about 11.10 when the company all broke up. Mr. John Mugridge was remarkably drunk, but I think all the others was very sober, some few excepted. There was in the even a very strong argument between Mr. Porter and Mr. Adams concerning the unhappy affair of poor Elizabeth Elless, when I think it must astonish almost any thinking person to see with what audaciousness the poor hardened wretch behaves, for he seems to glory in and give encouragement to crimes of the deepest dye, and his chiefest discourse consists in obscene words and oaths. Oh, may the poor unhappy creature think upon his vicious course of life and that the Almighty Disposer of all events may graciously grant him divine grace that he may repent and be converted and return unto the Lord his God. N.B. My family at home dined on the remains of yesterday's dinner. Mrs. Piper was this day about 1 o'clock delivered of a fine girl, but the poor old man her husband sat with us very easy both before and after the news were brought to him of his wife's delivery, contenting himself with saying he should get home soon enough to kiss the old woman, for they were no starters. Query: Did the old man's actions show he felt any inward pleasure at his wife's safe delivery or his being blessed with a fine daughter?

Sun. 7 Nov. In the morn Mr. Darby made me a present of a pig, and I in return gave him about a pound of sugar, value 8d. . . . We dined on the pig, given us by Mr. Darby, roasted, a piece of beef boiled, a plum rice pudding and turnips . . .

Mon. 8 Nov. . . . About 11.30 Thomas Durrant and I set out on our journey to Steyning. We called at Falmer and Patcham to bait our horses and arrived at Steyning in the evening . . .

Tues. 9 Nov. In the morn Mr. Burfield and I settled the accounts . . . After this we must walk up to Steyning Town with Mr. Burfield where he had us about from one of his friends's houses to another until we became not very sober. But, however, we got back to Mr. Burfield's and dined there . . . After dinner, knowing my wife would be very uneasy if I did not get home tonight, and also that my business would want me, and thinking myself capable to undertake such a journey, I came away, leaving Thomas Durrant there, who actually was past riding, or almost anything else. I arrived at home through the providence of God very safe and well about 7 o'clock. I spent in the whole journey 4s., and to give Mr. Burfield his just character in the light wherein he appears to me, he is a very good-tempered man, a kind and affectionate husband, an indulgent and tender parent, benevolent and humane to a great degree, and one who seems to have a great capacity and judgment in his business. Honesty seemingly is his innate principle: but after all a man very much given to drink. When I came home, Dame Durrant was like to tear me to pieces with words for leaving her son behind. But, there, poor woman, she and I were both in one pickle. But, however, Master Durrant and she and Henry Weller of Eastbourne and Fanny Weller spent the even at our house when it all came to rights with the assistance of 2 or 3 drams of her beloved Nantz.[55] Received of the gardener at Halland 12 bushels apples and ½ bushel potatoes. Steyning I think is but a small town, though both a borough and a market town, and also a free grammar school there.

Fri. 12 Nov. In the morn I got up and went over to Framfield and called at Tho. Farrant's, as I went along, for some money, but could get none . . . Came back and drank tea at my mother's in company with Mr. John Davenport and Mr. Soundy, the latter of whom has declined taking my mother's shop . . .

Weds. 24 Nov. . . . Mr. Snelling called on us and dined with us on the remains of Sunday's dinner with the addition of some pea broth. I also paid him 5s. in full for opening one of my temple arteries on Sunday the 10th October. Mr. Snelling stayed with us till about 4 o'clock . . .

Thurs. 2 Dec. . . . This day the child of Mr. Will. Piper was baptized; the sponsors was Mrs. Tealing, Mrs. French and Mr. Diplock. They made the poor old man's pocket suffer by the company's emptying as many of his old rusty bottles as they possibly could, and more than their behaviour could give sanction to (as I hear), supposing decency to be the standard for good behaviour.

Sat. 4 Dec. . . . In the even read to Tho. Davy an appeal to the public in behalf of Admiral Byng wherein he is clearly proved to be no ways guilty of what has been laid to his charge, nay, even so far from it that he behaved like a prudent and courageous commander in the Mediterranean; and his bad luck proceeded from an inferior fleet, and one which our treacherous or simple ministers, or the Lords of the Admiralty, or whoever the planners of the voyage were, could never

---

[55] Brandy: a corruption of the French place-name Nantes.

expect to have success, having but few men, not one hospital, nor fire-ship, nor never a tender sent with them; and every time they received a letter from him wherein he mentioned the want of fresh succours, or anything that might inform the public of the faults of the ministry, that never was published in the *Gazette*, and only such parts of his letters as might, for want of the whole, make him look odious to the world.[56] Tho. Davy supped with us and stayed near 3 hours with us. I also read Bally's poem on the wisdom of the Supreme Being, which I think is a very sublime piece of poetry and almost too much so for my mean capacity. But as I find the author's views are good, I do, as I am bound in duty, like it very much.

Fri. 10 Dec. . . . I was obliged to send James Marchant to Lewes for 2¼ yds cloth for Mr. Coates, and he, like a calf, stays all night. Oh such fools! By this piece of his folly I shall have the cloth left in my hands . . .

Sat. 11 Dec. . . . I this day received a letter from Mr. Sewell wherein he informs me he has given over all thoughts of taking my or my mother's shop . . .

Tues. 14 Dec. . . . My brother Moses came over (to take up a frock and pair of breeches for my nephew) in the afternoon; he stayed and drank tea with us, as did Master Piper. Received of my brother 8s. 8d. in full [for] 2 gallons grape brandy, exclusive of carriage. In the even Mr. French and Mr. Jos. Fuller coming into our house accidentally they stayed till about 10 o'clock, at which time Mrs. Fuller and her daughter came and called Jos. Fuller, thinking, as I must imagine, that by their superiority of wit they could best judge when it was proper for him to come home. But, however, to do the good woman justice, she, finding Mr. French at our house and her husband sober, so that she did not use much of that volubility of tongue which she is many times so unfortunate as to be afflicted with on such occasions. I would not, by this short digression, be thought to bear ill nature to the last, and therefore the most accomplished, part of the creation (I mean the fair sex), but would have they should employ themselves in all the soft and endearing ways that their nature seems framed for to delight man with, and not by assuming too much of the obstreperous and masculine behaviour render themselves (instead of the greatest happiness, which

[56] For the case of Admiral Byng see Julian S. Corbett, *England in the Seven Years' War* (1907), 1, pp. 96–138. See also above p. 58, note 39. It is interesting that Turner, who may have just been under the influence of the last pamphlet he had read, seemed convinced of Byng's innocence. For a description of Byng written by someone on board the *Monarque* see below, p. 85. For his death and Turner's opinion restated see below, p. 93.

the contrary behaviour must consequently yield to mankind) the greatest misery, trouble and misfortune that can I think accrue to man . . .

Weds. 15 Dec. In the forenoon went down to John Watford's and agreed to take of him one fat hog of about 10 or 12 stone, to be ready in the month of February, at 2*s.* 2*d.* per stone, and to weigh no feet and ears if I kill him at home . . .

Thurs. 16 Dec. . . . Read part of Locke's *Essay on Human Under-standing*, which I find to be a very abstruse book. I find trade very dull and money prodigious scarce and everything very dear, *viz.*, malt 4*s.* per bushel, wheat 5*s.* 8*d.* a bushel, oats 17*s.* per quarter, cheese 4*d.* per lb., butter 6*d.* per lb., and good beef 22*d.* and 2*s.* per stone. What will become of me I cannot imagine, my trade being so bad; and I know not what to do to earn a 1*d.* . . .

Fri. 17 Dec. After breakfast I designed to go down to Whyly to choose my hog and then to go to the keeper's and round by Laughton. But after I had been at Mr. French's and chose my hog, it rained, which prevented my proceeding in my journey any farther . . .

Sat. 18 Dec. . . . I can with truth say I have not been one pole from my house today. Oh, nothing to do, the thoughts of which almost drives me to despair!

Tues. 21 Dec. . . . This being St. Thomas's Day, I relieved the under-mentioned persons with one penny each:

1. Wid. Hope	18. Roger Vallow
2. Jn. Streeter	19. Will. Starks
3. Wid. Pilfold	20. Tho. Dallaway
4. Dame Burrage	21. Wid. Cornwell
5. Rd. Brazer	22. Rd. Vinal
6. Fran. Turner	23. Wid. Caine
7. Jn. Durrant	24. Sarah Vinal
8. Will. Henly	25. Tho. Ling
9. Jas. Emery	26. Jn. Dan
10. Lucy Mepham	27. Widow Day
11. Sam. Jenner	28. Wm. Eldridge
12. Tho. Tester	29. Robt. Baker
13. Tho. Heath	30. Rd. Trill
14. Rd. Heath	31. Ed. Babcock
15. Su. Swift	32. Ann Wood
16. Rd. Prall	33. Wid. Bristow
17. Jn. Elless	

. . . At home all day. In the even read 4 of Tillotson's sermons, and also in the even John Watford here a-beating out of some apples for some cider and who supped with us.

Weds. 22 Dec. . . . This day received . . . my pipe of cider which I had made by Mr. Heaver of Uckfield . . .

Sat. 25 Dec. At home all day. No churching here the whole day. James Marchant and the widow Caine dined with us on a sirloin of beef roasted in the oven with a batter pudding under it, a plum suet pudding, boiled potatoes and some bullace pies. In the even Tho. Davy sat with us about 3 hours and to whom and in the day I read 7 of Tillotson's sermons.

Sun. 26 Dec. In the morn the Rev. Mr. Hamlin of Waldron preached at our church and also gave the communion. My wife, self and maid at church (leaving my nephew at Richard Prall's). We all stayed the communion and gave 18*d.*, that is 6*d.* each, we giving our servant her 6*d.* I believe there might stop at the communion about 40 or more people which, I think, is more than I ever saw before. We had an excellent sermon, Mr. Hamlin in my opinion being the completest churchman of any clergyman in this neighbourhood, and one who seems to take a great deal of pains in the discharge of his duty . . .

Mon. 27 Dec. In the morn gave two women and 2 children with a pass 12*d.* Gave John Streeter the post 12*d.* to his box. Gave Tho. Hemsley 6*d.* to his box. Gave Robt. Hook 3*d.* to his box. About 12 o'clock I went down to Laughton to attend the funeral of Tho. Jones where there was, I believe, more than 80 people; his age 67 years. We had an indifferently good sermon, but I think very indifferently delivered to the audience. The text *Psalms* 39.14: 'For I am a stranger with thee, and a sojourner, as all my fathers were.' I came back about 5.40 when I went into John Jones's, the parish being met there at a public vestry to choose surveyors. The two surveyors chosen for the year ensuing are Mr. Will. Piper and Mr. Jos. Durrant: the electioners, Richd. Hope and John Vine Jr. The company were Mr. Jer. French, Messrs. Jos. and Tho. Fuller, Wm. Piper, Peter Adams, Jos. Burges, Jn. Potter, Edmd. Elphick, Fra. Turner, John Watford, Jn. Cayley, Robt. Hook and myself. I spent nothing. I came home about 7.40. I gave Eliz. Akehurst 6*d.* to her box; do. Tho. Turner 6*d.* to his box. Read 4 of Tillotson's sermons. Dined today on a piece of a shin of beef stewed. Just almost as we were going to bed, in came Tho. Fuller, Mr. Will. Piper and John Cayley, and as Mr. Will. Piper and T. Fuller are such (what shall I say?) spongers—no, only old Piper—that they must stay and smoke one pipe, they stayed smoking and drinking until they two was very drunk; and at last, poor fools, they must quarrel, and for no other reason that I can judge but because Tho. Fuller told that which in my opinion was really true, *viz.*, Master Piper, being lavish of his professions of kindness, and how much he loved his dear neighbour, which at last occasioned Tho. Fuller to tell him that he could never recollect any favour or kindness he ever showed him. But he did remember that once, on some emergent occasion, he wanted to borrow about £4 of him for a few days, but the poor old man would not let him have it, though he was well assured he had the money by him and could have spared it; and told him of many such-like mean actions, which made the poor old man at last so angry that he cried and bellowed about like a great calf. But, however, they all went away about 2 o'clock. Now let me shift the scene and meditate on the vice of drinking to see how despicable it makes a person look in the eyes of one that is sober. How often does it set the best friends at variance, and even incapacitates a man from acting in any respect like an human being because it totally deprives him of reason, and he is not capable of acting with reason, it is much to be doubted he will be guilty of that which is most vile and sinful. Oh, may the God of all mercy give me his grace always to detest and abhor this vice, and that the obnoxiousness of it may thoroughly work such an impression on my mind that I may never more be guilty of it, and that the many dangers which I have so often been exposed to when I have, as it were, rushed impetuous and headlong into destruction (by drunkenness), from which I have sometimes almost miraculously escaped by the undoubted providence of a all-wise Being! I say, that this may more thoroughly have its due effect upon my mind, let me never forget the goodness of God to his sinful creature, but may I meditate on it day and night that by so doing I may confirm my intentions of never giving myself the least liberty of doing anything that shall any ways tend to that heinous and ever-to-be-abhorred vice . . .

1757

Sat. 1 Jan. . . . Jos. Fuller smoked one pipe with us in the afternoon, and I got him to pay the poor tomorrow, I having promised to go to Framfield to serve a funeral there for my mother. But in the even received a letter from my sister that the person was not to be buried tomorrow as appointed, but is to be buried a-Monday . . .

Sun. 2 Jan. . . . After churchtime my brother Moses and cousin Moses Bennett came over to acquaint me that my mother wanted me to tie up 12 favours for her in order for me to carry tomorrow to the funeral, and accordingly my wife tied them up in the even . . .

Mon. 3 Jan. In the morn went over to Framfield where I arrived about 9.45. After eating a mouthful or 2 of buttered toast and drinking a few dishes tea, as also looking over and tallying and taking an account of the gloves, hatbands, favours etc., I set out for the funeral of Alice Stevens, otherwise Smith, natural daughter of Ben. Stevens (at whose house she died). I got to the funeral house about 11.50 where my brother Moses came to me soon after, in order to learn to serve a funeral and also to give me what assistance he could. The young woman's age was 28 years, and I think I never saw any person lament the death of anyone more than Ben. Stevens did for this poor girl, his daughter. After serving the funeral, as soon as it was possible we could, we set out for Buxted Church where she was to be buried, with a large company of people (she being carried on men's shoulders)[1] where we arrived about 4.20 and where we had an excellent sermon preached by the Rev. Mr. Lawson, Curate of Buxted, from *Hebrews* 9.27: 'And as it is appointed unto men once to die, but after this the judgment.' The young woman was laid in the ground about 5.50 after staying at the grave about 15 minutes, it being dug too small. When all was over, my brother, Charles Diggens and myself came to Framfield. I stayed at my mother's and baited; and then my brother came home with me and stayed and spent the evening with us, and stayed all night. We came home exactly at 8 o'clock. I received of my mother in cash 4*s.* and by old gold and silver 16*s.* which, with the £4 sent me the 30th December, makes together £5 which I have borrowed of her. My mother gave me 12*d.* for my gloves given me at the funeral and for silk used in tying up the favours. I gave away hatbands, favours, gloves etc. at the funeral for my mother's use as under:

---

[1] Turner may simply be remarking on the fact that the coffin was carried on shoulders rather than on a horse-drawn conveyance; alternatively, he may be referring specifically to the body being carried by men. It was often the case that women dying in their viriginity were borne by women. At such funerals also the favours distributed were often, as in this case, white rather than black. For examples of these practices in an earlier period see Clare St. Q. Gittings, 'Funerals in England 1580–1640: the evidence of probate accounts'. Unpublished Oxford M.Litt. thesis; Bodleian Library MS. M.Litt.d.1890; and for later survivals Anne Ruffell, 'To crown a virgin', *The Countryman*, 87, no. 3 (Autumn 1982), p. 33. See above, p. 7, note 5.

Hatbands in all 9, *viz.*,

Ben. Stevens Sr.	Sam. Smith
Henry Stevens Sr.	Arthur Knight
Will. Stevens Sr.	Ed. Willet
Richd. Stevens	Tho. Ware
John Newnham	

Favours in all 12, *viz.*,

1. Mary Ware	7. Henry Stevens Jr.
2. Jane Stevens	8. Ben. Stevens Jr.
3. Eliz. Willet	9. Ed. Willet
4. Mary Ware	10. Richd. Savage
5. Fran. Smith	11. John Staply
6. Sarah Colgate	12. Richd. Figg

2*d.* chamois gloves, 1 pair—*viz.*, Benj. Stevens Sr.

Women's white lamb gloves in all 24 prs:

1. Mrs. Willet	13. Mary Russell
2. Mary Ware	14. Sarah Colgate
3. Lucy Diggens	15. Mary Colgate
4. Mary Ware	16. Fran. Smith
5. Mrs. Attree	17. Mrs. Stevens
6. Mrs. Ware	18. Dame Newman
7. Eliz. Willet	19. Eliz. Ware
8. Sarah Willet	20. Dame Stevens
9. Mrs. Stevens	21. Dame Limpass
10. Mrs. Stevens	22. Hannah Simonds
11. Jane Stevens	23. Dame Belton
12. Eliz. Stevens	24. Mary Devonish

Men's white lamb gloves in all 35 pairs:

1. Tho. Attree	10. Charles Diggens
2. Ed. Willet Jr.	11. Robert Diggens
3. John Staply	12. Richd. Savage
4. John Buckwell	13. John Smith
5. Ben. Stevens Jr.	14. Hen. Stevens Jr.
6. Richd. Tealing	15. John Cornwell
7. John Stevens	16. Will. Cornwell
8. Tho. Starr	17. Will. Novice
9. John Cornwell	18. Ben. Stevens Jr.

19. Will. Stevens Jr.   28. Sam. Smith
20. Richard Figg   29. Arthur Knight
21. John Evans   30. Ed. Willet Sr.
22. John Shelly   31. Thomas Ware
23. Tho. Card   32. The Rev. Mr. Lawson
24. Henry Stevens Sr.   33. Will. Devonish
25. Will. Stevens Sr.   34. Tho. Turner
26. Richd Stevens   35. Moses Turner
27. John Newnham

John Newnham's 2 sons to have each a pair . . .

Sat. 8 Jan.   In the morn Tho. Fuller Jr. killed a hog for us, which I
am to have of Mr. French at 2*s.* 2*d.* per stone. He stayed and
breakfasted with us. We dined on the hog's sweet-bread fried with the
addition of some apple pie and bread and cheese. In the even Robt.
Hook got me to go down to Jones's in order for me to draw up his
account of surveyorship[2] to carry to the sitting a-Monday. But Mr.
French and I, seeing no reason but what he had already done was
sufficient, so that I did not do it; though in being so good-natured as to
go with him, it cost me 4*d.* Sure nothing is more despicable than to see
what slaves many people make themselves to that most detestable
poison called gin! . . .

Mon. 10 Jan.   In the morn Jos. Fuller cut out my hog, and we
weighed him, Mr. French being here, and he weighed 23 stone 7½
lbs.,[3] which, at 2*s.* 2*d.* per stone, amounts to £2 9*s.* 10½*d.* I did not pay
for him. At home all day . . . In the afternoon sent our servant to
Framfield to bring home some lump sugar, who acquainted me my
brother John was at Framfield . . .

Tues. 11 Jan.   After breakfast my brother Moses came over to our
house for my brother's little boy; so accordingly we went over to
Framfield and carried the little boy . . . My brother [John] was at my
mother's, where and with whom I stayed about 2 hours.

Fri. 14 Jan.   In the morn composed the following petition for Will.
Eldridge, which Mr. Elless wrote for him:
   'Whereas William Eldridge hath lately had the misfortune to
   lose a hog by it's getting under the ice and being drowned, and
   as it is at a time that corn and all other provisions are very dear it
   will render him incapable to get another which will be a great

[2] Hook had served the parish office of surveyor of the highways during 1756. This
office was held under the authority of the justices of the peace and an account had to be
rendered to them.
[3] See p.1, note 2.

loss in his family; he therefore assumes to ask the charity of his neighbours, hoping they will commiserate this his misfortune, and he will always (as bound in duty) gratefully acknowledge the favours they shall be pleased to confer on him.'

We dined on a hog's haslet roasted. At home all day. In the even went to see James Marchant and stayed about an hour. My wife very ill.

Mon. 17 Jan. . . . Gave Will. Eldridge 12*d.* on the petition I composed for him the 14th instant . . .

Sat. 22 Jan. . . . This day I took out of the draw[er], being the sum I have taken this week, £12—oh, but a small sum!

Sun. 23 Jan. In the morn about 5.30 Tho. Davy and I set out for Hartfield, it being a fine frosty morn. We arrived at Hartfield about 10.30. The reason of my going to Hartfield was on account of our hearing of Sam. and Ann Slater's being very ill, the former of which I think to be in a very dangerous way, but the latter is only ill with a cold. I should have mentioned we lost our road in the forest for near an hour or more. We dined at my father Slater's in company with Mr. Walis (officer of excise) on a sparerib roasted, a knuckle of pork, and part of a neck of mutton boiled, potatoes and apple-sauce . . . We spent the afternoon and even at my father Slater's, and who I think received us in a very civil and handsome manner and entertained us in a very free and generous manner, and I really think was sincerely glad to see me. But I wish I could say the same of my mother Slater, who seems, I think, to look upon me with an envious and scornful eye, and often-times in her talk seems to level a sharp satirical sentence or two at me; but there, let it pass, and God bless her. We lay at my father's all night. Mrs. French and her daughter Molly, with Mrs. Fuller and her husband, drank tea with my wife in my absence.

Mon. 24 Jan. About 10.30 we set out from Hartfield. In our road home I called at Mr. Courthope's at Uckfield to get a summons of him for Tho. Lewer and Eliz. Day (who have intruded themselves into our parish)[4] to bring them before him, or some other justice, to be examined and properly removed to their respective parishes. We came home about 3.45 . . . Paid for the summons on the parish account 1*s.* 0*d.* Gave my father Slater's maid and boy 6*d.* each. The reflections on this journey are no ways disagreeable to me, having been no ways the least concerned in liquor. In the even in reading the *Lewes Journal* I found the following remarkable character, which I admire not for the diction, but for the justness of it and for imitation: 'On Sunday the

---

[4] Without arming themselves with a certificate from their last place of legal settlement.

9th Jan: died Suddenly, the Rev. Mr. Lyddell, Rector of Ardingly, Sussex, aged 59; a Gentleman, who for his extensive Knowledge, unlimited Charity, genteel Behaviour, and other amiable Qualities, was an ornament to his profession; and yet so little Publickly taken notice of, that he never enjoy'd any Church Preferment, except a small Family living of one Hundred Pounds a Year. He was possessed of a good paternal estate, above one Thousand Pounds a Year; and though he lived in the most retired, Private manner, the Yearly income of it was disposed of, in assisting his friends in distress, and in Charity to the Poor. He died a Bachelor. The name is extinct. His estate devolves to Richard Clarke Esq. of Blake Hall in Essex' . . .

Tues. 25 Jan. After breakfast Mr. Jer. French came into our house, and we agreed to go and talk with Eliz. Day and Tho. Lewer, which we set off in order to do. But as we was going along, we heard that Eliz. Day was not in the way, upon which Mr. French asked me to go into Jones's, which we accordingly did and spent 2*d.* apiece. Just as we were coming out, we saw Dame Dan and her daughter Day pass by, whom we called into Jones's and talked to, and informed her of our intentions to remove her if she does not go out of our parish before Tuesday next. We then went down to Tho. Lewer's, but he was not at home; so we left word of our intentions with his wife. We came back again to Jones's (for it is, I think, impossible to get Mr. French by a sign post) where we had not sat long before we saw John Jones the overseer of Waldron ride by, who we called to and talked with about Eliz. Day. He says their parish no ways desires her to intrude on our parish, but had much rather she would come home. So that it's plain the fault of her living here is her own. We spent 12*d.* on the parish account . . . When I came home again, I dined on some sausages and hog's puddings. In the evening marked up 4 dozen hose.

Weds. 26 Jan. . . . We had very remarkable weather last night, *viz.,* about 9.30 very clear, and seemingly froze; about 11 a great shower of rain and hail, and about 12 a large clap of thunder and several flashes of lightning; and in the latter part of the night it froze hard and snowed.

Fri. 28 Jan. . . . In the afternoon posted my day book and wrote out Mr. Jn. Vine's bill. Oh, what a melancholy and dull time it is! No business, nor can I get in my debts; and everything very dear, *viz.,*

Wheat 7*s.* 4*d.* per bushel	Fat Hogs 2*s.* 3*d.* per stone
Barley 3*s.* 6*d.* *do.*	Beef 2*s.* *do.*
Oats 2*s.* 6*d.* *do.*	Mutton 3*d.* per lb. and 3½*d.*
Peas 3*s.* 9*d.* *do.*	Cheese 4*d.* *do.*

White peas 4s. 6d. per bushel   Cheshire cheese 5½d. per lb.
Brown Malt 4s. 6d.   Butter 6d. do.
Pale do. 5s. do.

What I shall do I cannot tell. I should never care how poor I lived in regard to eating and drinking if I could but make things keep in equilibrium, which I am afraid they cannot do as trade is so dull, and also as it hath pleased the all-wise Disposer of all events to afflict us with sickness, and then the many losses which I have almost incessantly had, some undoubtedly by our own misconduct, and some that could not be prevented, which altogether must have greatly hurt me. However, I will make it my greatest endeavour to be content with that station which it shall please God to appoint for me, and if it be my fortune to be poor and low in the world (as I can have no other hopes), I will endeavour to meet my fortune with pleasure. For thou, O Lord, knowest what is best for me . . .

Mon. 31 Jan.   Received by Eliz. Day a note from the churchwardens and overseers of the parish of Waldron wherein they acknowledge Eliz. Day and Ann her daughter to be legally settled and to belong to their parish[5] . . .

Weds. 2 Feb.   . . . Tho. Davy, who had this day been at Lewes, came into our house in the even [and] informed us that about 40 people out of several neighbouring parishes came there today in a kind of riotous manner in order to demand corn of a person who has lately bought up a large quantity, who very readily offered them any quantity they would have at 6s. a bushel, which prevented any further disturbance, though they did not accept of his offer on account of there being justices in town, to whom they made their complaint, and who very readily took their complaint into consideration and told them if they came again a-Saturday next, they would redress their grievances and they should have corn at a more moderate price; and also granted them several summonses for people that have quantities of corn by them but have [refused] to sell it at a market price. Oh! what a cruel thing it is to endeavour to grind the face of the poor, which many great farmers at this time do by keeping up their corn and not selling it at a time when it fetches so large a price as it does at this time.[6]

Thurs. 3 Feb.   At home all day but not busy. We dined on some

---

[5] See below, p. 92, note 15.

[6] 1756, the year that had seen the outbreak of the Seven Years' War, also saw a bad harvest. Prices of basic foodstuffs rose sharply, and outbreaks of civil disobedience were widespread. See George Rudé, *The crowd in history* (revised ed., 1981), chapter 2 'The English country riot of the eighteenth century'.

sausages baked in a batter pudding in Marchant's oven and some apple sauce. In the even Tho. Davy here and supped with us. He and I played at cribbage; I won 12d. If any person should by accident or curiosity peruse my several memorandums, they may think it somewhat odd and profuse or extravagant in me to entertain one person so often, which undoubtedly would be so, were there no reason for it, but I think there is. First he is a very sober man and one who has read a great deal, by which I oftentimes learn something. And then he is a man that is always ready and willing to do anything for me or to go anywhere for me and will never take anything for so doing; so that it is not altogether for the sake of company, but for the benefit of improvement and out of gratitude[7] . . .

Fri. 4 Feb. . . . I think this to have been as cold a day as I ever knew, and I believe the frost has not been totally out of the ground these 5 weeks.

Sat. 5 Feb. . . . This day the frost began to go away . . .

Tues. 8 Feb. Tho. Tester brought me the copy of a process which he had served on him by Thornton for Savage's debt, also a summons from George Courthope Esq. to appear before him either on Wednesday or Thursday next to show cause for not relieving him. In the even gave the poor creature 6d. At home all day. Nothing to do. Mr. Fran. Elless drank tea with us. Mr. French being at our house in the even, he appointed to go to Uckfield with Thomas Tester tomorrow . . .

Weds. 9 Feb. In the morn about 8.40 I went down to Whyly and called Mr. French and Tho. Tester (who was there at work), and we proceeded on our journey to Uckfield, and as we went, we called at my mother's and stayed about 30 minutes. When we came to Mr. Courthope's, he gave so much credit to Mr. French's fallacy that poor Tester was almost like to be hanged for saying of nothing. But, however, Mr. Courthope allowed him a grist more and then told us we should not be hard to him. But, however, Mr. French declared he should have no more on the parish account. We went with Mr. Thornton to *The Maiden Head* where I paid 18½d. for what we spent. Mr. Thornton offered to make up Savage's affair with Tester on condition that the parish should pay at Easter the debt to Savage and 10s. 6d. towards the expenses. But we could no way prevail on Mr. French to do it. As we came home, we called again at my mother's but did not stay. Mr. French and I both being pretty much in liquor, we

[7] Here Turner envisages his diary being read by others.

quarrelled very much, and the subject of our dispute was whether I should obey the justice's orders in giving Tester another grist or not. But, however, we went into Mr. French's and drank a bottle of beer. I came home very much in liquor. Oh! what an unfortunate wretch I am that I can drink but 2 glasses of liquor before I am drunk when it is a thing I am sure I despise and do try as much as possible to avoid it. Oh! may the ever abundant mercies of the Divine Goodness pardon this my weakness and imperfection and pour into my heart the grace of His Holy Spirit to strengthen my weak and frail resolutions that I may never more be guilty of this vice, but may always live in a constant state of virtue, temperance, justice, humility and charity. All this I humbly beg for JESUS CHRIST'S sake. Amen . . .

Thurs. 10 Feb. . . . A sad unpleasant day. Oh! the reflection of yesterday intolerable. Well, I am determined never to drink anything strong. Mr. Elless spent the even with us and read to us two of Torriano's sermons. The above-mentioned Torriano is curate of Hooe and Ninfield, has been in the linen drapery, and also has been clerk to some company, and is now a physician and divine.[8]

Fri. 11 Feb. . . . I called a public vestry today to consult about Tho. Daw . . . The result was as under, *viz.*, Mr. John Vine will try if Delves's blacksmith's shop etc. can be had for Tho. Daw[9] . . .

Sat. 12 Feb. . . . I have took no more than £8 this week. A terrible time . . .

Sun. 13 Feb. Just before churchtime my brother Will. came in, and he and I stayed at home. My wife and maid at church. My brother and Tho. Davy dined with us on a hog's cheek boiled, a plain batter pudding, turnips, parsnips and cabbage. My brother informed me that my uncle Hill had just received a letter from his son Charles on board the *Monarch* man-of-war in Portsmouth Harbour. I and Tho. Davy agreed to take a walk to see the contents; so accordingly, whilst they were at church in the afternoon, he and I set out for Horsted, leaving my brother to keep house, my wife and maid being at church. We got to my uncle's about 4.10, where we found my uncle at home and my brother Moses and Mr. Abra. Whapham and the Isfield miller, the two

---

[8] Nathanael Torriano.

[9] Thomas Daw, a one-legged ex-smuggler with a blind wife, posed great problems for the East Hoathly parish officers, and the diary reports continued attempts to set him up in trade as a blacksmith. Eventually the parish found him a house and the wherewithal to begin trade, but all in the parish of Waldron (see below, pp. 117, 122 and 125). The purchase of this property worth over £55—by East Hoathly parish but ostensibly by Daw—gave him settlement in Waldron parish and freed the East Hoathly rates from a considerable burden.

latter of which went away about 6.10. My uncle showed me my cousin's letter wherein he informed him of his health etc. and also that Admiral Byng was now confined under sentence of death on board their ship, that he was in high spirits for a man in his circumstances, and that he was about 5 feet 5 inches, somewhat corpulent, and had received a reprieve for 15 days on the 5th instant. But it was his and the officers' opinion that he would undoubtedly be shot. He also said it were computed he was worth £300,000[10] . . .

Tues. 15 Feb. . . . I agreed to take Philip Turner[11] of my mother at £5 a year to board and clothe him, that being the sum my father left by will to maintain him till he is 14 years of age. Dame Vinal here a-washing half the day. In the evening wrote to my cousin Cha. Hill on board the *Monarch* man-of-war at Portsmouth . . .

Thurs. 17 Feb. . . . Papered up 192 papers of tobacco weighing 43 lbs[12] . . . In the afternoon my brother Moses brought over Philip according to our agreement on Tuesday last. He stayed and drank tea with us. In the even read a sermon preached at this church on the 1st of August 1716 by the Rev. Mr. Richard Haworth on the wonders of providence in the defence of the reformation, and which in my opinion is an excellent discourse . . .

Fri. 18 Feb. . . . This day Ann Durrant was brought me (being big with child) by an order from Luke Spence and George Courthope Esqrs. dated today and delivered to me by one of the officers of Laughton Parish . . .

Sat. 19 Feb. . . . In the morn walked up to Jn. Durrant's and informed Ann Durrant (who yesterday declared to me and Mr. French she would swear the father at any time) that I would go with her today to Mr. Courthope's, to which she readily agreed. I came home and got Susan Swift to walk with her to Uckfield, and accordingly they set off about 9 o'clock and I after them. I called at my mother's, but did not stop, but went forward and informed Mr. Courthope there was a young woman a-coming to swear the father of a child she was then big of. I then went and stayed at Mr. Halland's at *The Maiden Head* until such time as she had had her examination and taken her oath, which was that she was actually with child and that the man who begot the said child was Geo. Hyland, a labourer and who now liveth in the parish of Laughton. When they came back to me, I went again to Mr.

[10] For Admiral Byng see above, pp. 41, 58 and 73 and below, p. 93.

[11] The illegitimate son of Turner's half-sister Elizabeth. He was some eight years old at this time. He died at the age of fifteen in 1764.

[12] Each packet thus contained 4 oz.

Courthope's in order to get a warrant to apprehend and take the said
Hyland, which was readily granted me by Mr. Courthope. I then
immediately came home, calling at my mother's (but did not stay) and
also at Mr. French's to acquaint him how I had made out. I came home
about 2.05 . . . Mr. Vine, Mr. French, Jos. Fuller and myself went
down to Jones's to consult of proper measures to take the man. We
agreed Mr. French and I in the even should go down to the man's
house and see if there were any light; if there was and he at home, to
take him. I did not stay, but I find the other three stayed and spent
11*d.*, which they charged to the parish account. In the even about 6.20
I went down and called Jn. Watford (he being headborough), and from
thence we went and called Mr. French and then proceeded on our
intended journey, where we could find no light nor any appearance
that the man was in his house. After staying and patrolling about the
house some time, we came home about 9.05 and, I think, never more
tired in my life. The money I expended today is:

Paid Mr. Courthope for Ann Durrant's		
examination and a warrant	2*s.*	0*d.*
Spent at Uckfield		8*d.*
Gave Susan Swift for going		6*d.*
Gave Jn. Dan	1*s.*	0*d.*
Do. Dame Burrage	1*s.*	0*d.*

The above all on the parish account.

Mon. 21 Feb. . . . In the afternoon went up to Mr. John Vine's
where I stayed some time in company with Mr. Martin, in partnership
with Mr. Tapsfield. I came home and called on John Watford, and we
went in the even to Laughton in order to take up Geo. Hyland again.
When we came to his house, we thought we perceived a light and also
heard somebody talk. We then went to the man whom we thought had
been the headborough of that Hundred, but he not being headborough
we were disappointed. It being somewhat late, we came home, where I
found Tho. Davy, who stayed about an hour . . .

Tues. 22 Feb. In the morn got up about 6 o'clock and went down to
Tho. Lewer's in order for him to go to Uckfield to swear his parish,
which he readily agreed to, assuring me he would be at my house by
8 o'clock, which he was; and then we both went together in order to call
Mr. John Vine and Eliz. Day, the former for company and latter in
order to remove to her parish of Waldron, and upon whom I served the
summons I had of Mr. Courthope the 24 January last. But she
absolutely denied going, saying that she would not go for anyone.

Whereupon going back to Mr. John Vine's (only Tho. Lewer being with me when I served the summons) we agreed to defer our journey till another day. Mr. Vine and I came down street together; he stayed and drank a bottle of beer with me . . . In the afternoon went down to Mr. French's in order to acquaint him with our proceedings. In the even Master Hook and I went down again to take up Hyland. When we came to his house, and after patrolling some time about the house, we found the man to be at home by hearing him talk. We then went immediately to Mr. Rabson, the headborough of Laughton, in order for him to serve the warrant, which he refused to do without more assistance, though there was two of us and the man we was a-going to take was a man with the use of one arm only. I then desired him to charge his son-in-law, who lived with him, and to whom I offered 12*d.* for his trouble, but he refused to charge him. I then proposed a 2nd person in our road to Hyland's, but he refused him. I then entreated him to go with us alone but could not persuade him. I then proposed his next neighbour, but he still rejecting my choice, but still continued protesting he would go and charge somebody, which he did, and I after him, entreating him to serve our warrant and not to use us ill, as I imagined he designed (and as I afterwards found true). For he went about ½ a mile the contrary road from which we was a-going under pretence to charge a man to go with him. But instead of doing that he went to one of the overseers and told them his business, *viz.*, that he had got a warrant to serve on Geo. Hyland and begged he would send word of it to the other overseer. And after staying at the overseer's about 20 minutes, he at last told him he should be glad of his company if he liked to go, but he should do as he liked about going. So that one may plainly judge of the baseness of the poor creature's principles. But we must, I think, say this of him: that he is a compound both of baseness and ignorance. But, however, we came back again by his house and called Master Hook where he had left him and then proceeded on our road to take the man (our company was Goad, the Laughton overseer; Rabson; myself and Hook). When we came to the house, the man and Ann Durrant were both in bed, but he soon came down and opened the door. And then we came away for Hoathly, leaving the poor old man so soon as we came into the park, and only G. Hyland, A. Durrant, R. Hook, Goad and myself came to Hoathly, where we arrived about 12 o'clock and sent for J. Watford. We went into Jones's and stayed till past 2. We spent 3*s.* 2*d.*, *viz.*, 3*s.* charged to the parish and 2*d.* I paid, though for the parish. A. Durrant, myself, Hook, Watford and Hyland came up to our house where we supped, or rather breakfasted on the remains of yesterday's dinner. We sat up all night . . .

Weds. 23 Feb. In the morn went down to Mr. French's to inform him what I had done. Tho. Cornwell and Tho. Durrant watched the man all the morn and about 8.30 Mr. Will. Shoesmith and Mr. Goad the Laughton overseers came and offered security for the child, *viz.*, 18*d.* a week and 40*s.* for her lying in. But our people, *viz.*, Mr. Porter, Mr. French, Mr. Vine and Jo. Fuller, thought it proper to have the man before Mr. Courthope before they took the security, thinking he might be persuaded to marry the girl, Mr. Porter endeavouring to persuade the man the justness of his so doing. About 10.20 Mr. French, Hyland and myself set out for Uckfield and when we come to Whyly, the man made some objection to go[ing] without a peace officer. We then sent for John Watford to go with us, who accordingly came, and we then proceeded on our journey. We called at my mother's but did not stay. When we came before Mr. Courthope, he was that very worthy man as never to ask the man whether he liked to marry, nor so much as once to reprimand him for bringing so much trouble on both parishes, but told us we should send to Mr. Shoesmith to let him be at Uckfield by tomorrow morning at 10 o'clock, and that we might leave the man in the care of our headborough till tomorrow (though he said the man must bear his own expense) and he would send the security by them. Spent 3*d.* at *The Maiden Head*, and seeing Mr. Thornton, he asked us to go into *The King's Head*, which we did and spent 9*d.* And in the meantime Master Watford, being loath (as I or anyone else should have been) to stay, went up to Mr. Courthope and asked his consent to go home, who readily granted him the favour, provided he would see that the man was brought there again tomorrow. As Mr. French and I came home, we called at my mother's but did not stay (Will. Bennett being there). We also called at Whyly where was John Jones the Waldron overseer, who told us we should certainly have a certificate with Eliz. Day on Sunday next without fail. I came home about 6.10 where I found all the people at my house. I then sent down for Mr. French, who came immediately and ordered them all down to Jones's, we both going with them; but I did not stay, being at home and abed by 8.30. My family at home dined on the remains of yesterday's dinner, but as for myself, I dined not anywhere. I think it is as near doing as I would be done unto if I charge the parish 6*s.* for the people's eating etc. at my house. In the even wrote a note to Mr. Shoesmith to be at Uckfield tomorrow by 10 o'clock and agreed to give John Hesman 6*d.*

1. East Hoathly village as shown on the tithe map of 1839: *E.S.R.O. TD/E 48*. Turner's house is numbered 328, The King's Head 324, the church and the rectory 371 and 372. Halland House had stood just off the bottom left hand corner of this map.

2a. Halland House from T. W. Horsefield, *History, antiquities and topography of the county of Sussex* (1835).

2b. Turner's house in East Hoathly. Photograph taken in 1984 by J. Meakin.

for carrying it, which he immediately did. My brother came over in the even for the mare but did not stay.

Thurs. 24 Feb. About 11.10 I was sent for down to Jones's but did not go. Again about 12.30 I was sent for, when I arose and went and then found the man, instead of going to Uckfield today, in the mind of marrying the girl (for Mr. French had agreed he should have £5, a ring value 10s., a wedding dinner, and married with a licence). I stayed about 30 minutes (finding most of 'em drunk) and came home to bed. About 6 o'clock, I arose and went down to see if the man was in the same mind, when I found he was not, then I again went home to bed, and about 7.30 I again was sent for. I accordingly arose and went, and the man then declared if I would get the licence, he would then be married. I then went up to Jos. Fuller and sent for Mr. French, and they both agreed I should go to Lewes to get a licence. I borrowed of Mr. French £1 16s. 0d. Jos. Fuller lent me his horse to go upon, and we agreed I should meet the people at Laughton Church on account they must be married there. So about 8.30 I set off for Lewes (but first I desired Mr. French in my absence to send Mr. Shoesmith word not to go to Uckfield) where I arrived about 9.20 and applied for a licence, but could get none on account the man was not there in presence. I breakfasted at *The White Horse* and then came away home by Laughton where I met Mr. Porter, Mr. French, Jos. Fuller, Jn. Durrant, Tho. Cornwell, Jn. Watford, Hyland and the girl, who all came away home again. I spent this journey as under:

To going through Peckham's	1d.
Turnpike	2d.
Breakfast	6d.
1 Pint Wine	6d.
Horse and ostler	3d.
	1s. 6d.

We dined today on the remains of Tuesday's dinner with the addition of some sausages. After dinner, the man being very sober and then in the mind of marrying, it was agreed we should go with him to Lewes to get a licence. We accordingly set off, he on a horse of Mr. French's, John Watford on one of Mr. Porter's and myself on one of Joseph Fuller's. We went very well till we came as far as Mr. Spence's, when the man declared he would go no farther, but would go in there and send for Mr. Dicker to be bound for him, but we persuaded him to go back to Hoathly and send for the overseers of Laughton, go to Mr. Courthope's and have no more trouble. Well, he agreed to it and was got on our road home as far as the turnpike when his mind altered, and he would marry if we would advance 40s. more. We then turned again and went to Lewes where we met with Jos. Fuller Jr., who gave him a note of hand for 30s. and 5 stone of beef to be paid on the day he was married to Ann Durrant. He then made all the solemn vows etc. that he would then marry her if we would take out a licence, which we did, and also bought a ring, and after smoking a pipe with Feron, John and Jos. Fuller and Aldersey at *The White Horse*, we came home about 11.30. I spent this journey as under:

To going through Peckham's			2d.
To the turnpike			6d.
Horses and ostler		1s.	1d.
Eating			6d.
Spent		2s.	0d.
Licence	£1	1s.	0d.
Ring		10s.	0d.
	£1	18s.	3d. . . .

Fri. 25 Feb. In the morn arose to go with the man and woman to be married, but was soon disappointed by the man's declaring he would not be married. We then concluded to go to Uckfield and put an end to so troublesome an affair. Our company was Mr. French, Mr. Vine, John Watford, Stephen Clinch, the man, and myself. We called at Mr. French's and my mother's in our road to Uckfield, but did not stay. When we came before Mr. Courthope, he discharged the man without examining him, but assured us he would take care that we should have proper security and that he would grant us some of our expenses since the man prevented our taking the parish security a-Thursday morning or rather a-Wednesday night; that is, as to the superfluous expense he has put us to except the licence and ring. We went to Mr. Halland's at *The Maiden Head* where we spent 18d. . . . We called at my mother's and Mr. French's as we came home, but did not stay. We came home thoroughly wet and fatigued about 6.20 and not a little overjoyed this troublesome affair is over . . .

Sat. 26 Feb. At home all day . . . Oh! a terrible time—took but £5 this week . . .

Mon. 28 Feb. . . . About 9.30 Geo. Hyland sent word to us that he would be married today if the parish stood to their agreement as before. Whereupon Mr. French, Joseph Fuller and Mr. Porter agreed to it. Whereupon I and Stephen Clinch and Tho. Fuller went to Laughton to get the parson ready, but he was not at home. Then Tho.

Fuller went back for Mr. Porter. Jos. Fuller walked to Laughton along with Geo. Hyland and Ann Durrant, and Mr. French went to call the clerk. After we had stayed at the church some time, Mr. Porter came and married them in the presence of Mr. French, Jos. Fuller, Robt. Durrant, myself and several people as ringers in the belfry. Stephen Clinch was father to give her away. Step. Clinch and Joseph Fuller signed the register. They was married by 11.35. I paid Mr. Porter for

Mr. Shenton's fees	10s.	6d.
do. the clerk	2s.	6d.
do. the ringers	2s.	11d. . . .

Tues. 1 Mar. In the morn Mr. French, John Watford, George Hyland and myself set out for Uckfield in order for him to swear his parish. We called at Mr. French's but did not stay. We then proceeded on our intended journey as far as my mother's, where I took my mare and rode to Uckfield, where we all met again and went to Mr. Courthope's, who examined the man and he swore his parish was Laughton. Mr. Courthope then granted us an order to carry him and Ann his wife home to their said parish of Laughton . . . we then parted, Mr. French, Hyland and Watford going to Hoathly . . . and I to Lewes to get the order signed, which I did by Luke Spence Esq . . . Then Mr. French and I went and delivered Hyland and his wife to Master Goad, one of the overseers of Laughton, who received them without any sort of dispute[13] . . .

Fri. 4 Mar. . . . This day Dame Vinal brought me a summons from Mr. Courthope to appear before him tomorrow morning by 10 o'clock to show cause why we use the poor so hardly.[14] In the even we had a public vestry when it was agreed by the consent of the majority of the said vestry to make a poor rate for the relief of the poor at the rate of 3s. 6d. to the pound . . .

Sat. 5 Mar. In the morning went up and called Master Piper and then proceeded on our journey to Uckfield to meet Dame Vinal . . . When we came before Mr. George Courthope, I represented the truth of Dame Vinal's affair to him and convinced him we were not hard to

[13] The frenzied activity over Ann Durrant and George Hyland which had occupied so much of Turner's time since 19 February illustrates well the enormous expenditure in the time of parish officers and in money from the rates necessary to ensure that a young unmarried pregnant girl should not become a charge on the parish. Any subsequent burden in this case would fall on the rates of Laughton parish. To achieve that end East Hoathly parish had had to pay the bridegroom £6 10s. 0d., give him five stone of beef, provide him with a marriage licence and a ten-shilling ring for his wife, and foot the bill for a wedding dinner.

[14] See below, p. 94, note 17.

our poor, or at least not in this point as Dame Vinal went about. Mr.
Courthope reprimanded her very much for coming to him with such a
heap of lies, and told us he would grant her nothing unless we liked to
give her daughter's child a pair of shoes, which I assured I would. We
spent at Uckfield 12*d.* and gave Mr. Courthope's clerk 12*d.* for his
signing our poor book . . . This day Mr. Sam. Durrant of Waldron
delivered a certificate to me for Elizabeth Day and Ann her
daughter[15]. . .

Mon. 7 Mar. . . . This day Tho. Daw came to me and informed me
if the parish would not find him a shop to work in or allow him 2*s.* 6*d.* a
week, he would make his complaint to a justice. I then assured him I
would call a vestry next Sunday and represent his complaint to the
parish . . .

Thurs. 10 Mar. In the morn two men from Willingdon came to my
house in order to take mine or my mother's shop . . .

Fri. 11 Mar. After breakfast the two men went away. The person
which wants to hire the shop, his name is Streton, about 22 years of
age, and I conject never in this trade nor any other, one who has seen
most parts of England and read a great deal and is extremely gay; but I
believe he is in reality a very indifferent scholar; though to do him
justice, I believe him to have a fine genius and an extensive capacity to
attain to learning, had he but an opportunity. I doubt he has drawn in
some of the Romanish principles by being intimate with a Jesuit. The
proposals I made him were that after my mother and I had took about
14 days to consult of it, I would inform him which of the shops we was
fixed on to part with, but be it which it would, I would part with it on no
other terms but by his taking the goods all at prime cost, and all the
fixtures, and the money to be paid down . . .

Mon. 14 Mar. . . . In the even went down to Jones's, there being a
public vestry, the result of which was that the parish should hire a
blacksmith's shop for Tho. Daw as soon as they can . . .

Weds. 16 Mar. . . . After breakfast I set out for Burwash on Tho.
Daw's mare in order to inquire after a blacksmith's shop for him and
which the parish had been informed there was one to let. But when I
came to Burwash, I found the mistake, there being none to let . . . I
heard at Burwash today that Admiral Byng was shot at Portsmouth on

[15] This settlement certificate, dated the previous day, is amongst the East Hoathly
parish records in E.S.R.O. PAR 378. It superseded the note obtained on 31 January.
The certificate, in the event, proved a burden for East Hoathly, since Elizabeth Day, a
widow, was soon to become pregnant again, by Robert Durrant. The pregnancy led East
Hoathly into a dispute with the parish of Waldron which Turner found unpleasant. See
below, pp. 108–9, 110, 114–6, 125.

Monday last, pursuant to the sentence of the court-martial held on him from the 27th December to the 27th January, though in my own private opinion I think him not that guilty person as many represent him, neither do I think it a prudent thing for him to be executed; but I suppose there was no calming a clamorous and enraged populace without taking away the life of this man, though if he is an innocent person, I think innocence should more than balance popular clamour.[16]

Sat. 19 Mar. . . . This day went down to Mr. Porter's to acquaint them I had got the raisins they gave me orders to get and also to inform them of the price, when Mrs. Porter, good woman, was so humane as to tell me she thought I ought in reason not to charge any profit for them, not, as she said, that it was her desire I should lose the carriage, but only my profit, which undoubtedly must be a very modest request at a time when money is so scarce and those raisins amounting to only £3 1s. 0d. and which I am assured I shall be paid for in about 10 months (though I must pay in 6 months). But all her humane dispositions and fallacious speech could not influence my obdurate heart to part with the raisins for less than the price I first asked . . .

Tues. 29 Mar. . . . In the afternoon Dame Martin came to our house and informed me that the man William Tull, whom Sarah Vinal has had a child by, was at their house today, upon which John Watford, Tho. and Joseph Fuller and myself went in pursuit of him; Tho. Fuller went round by Chiddingly and called at both the public houses there and then came to us at Edward Martin's. We then went and took a view of who there was at both the ale houses at Whitesmith, but found no one that answered to our description of him. We then proceeded to Mr. Will. Funnell's (where the man had formerly worked) who informed us that it was likely he might be at Mr. Guy's (a place where he had also worked) and accordingly we found him there in bed, whereupon, telling him our business, he very readily went along with us. In our road home we called at Mr. Funnell's and stayed about 1 hour when we proceeded on our journey home where we arrived about 11.15 and supped at our house, and Tho. Fuller, John Watford and my own family sat up all night. About 2 o'clock the man made an attempt to get away, but was disappointed. We then called up Thomas Durrant, who sat up with us the remaining part of the night. N.B. I spent 5d. in pursuit of the man.

Weds. 30 Mar. In the morn about 4.30 I went down to Whyly and

[16] Admiral Byng was executed on Mon. 14 March. For Turner's interest in the case see above, pp. 58, 73, 85.

called up Mr. French and informed him of what I had done. He and Sarah Vinal, the man etc. breakfasted with us. After breakfast, the man agreeing to marry Sarah, we then set off for Lewes in order to get a licence, *viz.*, the man, Jn. Watford, Jos. Fuller Jr. and myself and Mr. French, who met us at Mr. Sam. Gibbs's. We accordingly got the licence at Mr. Mitchell's, Mr. French being the bondsman for Tull. Also Mr. French was sworn that the girl had lived in this parish for 4 weeks last past, and the man also that his parish was Newport in the Isle of Wight. We dined at *The White Horse* on a shoulder of mutton roasted, a piece of pork boiled, greens, a butter pudding, cake and pickles . . .

Thurs. 31 Mar. In the morn Sarah and Mary Vinal breakfasted with us, and after breakfast the man and Sarah was married (I being father to give her away) in the presence of a great many people. Mr. French and myself signed the register book. After staying at Jones's some time, Mr. French, Mr. John Vine Jr. and Tho. Fuller came home with me and dined at our house on a piece of beef boiled, a piece of pork, a currant pond butter pudding and turnips. After dinner we went to Lewes with the man, for him to swear his parish. Mr. John Vine Jr., Jn. Watford, Tho. Fuller and Tho. Davy walked along with the man and I rode upon Mr. Vine's mare. We went to Luke Spence Esq, and before whom he swore himself on the parish of Thatcham in the county of Berks, and I accordingly took out an order to remove him and Sarah his wife to the said parish. Mr. Spence sent to Lewes for Mr. Fagg, who came and signed the said order, and while the order was a-making out, I rode up town, sending the foot people home. I called at Mr. Spence's as I came back and took the said order and examination . . .

Fri. 1 Apr. After breakfast I went down to Jones's where I stayed till near 12 o'clock, when we all went down to Mr. French's in order to set the people off to carry home the man and his wife. They set off about 1.20, *viz.*, the man and his wife and child, Mr. French, Jos. Fuller, Tho. Cornwell and John Durrant. I came home immediately and dined on the remains of yesterday's dinner . . . Gave Mr. French in cash £5 5s. 0d. to bear the expenses of the journey[17] . . .

Sat. 2 Apr. . . . In the afternoon a great quantity of snow fell, but did

---

[17] The Vinal family was a sore trial and a financial burden to the East Hoathly parish authorities. In March Dame Vinal had complained to the justices about the niggardliness of the parish's provisions for poor relief. Now one of her daughters put it to great expense organizing her marriage to the father of her illegitimate child and removing them both to Thatcham. Later in the year (see below, pp. 117–9) another daughter, Mary, already the mother of an illegitimate child, precipitated another parish crisis by declaring herself pregnant again.

not continue, it melting as it fell. Oh! how pleasant has this day been for what some of the last past have, they being spent in hurry and confusion, but this at home in my business and in reading! Oh! were I to choose my way of life, it should be a retired recluse life, making those about me happy if it were in my power.

Mon. 4 Apr. . . . Went up to Mr. Piper's and received of him £4 11s. 0d. for poor tax and paid him 14s. 1½d. on the parish account; *viz.*, 6s. for removing John Durrant, and 8s. 1½d. for 39 lbs of cheese for Dame Burrage. I stayed and supped at Mr. Piper's and smoked a pipe and came home about 9.20. As I came home, I found Tho. Cornwell and John Durrant came home from their journey and brought bad news, *viz.*, that they let the man make his escape, but they left Mr. French and Jos. Fuller to proceed on their journey with the woman . . .

Sat. 9 Apr. In the morn my brother and myself walked over to Framfield where we arrived about 7.15, where, after staying some time, and tying up some hat-bands, and telling and taking an account of the gloves and breakfasting, about 10.10 we set off in order to serve the funeral of Mr. John Cornwell, at which house we arrived about 11 o'clock. We had served the funeral about 3.40 when we set out for Buxted Church and buried him. Mr. John Cornwell was 77 years of age and a hearty man almost to the last, only afflicted with the stone. We gave away at the funeral 106 pairs of gloves and 8 hatbands[18] . . .

Easter Mon. 11 Apr. . . . I made up my accounts with the parish and there remains due to me £25 14s. 10½d. The officers chosen for the year 1757 were: myself, churchwarden, Jos. Fuller, electioner; Will. Piper, overseer, Ed. Hope, electioner—but as it was proved Will. Piper had served it very lately, it was agreed Ed. Hope should serve it.

Tues. 12 Apr. . . . Jos. Burges . . . agreed to be put in the poor book as electioner for overseer . . .

Fri. 15 Apr. . . . Read in the day part of Burkitt's *Poor Man's Help or Young Man's Guide*, which I think the best book I ever read of the size.

Sun. 17 Apr. . . . Called a vestry to consult about the overseers that were nominated on Monday last, they both declaring they will not serve it, but (as is the custom of our vestries) we came to no resolution concerning it[19] . . .

Fri. 22 Apr. In the morn I went down to Mr. French's in order to

[18] For Turner's activities at funerals see above, p. 7, note 5, and p. 77, note 1.

[19] In the event Turner went to Maresfield the next day to get his accounts as overseer for the previous year countersigned by the justices, and to receive the summons for his successor in office. He gave it to Joseph Burges who received it without objection and acted as overseer for the following year.

call him to go to Lewes to get two orders confirmed; *viz.*, Tull's and Hyland's, where after staying some time, we proceeded on our journey to Lewes. We dined at Mrs. Virgoe's on a roast loin of mutton . . . Bought for my mother 3 lb of gunpowder, which cost me 4*s.* 6*d.* . . . We got both our orders confirmed and set out for our road home about 7.20, both very much in liquor, and lost ourselves in the Broyle where we walked some time, though not without disputing whose fault it was that was the occasion of our mistaking the way. But we at last found our way to Will. Dicker's, where we found Dr. Stone and Richd. Savage, both very drunk; and we then fell out very much insomuch that I think Dr. Stone and I was a-going to fighting, but I cannot recollect on what account unless it must be that we were both drunk and fools . . .

Sat. 23 Apr. After I came away from Dicker's, I walked the Broyle until near 5 o'clock when I got out of it and went to Tho. Cushman's and lay down on their bed till about 10.25 when I arose and breakfasted with them and came away home. I called at Mr. Sam. Gibbs's and dined on a knuckle of veal, a piece of pork and greens, my family at home dining on the remains of Sunday's and Tuesday's dinners with the addition of some boiled tripe. I came sober about 3.35, and may I once more implore the most high God to give me grace to strengthen my weak resolutions that I may never again be guilty of this detestable sin. Oh! how doth the repetition of it make, as it were, my blood chill in my veins! I am quite distracted with anger at my own folly, but where can I run or go from the presence of a wounded conscience? But Oh! may I once more strive never, no, never to be guilty of this vice! I think, as I find my brains so weak, I will never drink anything stronger than small beer or water. In the afternoon Mrs. Fuller and her two daughters drank tea with us. I spent this journey, as near as I can recollect, about 4*s.* 6*d.* Oh, cruel is my misfortune (that I cannot bear the least matter of liquor, that is)!

Fri. 29 Apr. At home all day. My wife and I papered up about 140 papers of tobacco . . . This day there happened a very melancholy affair, *viz.*, one James Elless, a native of Chiddingly, but now kept a public house at Eastbourne, who, going to London on Sunday or Monday last, was there on Tuesday morn seized with violent fits (though what he had sometime been accustomed to); and Smith's team putting up at the same inn where he was, and coming from the same place, they took him into their wagon, and about half a mile before they came to our street, he was again taken with a fresh fit, and so he continued till they got him into John Jones's, where he expired in a few minutes. He was a young man of about 35 years of age, just a-going to be married. Oh! what a lesson is here, for mankind to prepare for death, when we have a scene now before our eyes of one cut off just in his full strength and almost, as it were, suddenly . . . A very melancholy time: corn and all other provisions being prodigious dear, *viz.*, wheat 10*s.* a bushel, barley 5*s. do.*, oats 3*s. do*, pale malt 5*s.* 3*d. do*, beef 2*s.* a stone, mutton 3½*d.* a pound, veal 3½*d. do.*, cheese 4*d. do.* . . . .

Mon. 2 May. . . . In the even my brother came over to our house in order to go to Lewes tomorrow to have Mr. Snelling's advice on his eyes, he being almost blind.

Tues. 3 May. . . . About 8.10 my brother and I set off for Lewes on foot where we arrived about 11.30. But as we found Mr. Snelling not at home, we was obliged to stay . . . About 7 o'clock Mr. Snelling came home and cut my brother an issue in his temple . . .

Weds. 4 May. We breakfasted at Mr. Thomas Scrase's and dined at Mr. Roase's on 2 pikes roasted and some veal cutlets . . . We came home about 7 o'clock, but not sober. Now am I resolved never more to exceed the bounds of moderation. Spent this journey, though I went entirely for company for my brother, 2*s.* 9*d.* My brother stayed all night in order for me to look after his issue.

Sat. 7 May. . . . Worked in my garden part of this day. In perusing an abridgment of the *Life of Madame de Maintenon* in *The Universal Magazine* for March, I find the following, being the last advice given her by her mother, Madame d'Aubigné: to act in such a manner as fearing all things from men and hoping all from God.[20] In the afternoon my brother went home and my nephew along with him, his mother being at Framfield and I understand has been there ever since Thursday. I must say I think it somewhat odd for her to send there for the boy without their either sending for me, my wife, self or Philip, or even her not coming over herself without sending word she would come one day next week if we would send word which day we would have her come. My relations seem all to look upon me with a coldness that I cannot but take notice of, and for a reason to attribute their so doing to, I am at a loss to find. I can only say this, that if I know my own heart (which I may not and therefore err, for as the prophet Jeremiah observes: 'She is deceitful above all things and desperately wicked'), I never think myself more happy than when I think myself a-serving any of my relations. Though undoubtedly by my so doing and their unkind

---

[20] *The universal magazine of knowledge and pleasure*, XX, p. 117. Turner is quoting (accurately) from an article entitled 'The Life of Madame de Maintenon. Extracted from a work, just published, intitled, Memoirs for a History of Madame de Maintenon, and of the last Age'.

retaliations, I am greatly injured and, I may say, impoverished. For to speak the truth, they seem to play at the game of outwit with me, or at least to take advantage of good nature or, if not that, of a simple and easy supineness or negligence. But their natural affection must, I think, so far bear sway that my ill-usage must proceed in part from their acting without thought or premeditated design, and from false and groundless chimeras formed in my own brain.

Sun. 8 May. ... This day have I taken up a resolution, GOD being my helper, to live a sober, virtuous and pious life, which God grant I may do, through the mercies and intercession of my blessed Saviour and Redeemer, Amen. At home all day ...

Weds. 11 May. ... This day there was a cock match at Jones's between Mr. Clinch and Watt. Dicker, but I did not go down.

Thurs. 12 May. ... This day our parish and Chiddingly played at cricket at Broad Oak, when ours was beat, Chiddingly having 8 wickets to go up.[21]

Tues. 17 May. ... This day a game of cricket was played on Hawkhurst Common between this parish and Chiddingly which was won by our parish, the two last men being in[22] ...

Weds. 18 May. ... This even as I stood in my court I see a star which I imagine to be a comet, it being a star of a different kind from any I ever saw, and as there is a comet daily expected, I conject this to be it.

Fri. 20 May. ... This day went down to Mr. Porter's to inform them that the livery lace was not come, when I think Mrs. Porter treated me with as much imperious and scornful usage as had she been what I think she is, that is, more of a Turk or infidel than a Christian, and I an abject slave. N.B. If Mrs. Porter is neither Turk nor infidel, I am sure her behaviour is not Christian, or at least not like that of a clergyman's wife.

Sun. 22 May. ... After dinner my wife and I went to the funeral of Master Marchant where I gave away the following glazed lamb gloves:

Mr. Porter—to be sent	John Vine Jr.
Master Baker	Dame Marchant, widow
Mr. Elless	Mrs. Marchant
James Marchant	Sarah Marchant
Will. Eldridge	Eliz. Marchant
Richd. Marchant	Hannah Marchant

[21] See above, p. 9, note 9. *The Cricket Quarterly* loc. cit. makes nonsense of this entry.
[22] See above, p. 9, note 9.

James Marchant	Dame Cornwell
James Marchant Sr.	Elizabeth Mepham and
Tho. Marchant	myself

Myself and wife only at church this afternoon. There was a funeral sermon for Master Marchant from part of *Numbers* 23.10: 'Let me die the death of the righteous and let my last end be like his', from which words we had a very good sermon, though whether it was a funeral sermon, they that preached it and they that pay for it alone must know. For I believe most of the audience must think it to be a sermon made before the death of Master Marchant. At home all the evening . . . N.B. Master Marchant's age was 67 years.[23]

Mon. 23 May. . . . This day read in the *Gazette* of the 20th instant that the King of Prussia[24] had on the 6th instant gained a complete victory over the whole combined forces of Austria (near Prague), taking their whole camp and 250 pieces of cannon and 6 or 7000 prisoners . . .

Thurs. 26 May. . . . At home all day; at work in my garden. Nothing to do in the shop; a prodigious melancholy time; what to do I know not . . .

Tues. 31 May. . . . Thomas Davy and I in the afternoon and even played, I conject, above 50 games of cribbage and then left off just as we begun, having neither won nor lost. A very melancholy time; nothing to do; provisions extreme dear . . .

Thurs. 2 June. . . . This day I was admitted a member of the Friendly Society[25] at Mayfield, being put up by John Streeter, John Vine and Robt. Hook.

Sun. 5 June. My whole family at church this morn, *viz.*, myself, wife, maid and two boys. There was a brief[26] read for damage done to

[23] For Turner's activities at funerals see above, p. 7, note 5, and p, 77, note 1.
[24] Frederick II.
[25] Friendly Societies were beginning to multiply in the second half of the eighteenth century for the purpose of mutual help in times of sickness or death. They generally met in a public house where subscriptions were paid, and some cheer was provided from Society funds. There was normally an annual dinner and sometimes an annual sermon, and this was the case at Mayfield where Turner attended the annual feast and sermon on 26 Apr. in 1758 and again in 1765. The Friendly Societies were the only 'working-class' organizations excluded from the ban on combinations during the Napoleonic Wars though they were subject to increasing governmental control from 1793 onwards. By 1801 there were reckoned to be 7000 such clubs in England and Wales. Turner's sponsors were a little lower in the social scale than Turner. All of them were, however, from the middling ranks of society: the annual subscription meant that membership of such Societies was not for the poorest and therefore itself conferred some status. For the subsequent growth of such Societies see P. H. J. H. Gosden, *The Friendly Societies in England, 1815–75* (1961).   [26] For briefs see below, p. 158, note 37.

corn, grass and hops etc. in some part of the counties of Kent and Sussex by a violent storm of thunder, hail, etc. on the 7 July last, which damage in the county of Kent is esteemed (by the brief) to amount to upwards of £1800 and the damage in Sussex to upwards of £1300; so that the damage sustained being so great, the said brief is to be collected from house to house . . .

Thurs. 9 June. . . . Gave 6*d.* to the brief which was read a-Sunday.

Tues. 14 June. About 4.20 in the morn Master Durrant and I set out on foot for Lewes (today being the visitation at Lewes) where I arrived about 6.30, Master Durrant going to Ashcomb. I breakfasted with my brother at his lodgings. About 11 o'clock we went to St. Michael's church, the visitation sermon being preached there. The service was read by the Rev. Mr. Bristed, clergyman at Slaugham, and the sermon preached by the Rev. Mr. Jefferies, vicar of Ditchling. The text in *Matthew* 11.5: 'The blind receive their sight, and the lame walk, the lepers are cleansed, and the dead hear, the dead are raised up, and the poor have the Gospel preached unto them.' After churchtime and the clergy were all called over, I was sworn into my office of churchwarden (with a great many more), for which I paid 4*s.* 6*d.* I dined at *The White Horse* with 16 persons more on a knuckle of veal, bacon and greens boiled, a leg of lamb boiled and spinach, a rib of beef roasted and green salad and two pond currant puddings (my family at home dining on the best end of the shoulder of veal roasted in the oven with a batter pudding under it). After dinner we sat and smoked one pipe, and then I met with Mr. Harraden with whom I smoked another at *The White Horse*. I drank tea with my brother in company with Mr. Tucker and Will. Bennett. I came home about 9.10, thank GOD very safe and sober. Mr. Tho. Scrase brought me going on my road home as far as Ringmer Green and my brother and Mr. Tucker as far as the Broyle gate. I spent today . . . half a crown of which I am to be allowed by the parish:

In the morn before breakfast	
1 pt. of beer	1½*d.*
At *The White Horse* with Mr. French	
and Mrs. Stemp	9*d.*
To my dinner	1*s.* 0*d.*
After dinner	7*d.*
With Mr. Harraden and Mr. Tanner	6*d.*
With my brother and Mr. Tucker at	
The White Horse	1½*d.*

To mending a pair buttons   3d.

	3s. 4d.
Allowed	2s. 6d.
Spent on my own account	10d.

I called to see Mr. Tho. Scrase and Will. Bennett. This visitation was held by the Rev. Dr. D'Oyly, archdeacon of this diocese, who tendered the oath to the churchwardens.[27]

Thurs. 16 June. . . . In the afternoon borrowed a horse of Mr. Burges to go to Lewes upon for things for the funeral of Mr. Richd. Goldsmith, widower, late of this parish, but died this morn at the house of Master Gladman at Laughton and was aged 80 years . . .

Fri. 17 June. In the morn at work in my garden. Drew off in the forenoon 2 barrels of cider, and a-corking of it broke a bottle and cut one of my fingers prodigiously with it . . . Mr. Richd. Comber and Charles Diggens drank tea with us. The finger which I cut in the morn bled to such a degree that I tried all possible means to staunch it but could not till I applied to Mrs. Porter for some styptic . . .

Sun. 19 June. . . . I and Jos. Fuller and Mr. Stone went to the funeral of Master Goldsmith (I riding on a horse of Master Fuller's) where I read the deceased's will to the relations and by which will be constituted Mr. Jos. Fuller and Mr. Rd. Stone sole executors, and gave to them all that was at his own disposal. I served the said funeral and gave . . . in all 53 pairs men's and women's gloves. About 1.20 we set out from the house, *viz.*, from Master Gladman's. I rode home and left my horse and walked with the corpse to Waldron where we arrived about 3.10 just as the people was gone to church. We had a funeral sermon preached by the Rev. Mr. Hamlin from *Job* 5.7: 'Man is born unto trouble, as the sparks fly upward' . . . N.B. I think this to have been the merriest funeral that ever I served, for I can safely say there was no crying.[28]

Mon. 20 June. This is my birthday and the day in which I enter into the twenty-ninth year of my age, and may I, as I grow in years, so continue to increase in goodness that, as my exit must every day draw nearer, so may I every day become more enamoured with the prospect and happiness of another world that I may be entirely dead to the follies and vanities of this transitory world. We dined on the remains of yesterday's dinner. In the afternoon I walked up to the common with

[27] For the attendance of newly-appointed churchwardens at the archdeacon's visitation see *Tate*, p. 95.

[28] For Turner's activities at funerals see above, p. 7, note 5, and p. 77, note 1.

an intent to see a cricket match played between an eleven of the Street quarter and an eleven of the Nursery quarter,[29] but when I came there, they not having enough to play, so that I was constrained to play for one, which I did, and we had the good fortune to beat the Nursery eleven 72 runs.[30] I went down to Jones's with the rest of the gamesters and stayed till 11.15. I spent only my shilling as a gamester . . .

Weds. 22 June.   Received of Mrs. Sarah Mary Marchant £3 12s. 6d. in full for the funeral expenses of her father's funeral . . .

Fri. 24 June.   After breakfast the boys and myself walked over to Framfield, it being the fair day at Framfield . . . I spent at the fair 6d. I also saw there the largest hog I think that I ever saw. His height was about 4 feet. He was 9 feet 4 inches from the tip of the nose to the tip of the tail. His ears was 12 inches by 9. [He was] about 3½ years. In my absence my wife received of Mr. Richd. Stone and Mr. Jos. Fuller £5 10s. 6d. in full for the funeral of Master Goldsmith . . .

Mon. 27 June.   . . . In the even I walked down in the park to see Mr. Elless and Tho. Durrant run, which they did, and Mr. Fran. Elless was beat. This day I subjected my good nature or ignorance to be imposed on by policy or villainy, *viz.*, some time since I looked upon a horse of Mr. Vine's of Heathfield with an intent to purchase him; and the price he had some time ago been offered at was £9 9s. 0d. Now I, knowing nothing of a horse myself, entreated the favour of Joseph Fuller as a friend to give me his opinion, which he did—that he did not know but the horse might suit me, but he thought it too much money, being more, he said, than he should choose to give. But, however, he having business near the house today, he promised me to call and treat with Mr. Vine for me about the horse. Now what we agreed on this morn before he went out was that if the horse appeared to him as if he would do, he was to buy him for me at the lowest he could, and for so doing I was to give him a treat. Now! my worthy friend accordingly went, and instead of buying the horse for me, raped[31] with Mr. Vine in some manner and got the horse for himself and when he came home sent for me down and offered me the horse at £11 and no otherwise. And by reason I seemed to think it ill usage to have the horse bought out of my hands in that manner when I had entrusted him to buy him for me as a friend, he only laughed at me and counted it as a piece of wit and a sharp look-out for a man to serve himself when he can and his neighbour next. But if this usage is consistent to honesty, religion or

[29] The village XI against the Halland hamlet XI.
[30] See above, p. 9, note 9.
[31] Bartered.

anything else that should be acted by Christians, I am utterly at a loss
to know what is right and wrong. But this I think of it: that if Joseph
Fuller could with as much impunity defraud a man in any way
whatsoever, I shall make it my opinion he would do it. For I think
robbing a man on the highway is not a baser action in proportion to the
consideration than this. But still I do not envy him his talent of having
wit and a sharp look-out without honesty. Today Dame Cornwell
made us a present of some eels.

Tues. 28 June. . . . This day I was to give my answer in to Honest
Joseph Fuller about having the horse, for provided I had him not today,
I should not have him tomorrow under £11 11s. 0d. So in the morn I
had concluded to have the horse, but in the even, instead of the horse,
I bought a young colt of him at about 30 per cent too much. It is a
gelding, about 3 years old, 12 hands 3 inches high, never ridden, and
for which I gave him £5 6s. 0d. . . .

Weds. 29 June. . . . Just before dinner my brother came in, who
dined with us on some boiled mackerel, a kidney and some pork fried
and green salad. My brother's eyes being very bad, he wanted me to
walk with him to Mr. Snelling's at Alfriston to have his farther advice.
And accordingly about 2 o'clock we set off. We called at Crowhurst's
at Bayley's Lane and bought some earthenware; from thence through
Wilmington to Alfriston, where we arrived about 5.10 and was not so
unfortunate as to find him gone from home. We stayed and drank
1 glass wine and then we went to *The Star* and had a mug of beer and a
piece of bread. We then came home through Berwick, Selmeston,
Ripe, and so to Whitesmith where we went in and had a mug of beer.
We came home about 9.10. We spent in our journey 4½d. apiece. My
brother stayed all night at our house . . .

Thurs. 30 June. . . . Today in reading *The London Magazine* for May,
I find the following description of the comet that is shortly expected to
appear, *viz.,* that it has appeared 6 times already, *viz.,* in the years
1305, 1380, 1456, 1531, 1607 and 1682, and that it revolves about the
sun at the intervals of 75 and 76 years alternately, and since the last
period, *viz.,* in 1682, was 75 years, it is presumed the present period
will contain 76 years, and therefore its next appearance will probably
be in 1758. But the time of its appearing is uncertain, and it may
happen the latter end of the present year 1757, or the beginning,
middle or latter end of the next year.[32] After 85 days it will attain to its

[32] This was Halley's comet which in 1705 Edmund Halley had predicted would
return in 1758. His prediction was justified when the comet reappeared on Christmas
Day in that year. The article which Turner read—'An account of the remarkable Comet

perihelion, or be nearest of all to the sun, and after 130 days it will come to its descending node, at which time it will be very near the earth's orbit; and should that happen the 12th of May, we should then be in a dangerous situation as the denser part of its blazing tail would envelop the earth. It seems to be of those that rise to the least height from the sun, its greatest distance being only 35 times greater than the distance of the earth from the sun, so that at the farthest it does not run out four times farther from us than Saturn . . .

Fri. 1 July. . . . Mr. Snelling ordered my brother to be entirely debarred from beer, brandy (or any kind of spirits), and meat, and to drink the following for his constant drink, *viz.*, take 1 ounce of cream of tartar, ½ lb of lump sugar, the peel of a lemon; pour a gallon of boiling water on them and let it stand all night, then strain it off and bottle it for use. He also ordered him the cold bath, and blisters behind the ears to be perpetual, notwithstanding he has an issue both in the temple and arm.

Sat. 2 July. . . . This day Mr. Fran. Elless and Tho. Durrant ran in Halland garden a foot race 20 rods for 2*s.* 6*d.* each, which was won by Tho. Durrant, I suppose with ease (I not being there). I went half the bet with Tho. Durrant so that of consequence I won 15*d.*

Tues. 5 July. . . . After breakfast Mr. Burges and myself went to talk to Anne Jeffrey and the servant of Thomas Osborne,[33] who we had heard were both with child. But upon talking to them we have no reason to think so . . .

Thurs. 7 July. My mother's servant came over in the morn and breakfasted with us . . . James Marchant drank coffee with us in the afternoon on account it was my brother's birthday, who treated us with it, and entered into the 25th year of his age . . .

Sun. 10 July. . . . The Rt. Hon. Geo. Cholmondeley, Earl Cholmondeley, Viscount Malpas, Joint Vice-Treasurer of Ireland, Lord Lieutenant, Custos Rotulorum and Vice-Admiral of Cheshire, Governor of Chester Castle, Lord Lieutenant of Anglesey, Carnaervon, Flint, Merioneth and Montgomery, steward of the royal manor of Sheen in Surrey, and Knight of the Bath, being a-visiting at Mr. Coates's, was at church this morn.[34] We dined on a lamb's heart

---

whose appearance is expected . . . '—opened the May issue of *The London magazine or gentleman's monthly intelligencer*, XXVI (1757), p. 211.

[33] Thomas Osborne's servant's name was Mary Hubbard. In her case the parish officer's suspicions were well-founded. See below, p. 134, note 9.

[34] George Cholmondeley, third Earl of Cholmondeley (1703–1770). Turner seems to have copied his impressive collection of offices from some work of reference, perhaps Collins's *Peerage* of which he possessed a copy.

pudding, a piece of bacon, a lamb's head and the remains (which the cat left us) of a lamb's lights and the lamb's tongue and brains, carrots, green salad, gooseberry pie and custard . . .

Mon. 11 July. . . . My brother came over in the even in order to have 2 blisters laid on behind his ears . . . Read part of *The Universal Magazine*[35] for June wherein I find the following receipt recommended (in an extract from Dr. Lind's essay on the most effectual means of preserving the health of the seamen in the Royal Navy) as a specific against all epidemical and bilious fevers and also against endemic disorders. He first recommends a regular course of life and to abstain as much as possible from animal food, and to confine as much as possible to a vegetable diet. Then he orders about 2 ounces of the following tincture to be taken every day upon an empty stomach (and better if taken at twice):

8 ounces of bark,

4 ounces dried orange peel, which infuse in a gallon of spirits. But in my opinion it might be rendered a more grateful bitter if instead of the 4 ounces of dried orange peel there was put 1½ ounces of it and 3 ounces of undried lemon peel cleared from the white.

I also found an account of the following powder, called the Duke of Portland's gout powder:

For the gout or rheumatism:

Aristolochia Rotunda, or Birthwort Root
Gentian Root
Germander ⎫
Ground Pine ⎬ tops and leaves
Centaury ⎭

Take all of these well dried, powdered and sifted as fine as you can, equal weight; mix them well together and take one drachm of this mixed powder every morning, fasting, in a cup of wine and water, broth, tea or any other vehicle you like best; keep fasting an hour and half after it; continue this for three months without interruption. Then diminish the dose to ¾ of a drachm for 3 months longer; then to ½ a drachm for 6 months more, taking it regularly every morning if possible. After the first year it will be sufficient to take ½ a drachm every other day. As this medicine operates insensibly, it will take perhaps two years before you receive any great benefit; so you must not be discouraged though you do not perceive at first any great

[35] Turner quotes from two separate articles in *The universal magazine of knowledge and pleasure*, XX, pp. 250–54 and pp. 263–66.

amendment. It works slow but sure. It doth not confine the patient to any particular diet, so one lives soberly and abstains for those meats and liquors that have been always accounted pernicious in the gout, as: champagne, drams, high sauces etc. N.B.: In the rheumatism that is only accidental and not habitual, or that has been of long duration, then you must take it as for the gout. The remedy requires patience as it operates but slow in both distempers.

Mon. 18 July. . . . At home all day. In the even my brother came over to have 2 more blisters laid on behind his ears, which I accordingly laid on. In the afternoon drawed out part of a genealogical table of the royal family for Tho. Davy.

Tues. 19 July. . . . In reading Josephus's *Jewish Antiquities* I find his opinion was (or at least it was a prevailing notion in his time) that the earth was the centre of the planetary system . . . My wife very ill. Halland gardener made me a present of 2 melons.

Thurs. 28 July. About 6.20 in the morning I set out on my journey to Hartfield where I arrived about 9.20. After staying there and break-fasting, my brother, Sam. Slater and I rode to Broadstone to weigh up Mr. Martin's wool for my father Slater, which we accordingly did . . . Though I went to Hartfield purely to serve my father—for he knew nothing about wool, not even how many pounds there was either in a peck or tod of wool—but notwithstanding that—during the little stay I made at Hartfield, I think Mrs. Slater (for I am sure her usage to me was never like that of a mother) used me with abundance of reflecting scurrilous language, though what for I know not, but I think if there can be any parallel in this life with the miserable state in the future, it must be in living with a person of Mrs. Slater's unhappy temper, and more particularly so if she happens to be any ways related. Mrs. Slater might do well to sell oysters at Billingsgate, but to live amongst civilized people she must be an obstruction to theirs and her own happiness.

Sat. 30 July. In the morn my brother came over and breakfasted with us, and after breakfast he carried my wife to Lewes . . . At home all day, busy. Mr. Gibbs being a-drawing of the pond in the Street, he made me a present of 2 carp. I cannot say I think it prudent of my wife to go to Lewes now as I look for the Duke of Newcastle down at Halland next week and as I have several journeys to go which I must postpone on account of her going, whereby by business sustains a disadvantage. But, alas! what can be said of a woman's temper and thought? Business and family advantage must submit to their pride and pleasure. But though I mention this of woman, it may perhaps be as justly applied to men, but most people are blind to their own follies.

Sat. 6 Aug. . . . Down at Halland today 6 times. This day the Duke of Newcastle came to Halland . . .

Sun. 7 Aug. . . . In the morn my wife, self, nephew, Mr. and Mrs. Beard, Dame Weller and Master Weller and their daughter went down to Halland to see some turtles, and where we see 2, one of which weighed 367 lbs. . . . This being a public day at Halland, I spent about 3 hours there in the afternoon, *viz.*, from about 5 to 8 in company with several of our neighbours. There was a great company of people of all denominations from a duke to a beggar. Among the rest of the nobility were his Grace the Duke of Newcastle, the Hon. Lord Cholmondeley, Lord Gage, the Earl of Ashburnham, Lord Chief Justice Mansfield, Mr. Justice Denison and a great number of gentlemen.[36] My cousin Potter drank tea with my wife, as did my mother. My mother and brother went home about 7.30. I was down at Halland three times today. What a small pleasure it is to be in such a concourse of people, 1 hour spent in solitude being worth more, in my private opinion, than a whole day in such a tumult, there being nothing but vanity and tumult in such public assemblies, and their mirth being rather obstreperous than serious and agreeable! O, how silly is mankind to delight so much in vanity and transitory joys, which, instead of leaving a lasting joy behind, must only leave a sting to continually gnaw the conscience of us poor deluded mortals.

Mon. 8 Aug. . . . Down at Halland once. This day the assizes were held at Lewes before the Rt. Hon. Lord Mansfield, Lord Chief Justice of the King's Bench, and Mr. Justice Denison when 8 prisoners was tried, *viz.*, Tho. Sheather and Robt. Rutley for violently assaulting, beating and abusing Robert Dunk and Anthony Goddard on the King's highway, and taking from them 25 sacks of wheat, value £40; Sheather was acquitted and Rutley found guilty—death, the judge reprieved him before he left the town; Henry Boxall, concerned in a riot—to be transported for 7 years; Jn. Smith, Richd. Hopkins and Jn. Hawkins, concerned in riots—acquitted, nobody appearing against them; Jonathan Evans for stealing a bed-tick, bolster etc. was discharged by proclamation; John Ayling, attained at the assizes before, was ordered to be transported for 14 years.

[36] The Duke of Newcastle's guests were Lord Cholmondeley (for whom see above, p. 104, note 34); William, Viscount Gage, who subsequently became Baron Gage of Firle in Sussex (1718–1791); John, Earl of Ashburnham (1724–1812) who was Lord Lieutenant of Sussex; William Murray, Earl of Mansfield (1705–1793) who was Lord Chief Justice of the King's Bench; and Sir Thomas Denison, Judge in the King's Bench (1699–1765).

Fri. 12 Aug. . . . This day being the first race-day at Lewes, about 4.25 my sister Ann Slater and I (upon a horse borrowed of Mr. French) rode to Lewes where we arrived just as the people came from the hill. We went in to see the ball, which, in my opinion, was an extreme pretty sight. I went down and smoked one pipe with Mr. Geo. Verral in company with my brother Moses and Will. Bennett. My sister and I came home about 2 o'clock . . . This day the King's Plate of £100 was run for on Lewes Hill by Mr. Warren's horse Careless and Mr. Rogers's horse Newcastle Jack. It was won by Careless, the other being [with]drawn after the first heat. 'Tis said there were £100 laid by the grooms that Careless beat the other six-score yards, which he did . . .

Sat. 13 Aug. . . . This day the Subscription Plate of £50 was run for on Lewes Downs when 4 horses started, *viz.*, Lord Eglinton's bay horse Nothing, Robert Smith's black mare Frolic, Mr. Dye's bay horse Looby, and George Rogers Esq's grey mare (or hermaphrodite) Chastity; which was won by Chastity, she winning the two first heats. The two first were distanced, and the heats were very good between the other two.

Mon. 15 Aug. In the morn my brother came over and breakfasted with us, and after that we packed up my wool, with the assistance of James Marchant's man . . . My brother and James Marchant's man dined with us on a loin of mutton roasted in the oven with a plain batter pudding under it, with the remains of yesterday's dinner. After dinner my brother and I rode over to Framfield, and, with the help of Charles Diggens, packed up my mother's wool . . .

Tues. 16 Aug. . . . Sent the following wool to Messrs. Margesson and Collison for them to sell for me:

	My own wool Fleeces	Tod	lbs.
No. 1	118	7	29
2	192	13	--
	My mother's		
3	179	13	--
4	115	7	7
5	187	13	16
	791	54	20

with 5 cloths at 23*s.* 6*d.* on my own account . . .

Weds. 17 Aug. . . . In the even Mr. Burges and I walked over to Uckfield to ask Mr. Courthope's opinion whether we could remove Elizabeth Day, who we understand is with child, but at the same time we have a certificate with her from Waldron. His opinion was that we could not remove her without she first asked for relief . . .

Mon. 22 Aug. . . . About 11 o'clock I went to Framfield where I dined with my mother on some cold beef and French beans . . . From my mother's I set off to Piltdown where I saw Charles Diggens and Jn. Fowle run 20 rods for 1 guinea each, which was won by Diggens with ease. I got never a bet, but very drunk. I lay at my uncle Hill's all night. I find one glass of liquor disorders me so that I must and will, I am determined, leave off drinking anything strong . . . I was so much in liquor I have forgot what I spent.

Tues. 23 Aug. Came home in the forenoon, not quite sober. Mr. Evans dined with us on a coast of lamb roasted and cucumbers. At home all day, and I know I behaved more like an ass than any human being, so doubtless not like one that calls himself a Christian. Oh, how unworthy are I of that name! But let me hope and endeavour with the grace of God to get the better of this and that I may every day become a better Christian. The uneasiness that have so long subsisted between me and my wife now, I believe, are ended[37] . . .

Sat. 27 Aug. . . . In the afternoon Mr. Elless and I walked down to Peter Adams's in order to see Mr. Relfe the surveyor at Lewes[38] measure and map the said land, but could not find him. We was wet through a-going our journey. Dame and Master Durrant and Tho. Durrant at our house in the evening drinking what they call colt ale; that is, liquor for shoeing a colt the first time. This day, I think, has been as wet a day as I ever knew in my life at this season of the year. There have been today several very loud strokes of thunder.

Sun. 28 Aug. . . . My whole family at church this afternoon; the text in *Jeremiah* 17.7: 'Blessed is the man that trusteth in the Lord, and whose hope the Lord is.' And I think we had two as good sermons today as I ever heard Mr. Porter preach. Tho. Davy at our house in the

[37] A leaf of the original manuscript is missing at this point.
[38] F. W. Steer lists maps of Ticehurst in 1736 by John Relf, and of South Malling *c*.1750 by Relfe of Lewes. *A catalogue of Sussex estate and tithe award maps.* (*S.R.S.*,LXI, 1962), p. 42; and *A catalogue of Sussex estate maps, West Sussex inclosure maps, West Sussex deposited plans, miscellaneous and printed Sussex maps* (*S.R.S.*, LXVI, 1968), p. 12. I am informed by Mr. Christopher Whittick, however, that the latter map must, on internal evidence, be after 1760 and, intriguingly, the hand is remarkably similar to Turner's. John Relf is known as a Lewes plumber, glazier and painter (*Sussex Weekly Advertiser*, 19 Sept. 1744) who lived next door to the Rands who were schoolmasters and surveyors. See W. H. Godfrey, 'The seatholders of St. Michael's Church, Lewes, in 1753 . . . ' *S.N.Q.*, 1 (1927) and J. H. Farrant, 'Civil Engineering in Sussex around 1800, and the career of Cater Rand', *Sussex Industrial History*, 6 (1973/4), p. 4. I am grateful to Mr. Farrant for providing me with these references.

even, to whom, and in the day, I read 4 of Tillotson's sermons. This day while we were at church in the forenoon, we had the highest wind that I ever knew.

Fri. 2 Sept. . . . In the even wrote out a list for John Watford of people in this parish properly qualified to serve in the militia, *viz.*, all such as are between the ages of 18 and 50, parish and peace officers and apprentices excepted, which in our parish, according to the said list amounts [to] 45 persons. I also wrote him out a list of the persons properly qualified to serve on juries in this parish, which are but 6.[39] In the evening read one of Tillotson's sermons.

Sat. 3 Sept. About 6.45 Mr. Jos. Burges and I set out for Lewes on foot to ask advice whether we could remove Eliz. Day, a certificate person to our parish (but now big with child), but none of the justices being in town, their clerk, Mr. Ed. Verral, informed us that we could not remove her. But the child, though born a bastard, would not belong to this parish . . . In the even Tho. Davy at our house, to whom I read a sermon preached by the Rev. Mr. James Hervey, A.M., rector of Weston-Favell in Northamptonshire, being preached on some of the late fast days. I bought 3 of them today at Lewes, being lately published and stitched all together . . .

Weds. 7 Sept. In the forenoon I set out with Mrs. Virgoe on one of her brother's horses to Mr. John Burges's at Brook House in Rotherfield in order for her to ask his advice in regard to selling and letting her house at Lewes . . . Mr. Burges has a sister to keep his house, whose name is Jael, and I think the greatest oddity I ever saw. It's her misfortune to be deformed by nature; i.e., squint-eyed, a great stammering in her speech, and very much on one side. But yet her greatest misfortune is in that of her unhappy temper, for she even appears to [be] so miserable that we may justly say she is poor in the abundance of riches . . .

Mon. 12 Sept. . . . After breakfast Mr. Burges and I went down again to talk with Osborne's servant about her being big with child, but she would give us no satisfactory answer . . . Very busy all the afternoon a-gathering of pears, gathering and shaking down near 20 bushels . . .

Weds. 14 Sept. In the morn Mr. Burges and myself went down to Bean's at Whitesmith with about 20 bushels pears, of which I had made 50 gallons of perry, which at 1*d.* a gallon amounts to 4*s.* 2*d.*, but

---

[39] The principal qualification for jurors was to possess freehold, copyhold or customary land to the value of £10 *per annum* or leasehold land to the value of £20 *per annum*.

having no silver, I could not pay Mr. Bean. I also received of Mr. Bull
3 plugs, 3*d*. . . . Mr. Burges, in gratitude for favours he says I have
conferred on him, gave me the carriage of the pears . . .

Fri. 16 Sept. In the morn my brother came over and breakfasted
with us, and after breakfast my wife and I set out for Pevensey where
we arrived about 11.05 in order to consult Mr. Breeden about hiring of
a shop at Ninfield, in this county. My reason for being so anxious to
leave this shop is its being an extreme dear rent, *viz.*, £8 per year, and
so very bad in repair that my goods sustain a great deal of damage
thereby. And though I mention but £8 per year, the rent, I am obliged to
hire a stable, warehouse, cellar and 2 seats in the church. And I
observe that almost all the people in the parish seem to be growing
poor and are so long [to] pay that no tradesman I am assured can bear
it, for even the best will not pay above once a year. Then living so near
my friends is I think a very great disadvantage, though I should be
willing to lend them all the assistance that is in my power and help
them in anything that I can. But they seem to study nothing but self-
interest, or if they have no design to take advantage, it turns out very
much to my disadvantage. Whether it is with a premeditated design I
cannot tell, but I must rather think from a natural covetous disposition
and want of proper thought; and another motive, which I think none of
the least, is I shall never be able to get in my debts so long as I tarry at
this place . . .

Sat. 17 Sept. . . . I am come to a resolution, as I am so continually
almost troubled with the inflammation in my eyes, to leave off during
life (unless anything very material should intervene) eating any sort of
meat, unless sometimes a bit of boiled lamb, mutton, or veal, or chick,
or any such harmless diet; as also to refrain from all sorts of strong
liquor; and to continue not eating any supper, from eating of which I
have debarred myself for the general part this 12 months.

Sun. 18 Sept. My whole family at church in the morn, that is,
myself, wife, maid and 2 boys . . . We dined on a piece of boiled beef, a
currant suet pudding and carrots . . . We have had I think two extreme
good sermons this day preached unto us . . .

Mon. 19 Sept. In the morn pretty busy. After breakfast John
Watford and myself set out for Selmeston Fair. I went in order to talk
to Dr. Snelling concerning my eyes being very bad. We arrived at the
fair about 12.30, where, after walking about till near 6 o'clock, Mr.
Piper, myself and Jn. Watford came home; we arrived about 7.20. I
paid Thomas Bean at the fair 4*s*. for making out my perry the 13th
instant. I think this to be as pleasant and large a fair as I ever saw in the

county of Sussex, there being great numbers of sheep and cattle and a great number of people . . .

Mon. 26 Sept.   In the morn Master Piper and Master Durrant came along with an 'Here, we have got a job of writing we want you to do.' Which I readily did, and when I had wrote out a sheet of paper and wrote about an hour, they sneaked away without so much as ever offering to pay for the paper, or I think to say they were obliged to me for doing it. But what can we expect from two such covetous and miserable wretches, whose whole delight is self-interest? The only thing that I know of is that they must remain as they are, and God bless them . . . In the afternoon I rode out a-money-catching, but got none. I came home through the park where I stayed and saw part of a game of cricket.[40] Gave some seamen with a pass 6*d.* At home all the even.

Tues. 27 Sept.   . . . In the forenoon busy a-writing for John Watford, but he pays for paper and is almost ready to smother me with thanks and does promise to serve me as far if opportunity offers . . .

Weds. 28 Sept.   In the morn I set out for Pevensey where I arrived about 10.20, and after staying about half an hour, Mr. John Breeden and I rode to Ninfield in order to hire the shop there, but when we came to Ninfield, we could not hire it, the shop not being to be parted with as yet. Mr. Breeden and I dined at *The Spread Eagle* on some mutton chaps (as did my family at home on a duck pudding and turnips). We stayed at Ninfield till about 3.40 when we parted, and thanks be to God I got home very safe and sober about 6.55. I spent this day 2*s.*, *viz.*, the turnpike 2*d.*, our dinner (i.e. both) 9*d.*, horses and beer 1*s.* 1*d.* . . .

Fri. 30 Sept.   In the morn walked up to a sale of the late Master Goldsmith's, where, after staying an hour or two and buying an handbill and a pot-hook for 8*d.*, I came home . . . In the evening my brother came over who stayed all night. Paid Mr. Burges 16*s.* 6*d.* for 1 brass pot which he bought at the sale, but afterwards had a dislike to it. I thought I had bought it at the price of old bell brass, but I doubt I am deceived, which if I am confirms the old proverb that covetousness never brings anything home . . .

Tues. 4 Oct.   . . . Paid Mr. Will. Kemp[41] for a silver seal, the 2 initials of my name ciphered on it, 7*s.* 0*d.* . . .

Weds. 5 Oct.   After breakfast Joseph Fuller called me in order to go to the sale at Bentley, and we accordingly went. As we went by Peckham's at Terrible Down, there we met with Mr. French and Mr.

---

[40] See above, p. 9, note 9.   [41] William Kemp was a watchmaker at Lewes.

Sam. Gibbs, both as seeming happy as could be over a dram of gin. We stayed there and spent 4½*d.* though I drank nothing but a little mild beer. We then went to the sale where I bought a brass candlestick which cost me 8*d.* We came home by Peckham's, where we stayed and spent 1½*d.* in company with Mr. James Shoesmith. I came home and dined on the remains of what my family left, *viz.*, some boiled plaice. Dame Dallaway, buying many things, went in and drank tea with us. In the even read part of the 5th volume of *Medical Essays and Observations*, published at Edinburgh by a society of physicians.

Thurs. 6 Oct. . . . After dinner our servant went to Blackboys Fair and stayed all night. There was a public vestry held at Jones's, to which I went; but not being very well I did not stay to conclude upon anything, but came home immediately. Tho. Davy at our house in the even and he and I played a few games of cribbage and I won 1*d.* This day how are my most sanguine hopes of happiness frustrated! I mean in the happiness between myself and wife, which hath now some time been continued between us, but, Oh, this day become the contra[ry]! The unhappiness which has, almost ever since we were married, been between us has raised such numberless animosities and disturbances—and amongst our friends—that I think it hath almost brought me to ruin. What the causes of it is I cannot judge. I cannot judge so ill of my wife as to think she is only to blame, and I think I have tried all experiments to make our lives happy, but they all have hitherto failed of their end. I can see nothing that so much contributes to our unhappiness as an opposition that proceeds from a contrariness or at least spitefulness of temper, but an opposition that seems indicated by our very make and constitution . . .

Sun. 9 Oct. My wife, self and maid at church in the morn (the 2 boys staying at home) . . . This day the holy sacrament being administered, my wife, self and maid all stayed. We gave 6*d.* each, we paying for our servant. This day have my wife and I taken up a resolution in the presence of our almighty God and Saviour, with His divine grace and Holy Spirit, to forsake our sins and to become better Christians. And, Oh, may the God of all goodness and perfection pour into our hearts His Holy Spirit that we may live together in true unity, love and peace with each other, bearing with each other's infirmities and weakness, and that we may also live in peace with all mankind. We dined on beef pudding, carrots and some cold raisin suet pudding. My whole family at church in the afternoon . . . Tho. Davy at our house in the even, to whom, and in the day, I read 6 of Tillotson's sermons.

Tues. 11 Oct. About 6 o'clock my brother and I set out for Battle Market where we arrived about 10.10. I went in order to see Mr. Breeden, but he was not at the market. I bought of Mr. John Hammond 50 lbs of powder, which cost me £3 10s. 0d. We dined at *The Bull's Head* on a rump of beef boiled, a loin of veal roasted, a roast goose and a currant suet pudding and an apple-pie . . . We came home very sober about 7.10. My brother stayed all night. I think Battle to be a pleasant situated town, and there seems to be a considerable market for stock etc. on the 2nd Tuesday in every month. The abbey, which belongs to the family of the Websters (and which was built just after the conquest in memory of that battle fought near that place between the Conqueror and Harold the then King of England, in which the latter, his 2 brothers, most of the English nobility, and 97,974 common men was slain), is the remains of a fine Gothic structure. There is also a deanery and a large low-built church. There is also in the town a free school. We spent as under:

By being decoyed in, by an acquaintance, to an ordinary, it cost for dinner only and about 2 glasses wine	3s.	0d.
Our horses and ostler		11d.
Spent		2d.
	4s.	1d.

The half of which my share is	2s.	0½d.

Paid for stuffing my saddle 6d. . . .

Weds. 12 Oct. My brother stayed and breakfasted with us and then went home. In the forenoon Mr. Jer. French and Mr. Jos. Fuller came and informed me that Robt. Durrant was in custody of the parish officers of Waldron on account that Eliz. Day, belonging to Waldron, but a certificate person to this parish,[42] being big with child, and which she hath sworn upon Robert Durrant, they accordingly desired Mr. Burges and I would go and give the officers security on account of this parish in behalf of the man, and we both accordingly went. And when we came there the man being so ill that he could not go before a justice, and they seemed not willing to take anything less than an indemnifying bond, so that I came back again to consult the people of this parish (leaving Mr. Burges there). And it was agreed I should go and ask Mr. Courthope the most they could oblige us to pay per week which I accordingly did. But when I came to Uckfield, Mr. Courthope

[42] See above, p. 92.

was not at home so that I could have no advice from him. So that after staying at Uckfield and getting some mutton chaps (having had no dinner), I went to Waldron again where, to my great surprise, I found Mr. French, Mr. Piper and Tho. Fuller come to treat with the officers. They then offered to take our bond for 15*d.* per week for the maintenance of the child and 40*s.* for her lying-in, but Mr. French was obstinate and would not permit us to give above 12*d.* per week and insisted for it to be left for Mr. Geo. Courthope to decide. Mr. Burges and I accordingly gave Mr. Nich. Attwood our note of hand to forfeit £20 if we do not meet him or some other of their parish officers on Monday next at Uckfield in order to have Mr. Courthope's determination, and if he will not determinate it, then to deliver them up the body of Robert Durrant again. I come home, but wet enough, about 10.15, and very sober. Spent this day on the parish account 23½*d.* ...

Sat. 15 Oct. ... This day received a letter by the post from Mr. Hodges the keeper of the new state lottery office in Cornhill that No. 66612, being the ticket which belongs to Mrs. Atkins, and one of the tickets which Mrs. Atkins and my wife went partners in, was drawn a blank on the 12th instant. I also received a letter from J. Hazard's lottery office that the ticket No. 38567 which I registered for Master Hook was drawn a blank the 11th instant[43] ...

Mon. 17 Oct. In the morn Mr. Burges and I went to Waldron in order to accompany the parish officers to Uckfield previous to our agreement of the 12th instant to give security for Robt. Durrant. We called at Mr. Wood's and breakfasted, and Mr. Attwood and both of us went to Uckfield where we met Mr. Bonwick the overseer of Waldron, but found Mr. Courthope was not at home; so that we could not have his determination in this affair. We then went to Mr. Halland's to endeavour to make it up between ourselves, but could not on account that some of the cunning people of our parish think 18*d.* per week too much and sent us with orders to give but 15*d.*, and they insist on 18*d.* per week and 40*s.* for her lying-in and what in reality we must come to pay, though we might have made it up the 12th instant for 15*d.* per week. And then Mr. French would give but 12*d.* per week. But, Oh! all this trouble and charge proceeds from the ignorance of that man; for in the first place he obliged the man to be sent away with a view to

---

[43] Throughout the eighteenth century the lottery was a regular feature of English government finance. There were 170 state lotteries between 1694 and 1826. This one was the Guinea lottery of 1757 and was not as great a success in raising money as previous ones had been. It was, in fact, undersubscribed, and, as a result, the lottery as a source of revenue was discontinued until 1769. See R. D. Richards, 'The lottery in the history of England government finance', *Economic History*, III (1934–7), pp. 57–76.

defraud the parish of Waldron of having any security (which cost us
15s.). And now to dally with an affair that almost a child must know the
end of! But, however, in complaisance to Mr. French we did not
comply with the 18d. per week, but took up the note we gave them the
12th instant and gave them another to forfeit £20 if [we] did not meet
them the 24th instant at Uckfield to give such security as shall be
adjudged proper by Mr. Courthope, or otherwise to deliver up to them
the body of Robert Durrant. Now how black and unjust must it appear
before the justice to think we should send the man off with an intent to
evade justice and to put the parish of Waldron to an unnecessary
expense in taking him. As we came home we met Mr. French, whom
we informed of what we had done, but he blamed us for not (instead of
leaving it to Mr. Courthope) entering into a recognizance to have it
tried at the sessions . . . Spent on the parish account this day as under,
viz.,

1 pint wine and beer	1s.	6d.
Horses and ostler		6½d.
Turnpike		2d.
	2s.	2½d.

Received a letter today from the new state lottery office in Cornhill that
the lottery ticket between my brother and I, viz., No. 66643, was drawn
a blank the 14th instant so that I have had no fortune in this lottery . . .
The gardener at Halland made us a present of a fine cauliflower and
some pears.

Fri. 21 Oct. . . . Tho. Durrant came in the even to cut my wife's hair;
he stayed and smoked a pipe with me and stayed till about 8 o'clock. At
home all day and very ill with a cold.

Sun. 23 Oct. . . . My whole family at church in the morn . . . This
being the second time of the asking between Richd. Parkes and Mary
Vinal, it was publicly forbid, and Mr. Porter ordered the woman to
come to him after churchtime to show him just cause for her so doing.
After we came out of church, Mr. Porter sent for me to come in to hear
what the woman had to say, and I accordingly went into Mr. Porter's.
The poor girl, whose name was Anne Stevenson, declared that about
3 years ago she had a child by him, and that he had many times promised
her marriage, which she was ready to make oath of at any time (if ever
she should be called upon to do it), and that he had kept her company
so lately as Michaelmas last, and farther that she would be glad to be
married to him at any time. Upon which Mr. Porter assured her he
would postpone asking them any more till such time as he had satisfied
and agreed with her . . .

Mon. 24 Oct. In the morn I went up to Mr. Burges's in order to call him to go to Waldron according to appointment, but whilst I was a-staying for his getting himself ready, Mary Vinal came in there and informed us she was with child by the man to whom she had been twice asked and yesterday forbid, upon which I ordered her to go to Mr. Courthope's to swear the father, which she readily agreed to. We then proceeded to Waldron and went to Mr. Nich. Attwood's, where we could not agree up our affair upon any other terms than what they offered us the 17th (and what in reality was just). Then Mr. Attwood and we set out for Uckfield where we arrived about 12.40. We dined at *The Maiden Head* on some mutton chaps . . . After dinner we laid our affair before Mr. Courthope (in order to abide by his arbitration), and he adjudged that we must give security to the parish of Waldron to pay them 18*d.* per week so long as the child shall be chargeable to the said parish of Waldron, or be by us otherwise provided for, and also to pay the sum of forty shillings in part to defray the expenses of her lying-in. We then applied to Mr. Charles Thornton, attorney, to make out a bond for us which he did upon the aforesaid terms and Mr. Burges and I signed the same and took up our note that we gave to Mr. Bonwick and Mr. Attwood the 17th instant. I found Mary Vinal had, according to her promise, been with Mr. Courthope and sworn that the child she was now pregnant of was begot on her by Richard Parkes, husbandman of the parish of Ringmer, upon which I took out a warrant to apprehend the said Parkes. We came home about 6 o'clock and I immediately went down to Jones's, there being a public vestry (notice thereof being given yesterday). It was the unanimous consent of all that was at the vestry, *viz.,* Tho. Turner, Joseph Burges, Joseph Fuller, Will. Piper, Joseph Durrant and Jer. French, to give to Thomas Daw, upon condition that he should buy the house in the parish of Waldron and which he hath some time been treating for (by reason he then would be an inhabitiant of Waldron and clear of our parish, it being a purchase of £55),

half a ton of iron	£10
1 chaldron coals etc.	£ 2
in cash	£ 8
	£20

and find him the sum of £20, which he is to pay interest for; and also for Mr. Burges to go tomorrow with him to Mrs. Browne's at Pemberry to buy the said house, a fine present for a man that has already about the value of £80. But yet I believe it a very prudent step in the parish, for he being a man with but one leg (and very contrary withal), and his wife being entirely deprived of that great blessing, eyesight, there is great room to suspect there would, one time or other, happen a great charge to the parish, there being a very increasing family. And I doubt the man none of the most prudent ones, he having followed smuggling very much in time past, which I doubt has brought him into a trifling, lazy way of life.[44] Mr. Burges and I spent on the parish account as under (all of which he paid except 14½d.), viz.,

To half the charge of the bond	3s. 10d.
To half the dinner, drinking etc.	1s. 11¼d.
To half the horses	2¾d.
Ostler	1½d.
Turnpike	4d.
Spent at Waldron	7d.
Mary Vinal's examination and warrant	2s. 0d.
	9s. 0½d.

My wife at home very busy today in putting up the goods for the audit.

Tues. 25 Oct.   In the morn about 2 o'clock Mr. French called me to go and take up Parkes, upon which I arose, and we called Tho. Fuller Jr. and all set out together on our intended journey. We called and stayed at Martin's at the Broyle Gate about one hour. The reason of our calling was to know where their headborough lived; who, he informed, was one Nich. Pocock, upon which we went forward on our expedition. We called up the headborough and then proceeded to the house of Mr. Will. Barnett where the said Parkes lived, and where we arrived about 5.40. The people of the house being just up, we went in and found the man sitting by the fire, whom we took without any resistance. We brought him back to Martin's where after staying about 15 minutes, and the man declaring he was willing to marry the girl, Tho. Fuller and I went with him to Lewes to take out a licence (Mr. French going home on foot to get the bride etc. ready for the celebration of the nuptials). We got to Lewes about 7.15 and breakfasted at *The White Horse* and took out a licence, I being a bondsman for the poor creature. We came back just by 11 o'clock when we immediately repaired to church and Mr. Porter married them in Mr. French's, the clerk's, and my presence, I being what is commonly called 'father'.[45] Mr. French and I both signed the register book as witnesses to the said marriage. Then the man, John Durrant and myself set out on foot for Uckfield for the man to swear his parish (Thos. Prall going on horseback to get the order signed). When we

---

[44] See above, p. 84, note 9.   [45] That is to say Turner gave the bride away.

came to Uckfield, the man swore himself upon the parish of Hellingly; and finding Mr. Bridger to be at Mr. Courthope's, we sent T. Prall home to carry the woman to Hellingly, we proposing to come after on foot. We accordingly took out an order and got it signed by Mr. Courthope and Mr. Bridger, we then making the best of our way home, though when we got there we did not stay above 10 minutes, but we set forward for Hellingly (Thomas Prall and the bride being gone before). When we came to Horsebridge, we found the nearest parish officer lived at the Dicker, upon which we went on till we came to the churchwarden's house, whose name was John Acton, and to whom I delivered them both about 8 o'clock, with a true copy of our order, in the presence of Tho. Prall and Jn. Durrant. I came home about 9.05 and I think made a good day's work of it[46] . . . I spent on the parish account today as under, *viz.*,

At Martin's	2*s.*	10½*d.*
Turnpike		6*d.*
At Lewes for eating, horses etc.	3*s.*	3*d.*
The licence	£1 6*s.*	0*d.*
Paid Mr. Porter	10*s.*	6*d.*
Do. the clerk	2*s.*	6*d.*
Spent at Uckfield	2*s.*	1*d.*
Ring	1*s.*	1*d.*
Spent at Horsebridge		8½*d.*
The order and signing	5*s.*	0*d.*
Paid Jn. Durrant	1*s.*	6*d.*
Paid Tho. Prall	2*s.*	0*d.*
Liquor etc. at my house	1*s.*	6*d.*
	£2 19*s.*	6*d.* . . .

Weds. 26 Oct. . . . In the morning wrote a list of the inhabitants of this parish for John Watford to carry in to the court today for the steward to choose a headborough out of.[47]

---

[46] The feverish activity of the hours between 2 o'clock in the morning and 9 o'clock at night shows just how ruthlessly efficient the parochial machinery could be. In one day Parkes had been arrested in Ringmer, taken to Lewes to get a marriage licence, brought back to East Hoathly and married, taken before a justice at Uckfield to swear his parish, and removed with his wife to Hellingly *via* the Dicker. Mary Vinal was already the mother of one illegitimate child, Anne Thomas Vinal: the parish therefore had an interest in seeing her settled as someone else's responsibility. What became of the jilted Anne Stevenson, Turner did not record.

[47] The Laughton manorial court. The headborough or constable was a manorial office.

Thurs. 27 Oct. Very busy all the day. We dined on a piece of beef boiled, a light pudding, turnips and cauliflower. In the even one Eliz. Blackman came to us with an order from the justices of the town and port of Seaford, but as she came of herself and was not brought by any parish officer, neither was her examination upon oath, so that we did not receive her (on account we was not obliged to it). Mr. Burges and I went and informed Mr. French and Mr. Coates of what we had done, who acquiesced in what we had done. Robt. Hook, Tho. Davy and myself in the even went and asked Osborne's servant to swear her parish tomorrow, which she agreed to . . .

Fri. 28 Oct. In the morn Tho. Prall and I went down to Osborne's and took their servant and carried to Uckfield when she swore herself upon our parish, but could not be persuaded to swear her great belly . . .

Sat. 29 Oct. At home all day and busy. We dined on the remains of yesterday's dinner with the addition of a light pudding and some turnips. In the afternoon I sawed some wood. Oh, how pleasant is it to be at home in business, rather than hurrying about in the manner I have this last week; so I think no life can be so pleasant as a retired country life with a small competency . . .

Weds. 2 Nov. . . . Oh, how transient is all mundane bliss! I, who a-Sunday last was all calm and serenity in my breast and seemed desirous of nothing so much as my eternal and immortal happiness, am now nought but storm and tempest occasioned by the unhappiness that subsists between myself and wife. Well might the wise man say, 'It were better to dwell in a corner of the house-top than with a contentious woman in a wide house.' The thought of such disturbances almost distracts and weakens my reason, and when I think there is no hope of their end, as undoubtedly there is not so long as we both live, it almost drives me to despair. Oh, how could I apply myself with the assistance of God's grace to work out my salvation! But now by such tumults I am as it were rendered incapable to lift up so much as an eye or hand to that great Benefactor of mankind without whose blessings the endeavours of frail man is but foolishness . . . In the even read the play of *Tamerlane*, wrote by Rowe, which I think a very good play; the character of Tamerlane is such as I think should be the character of all mankind.

Thurs. 3 Nov. . . . We dined on the remains of Sunday, Monday, Tuesday and yesterday's dinners with the addition of a light pudding and some turnips. At home all day . . .

Sat. 5 Nov. In the morn Mr. Burges brought me a summons which

Wednesday 14 July
Whom all the foresaid for
Setter for a Bullock or
Heart B or Dined on the
Heart Baked in the Oven
and Suphant a pudding
under it about 4 OClock
Mr Porter come to me &
told me he thought it was
the parishes Duty to
Examine into the Death
of this poor Creature who
Died Yesterday and have
her Opened for there was
according to all Circumstances

reason to suspect the heat
or some other person her
administered something
to her or put or cho...
of life in they hard of men
with a mind to come thee
Oh wick the is very ugly
Oct and many too
with so ugly a Black and a
spoon or intehs stamp
from the Creature was very
well till the say morning
and Baked and after she
had taking the Bread out
of the Oven she took a
Walk in Garton about 3
8 OClock and about 10
Clock or between 9 & 10

Expences in Decemb.r Continued

Expences for January Continued

4. East Hoathly overseers' account book, 1762–79: *E.S.R.O. PAR 378/31/1/1*. The entire volume is in the hand of Thomas Turner. The payments between 2 and 14 January 1764 include those incurred over the marriage of William Williams and Lucy Mepham (see pp. 283–4) and the death of William Lidlow (see p. 284).

Richd. Brazer hath lately been at Mr. Courthope's for, and he desired I would go and appear at Uckfield in his room, he being obliged to go to the funeral of Jn. Carter, which was uncle to Mrs. Burges, to which I did. The reason of Brazer's complaint was: the parish having for several years past paid for keeping 2 of his girls, one about 10 and the other about 12 years of age, one of which was kept by Richd. Trill, the other by Will. Elphick, both of which girls have for some time past been very saucy and impertinent, and Elphick, upon account thereof, corrected her by the order of a public vestry at Easter last, which so affronted Brazer that he declared she should live there no longer, pretending she was abused, and accordingly took her home where she has been near 3 months and has lately demanded pay or relief for keeping her, which we refused so he thereupon got a summons. About 9.20 I set out for Uckfield, calling at the Nursery and taking Edmd. Elphick along with me. I laid the story impartially before Mr. Courthope, who severely reprimanded Brazer and told him all that he could do was to advise him to beg of the parish officers to put her out again to the same place. I came home about 1.50 when, after staying a short time, I went down to Mr. Porter's, this being the day of his tithe feast, and dined there on a buttock and piece of brisket beef boiled, a sirloin of beef roasted, turnips, carrots and a fine raisin bread suet pudding (my family at home dining on the remains of yesterday's dinner) in company with Mr. Porter, Mr. Coates, Mr. Jer. French, Mr. Jos. Fuller, Mr. Will. Piper, John Vine Jr, Joseph Durrant, John Cayley, Ed. Hope and Richd. Hope. I paid Mr. Porter 8*s.* in full for 1 year's tithe, due at St. Michael last. I stayed and smoked one pipe after dinner and then came home. About 5.40 I set out to the house from which John Carter was this day buried in order to read the will of the deceased (by desire of Mr. Burges) to his relations, they being all met to hear the same . . .

Sun. 6 Nov. . . . This day and in the night there has been a great deal of rain, which is the first appearance we have had of winter in regard to the weather, for the roads till now have been almost as good as in summer, the dust at most places streaming as in the summer months; so that has been the finest season for the farmer to sow in as I believe was ever known.

Tues. 8 Nov. . . . A very cold frosty day. In the even examined Eliz. Blackman in regard to her settlement and I doubt she will belong to this parish.

Thurs. 10 Nov. . . . In the even Master Paris came to me and begged I would draw up a petition in behalf of his sons for them to endeavour to ask some relief of their neighbours, upon which I drawed up the following petition, *viz.*,

'Whereas Francis and John Paris having formerly through mistaken notions followed that unwarrantable practice of smuggling though for a considerable time past being convinced of the mischievous consequences of such a practice (to the British nation in general as well as to every individual thereof), have entirely refrained from the said practice, but as they who have once ventured on such an illegal course may years after become subject to the law (as many unhappy instances too justly testify) and which is now become the case of those unhappy men who have lately been sworn against in his majesty's court of Exchequer for a very considerable sum. Which if the law is executed against them in the most rigorous manner they must be obliged to abandon their native country and—that which is still more dear to them—their family and relations, but as they have some prospect and hopes to believe the said affair may be made up for a small sum in proportion to so great a one which they are sworn against for, though still so great that they are unable to raise the same from their effects. They therefore humbly implore the assistance of their neighbours and acquaintance hoping they will commiserate their unhappy affair and yield them some relief and succour in this their day of adversity and trouble and they will ever (as bound in duty) thankfully acknowledge the favours they shall be pleased to confer upon them.'

I gave the man 2*s*. 6*d*. for his son—not that I did it so much from principle of charity as self-interest, having formerly bought some brandy of them. I could not tell but their poverty might induce them to do that for me which another has done for them, in order to clear themselves.

Fri. 11 Nov. . . . In the even went down to Jones's, there being a vestry to consult further about Tho. Daw's affair. It was agreed that the parish should borrow of John Watford the sum of £20 and pay interest for the same after the rate of £4 per cent per annum, and for which Mr. Jos. Burges and myself should give our notes of hand for in behalf of the rest of the parishioners of this parish; and that we should lend the said money upon mortgage to Tho. Daw, to bear interest at £4 per cent per annum . . .

Sat. 12 Nov. . . . About 11.20 Mr. John Box, attorney-at-law at Hailsham, called on me in order to settle the affairs about Tho. Daw. We then went down to Mr. Cayley's at the Moat where Mr. Burges was already and John Watford and Tho. Daw came soon after. We then borrowed of John Watford the sum of £20 and gave him the following note written by Mr. Box:

> 'We Thomas Turner, churchwarden of the parish of East
> Hoathly, and Joseph Burges, overseer of the poor of the
> said parish, do for ourselves jointly and severally and for
> the rest of the parishioners of the said parish promise to
> pay Mr. John Watford or order the sum of £20 for value
> received with interest for the same after the rate of £4 per
> centum per annum within three months notice thereof.
> Witness our hands this twelfth day of November 1757.'

We then lent the same to Tho. Daw and took his note of hand till such time as Mr. Box can draw up the mortgage. Tho. Daw then paid to Tho. Price, steward or servant to Mrs. Browne, the sum of £55 for a tenement etc. in the parish of Waldron, and had proper writings delivered to him, which Mr. Box had perused and declared the title to be very good. I drank there some beer which they informed us was above 60 years old . . . Tonight there has been the greatest light in the air that I think I ever observed, being, I imagine, what is called the *Aurora Borealis.*

Fri. 18 Nov. In the morning walked up to Mr. Piper's and persuaded him to let Tho. Daw's mortgage be made to him as mortgagee in the stead of Mr. Burges and myself . . .

Mon. 21 Nov. In the morn wrote the following petition for Mrs. Virgoe, which Mr. Porter composed:

> 'Madam,
> I humbly hope your goodness will excuse my boldness in
> presuming to trouble you with this. My late husband Isaac
> Virgoe had the honour to serve you as a butcher but he
> dying about five years past left me a widow with two small
> children in distressed circumstances. He left only the
> house we lived in to support us and that deeply mortgaged.
> John Fuller hired this house and shop of me and you have
> been so kind as to continue your custom to him. But he has
> lately left my house and shop and shut it up, by which
> means all custom is stopped from it. A brother of Mr.
> Fuller's has offered to hire this house of me if you will be
> so kind as to let him have part of your custom. He is a very

sober and diligent young man and will do his utmost to oblige you. This is the favour I humbly beg of you which if you will please to grant I shall have the usual rent of my house to support me and my little ones and you will save us from the greatest distress, and this your great kindness to me shall ever be gratefully acknowledged by your most dutiful servant, Mary Virgoe

To Mrs. Medley'[48] . . .

Tues. 22 Nov. In the morn after I had breakfasted, I carried Mrs. Virgoe to Lewes on my mare in order for her to present the petition I wrote for her yesterday to Mrs. Medley. We arrived at Lewes about 11.30. Mrs. Virgoe accordingly presented her petition, but could not have any direct answer thereto on account Mrs. Medley's being just a-going out. I dined at *The White Horse* today on a boiled hog's cheek, a piece of boiled beef, an apple pudding, greens and potatoes . . . Spent today as under (which I think may be justly charged to Mrs. Virgoe, as I never charge her anything for myself or horse) *viz.*,

My dinner	6*d.*
Spent	5*d.*
Horse	5*d.*
Turnpike	2*d.*
Ostler	1*d.*
	1*s.*7*d.*

Weds. 23 Nov. After breakfast Mr. Burges and I once more went down to talk with Osborne's servant when she promised that, provided the man she was now big with child by did not come and give this parish proper security to their liking, she would certainly swear the same on Monday next . . .

Thurs. 24 Nov. . . . Sent by John Streeter to Mayfield 3*s.* 9*d.* for my club arrears, due today, it being at quarterly club night[49] . . . At home all day . . .

Fri. 25 Nov. . . . In reading *The Gazette* for the 22nd instant I find the King of Prussia, with about 20,000, has beat the combined forces of the empire and France, which were 60,000;[50] he having totally routed them and taken almost or quite all their cannon, baggage, etc., taking and killing in the field of battle and the pursuit 10,000 men. Oh, could England boast of such a warlike P----- and to have such a

---

[48] The wife of George Medley. See above, p. 30, note 9.

[49] See above, p. 99, note 25.

[50] Frederick II defeated the Franco-Saxon army at Rossbach on 5 November.

true heroic courage diffused through all its martial men (as must evidently appear to have been in this small host)! But how can we expect to find such courage in the poor degenerated people of England, when it is virtue, and that alone, that constitutes the true hero and inspires courage into the breast of the warrior (which at this time seems almost to be extinguished in the British nation). For dissolute-ness of manners, a spirit of effeminacy and self-interest, together with an intolerable share of pride and luxury, seem almost to over-spread the whole face of this kingdom. And I presume when such are the vices of a nation, they must inevitably be ruined without a speedy reformation.

Sat. 26 Nov. . . . In the day read part of several new almanacs[51] which came down today, and I doubt but few will be sold by reason of the additional duty of one penny on the sheets, and two pence on the stitched.

Sun. 27 Nov. . . . This afternoon in service-time was baptized the natural daughter of Eliz. Day and which made me remark it was baptized by the name of Sarah Durrant Day. In the even and the day read 2 of Tillotson's sermons and part of Sherlock upon death. I this day completed reading of Tillotson's sermons over the second time, and so far as I am a judge I think them to be a complete body of divinity, they being wrote in a plain familiar style, but far from what may be deemed low.

Mon. 28 Nov. . . . This day Tho. Osborne came to me and offered me his bond to indemnify and save harmless this parish from any charge that shall ever arise touching or concerning his servant being now pregnant, and from all charges that shall or may arise there-from . . .

Fri. 2 Dec. . . . We dined on the remains of Sunday, Monday, Tuesday, Wednesday, and yesterday's dinners with the addition of some mutton broth and cold rice pie. Drew out a barrel of perry today. Paid 4*d.* for 2 lb. tripe. Gave a man and a woman with a pass 6*d.* on the parish account. At home all day . . .

Sat. 3 Dec. . . . Very busy all day; marked up a parcel of hose, and in the even read part of Sherlock on death. Oh, how pleasant has this day been—to have been almost continual busy, and then after the hurry of the day to enjoy the pleasure of the even in reading.

Sun. 4 Dec. . . . Myself and the two boys at church this afternoon; the text in the former part of *Hebrews* 10.25: 'Not forsaking the

[51] For the popularity and content of almanacs in the eighteenth century see Bernard Capp, *Astrology and the popular press* (1979), Ch. 8.

assembling of ourselves together, as the manner of some is; but exhorting one another.' From which words we had in fact a very good sermon, though it appeared to be delivered with a warmth that had a tincture of anger in it, or at least seemed to point at some particulars as if the deliverer had taken a pique at some one or more of the hearers. But that he alone must be best judge of . . . This day completed the reading of Sherlock on death and which I esteem a very plain, good book, proper for every Christian to read; that is, rich and poor, men and women, young and old . . .

Mon. 5 Dec. . . . Received of Mr. Sam. Gibbs £1 3*s.* 9*d.* in full. When Mr. Gibbs paid me his bill, Mr. French and Tho. Fuller was in company with him; so that common civility and gratitude obliged me to ask them all to walk in, which they did and stayed with me till near 7 o'clock. But I think nothing can be more frothy than those men's discourse, for let us only think that they are all masters of families and fathers of many children, and yet their whole discourse seems turned to obscenity, oaths, gaming and hunting—nothing to the improvement of the mind nor the honour of God or man, which still confirms me in the opinion I have of the degeneracy of this present age. Not that I would be thought to a-draw a conclusion from a single instance or a few people, but it is what I see daily and almost hourly . . .

Tues. 6 Dec. . . . At home all day and indifferently busy. In the even Peter Cornwell made me a present of 6 trout, for which I gave him 6*d.*, though in gratitude I must say it was not enough. But the smallness of my finances would not admit of more. My wife very lame all day with an ulcer lately broke out on her left leg . . .

Weds. 7 Dec. . . . In the day read part of some *Monitors* lent me by Mr. Calverley, but which paper the author endeavours to point out the only way to restore this nation to its former strength and dignity, which is by suppressing vice and immorality and encouraging virtue and merit . . .

Thurs. 8 Dec. . . . My wife continues very lame with her leg. Oh, the misfortunes that attend me! But still they are small to what we deserve, considering the many animosities that have subsisted amongst us since I have been married. And many doubtless, ah, too, too many have the faults been that I have committed, though many and great are the difficulties I labour under. But yet why, why should I palliate my own transgressions by laying them on other people, though sure I and my friends have all contributed to forward my ruin. The ill usage and unjust treatment I think I may say I have received has been great, but then how silly and thoughtless as a Christian (if I deserve that name) have I been to mind it and to make it turn out so much to my disadvantage. But I heartily forgive it all, hoping for the same of the great Disposer of all events . . .

Sun. 11 Dec. . . . After churchtime Mr. Stone paid a visit to my wife and assured us the ulcer on her leg was a scurvy . . .

Tues. 13 Dec. . . . This day about 2.10 died Mary Shoesmith, child maid at the Rev. Mr. Porter's, after about 10 days' illness. How should such instances of mortality awaken mankind to provide and think of eternity! This poor girl was cut off in the prime of her youth, nay, even before she arrived to years of mature understanding, not being 17. Oh, let mankind consider that no age nor sex is exempt from death. No, nor the most sanguine and florid constitution but must sooner or later be subject to it. But what is it that makes mankind so tremble at the approach of death? Only their vices which darkens and casts a cloud on the understandings of poor deluded mortals, for by sin death entered into the world and entailed that curse on all the posterity of Adam: 'Dust thou art, and unto dust thou shalt return.' Would men but seriously reflect upon this and refrain from evil and do good and return unto the Lord their God, who hath promised mercy and forgiveness (unto them that truly and sincerely repent them of their sins and perform works meet for repentance) in and through the merits and redemption of our blessed Saviour Jesus Christ; I say, would mankind but do this, then the prospect of death would be but as a translation from a life of misery to an eternal and endless state of bliss and happiness . . .

Weds. 14 Dec. . . . At home all day. My wife continues very lame . . .

Thurs. 15 Dec. . . . In the evening gave my wife a vomit . . .

Weds. 21 Dec. . . . This being St. Thomas's Day I relieved [32] poor with one penny each and a draught of beer . . .

Sun. 25 Dec. In the morn, myself, two boys, and servant at church. Myself and servant stayed the communion. I gave 6*d*. . . . This being Christmas Day the widow Marchant, Hannah Marchant and James Marchant dined with us on a buttock of beef boiled, a plum suet pudding and a pearl barley pudding, turnips, potatoes and boullis pie.[52] Myself, two boys and servant at church again in the afternoon . . . The widow, James and Hannah Marchant drank tea with us. Tho. Davy at our house in the evening, to whom I read two nights of *The Complaint*, one of which was the Christian triumph against the fear of death, which must be allowed by all Christians a noble subject, it being the redemption of mankind by Jesus Christ, and I think the author has treated it in a very moving and pathetic manner.

[52] Probably bullace pie.

Mon. 26 Dec. . . . About 3.20 I was sent for down to John Jones's, there being a public vestry for the choosing surveyors of the highways for the year ensuing. The present surveyors, *viz.*, Will. Piper and Joseph Durrant, wanted me to draw up an account for them to carry to the sitting, which I did, *viz.*,

By sundry disbursements	£3 4s. 0d.
By cash received of several defaulters and others	£3 4s. 0d.

Now they had disbursed in reality but £2 18s. 0d. so there remains in the surveyors' hands 6s., with which they are to secure the arch of Scallow Bridge and put up some posts etc. I then wrote the nomination of the surveyors for the ensuing year, which were Richard Hope and John Vine Jr. The electioners were Jeremiah French and Joseph Burges. I came home about 6.10. Master Bull at Whitesmith stayed about an hour or two with me in the first of the even. And when the vestry broke up, Will. Piper, Tho. Fuller and Mr. Hutson called at our house in their way home, according to their annual custom (that is, of Mr. Piper and Tho. Fuller). They stayed till 12.20 and, I may say, still continued encroaching for liquor till they two were as drunk as they could walk. Oh, that mankind should give way so much to this pernicious vice which incapacitates his reason so that he may more justly be ranked amongst the brute creation . . . Gave the following boys (as box-money): the post 1s. 0d., Thomas Turner 6d., Sam. Tooth 3d., Thomas Smith 3d. Very busy (thank God) all day. Gave some boys who came a-singing 3d.

Weds. 28 Dec. . . . This even Mrs. Porter was safely delivered of a fine boy . . .

Fri. 30 Dec. In the even Tho. Durrant came in to cut my wife's hair . . . This day my wife gave our servant Mary Martin warning to provide herself with another service at new Lady Day next.[53]

1758

Sun. 1 Jan. . . . I gave my cousin John Bennett 12d. in order for him to buy for me at Edm. Baker's circulating library Budgen's account of the hurricane in Sussex in 1729[1] . . .

Weds. 4 Jan. . . . In the afternoon James Marchant and I walked over to Framfield in order to get some blue cloth of my mother for a

---

[53] See above, p. 7, note 6.

[1] Edmund Baker's bookshop in Tunbridge Wells was destroyed by fire in 1761. *Sussex Weekly Advertiser*, 19 Jan. 1761.

coat for Master Richd. Porter, by Mr. Coates's order[2] . . . This day the son of John Browne, an infant of about 3 months, was buried. Oh, what pleasure would it be to me were there a good understanding between my mother and self, but why do I say myself, when I am assured if I know my own heart, that I am so far from having any ill against her that I have almost undone myself to serve her. Now this assertion is no more than what every unprejudiced person must allow were they thoroughly acquainted with the whole state of our affairs. But the reason of my mentioning this in so abrupt a manner was the seeming distant behaviour with which my mother treated me today, seeming so mistrustful that I should cheat her, and even appeared almost to be loath to trust me with a few remnants of cloth for fear I should cut a nail[3] or 2 more than I inform her of. I will stop the progress of my pen and thoughts and only conclude with saying and thinking she is my mother. Dr. Stone paid my wife a visit and cut her an issue.

Thurs. 5 Jan. . . . In the forenoon I went down to John Watford's to borrow his horse to carry my wife to see Tho. Fuller. Thomas Davy dined at our house on the remains of Sunday's dinner with the addition of some pear pie and potatoes. After dinner Thomas Davy carried my wife up to Tho. Fuller's . . . About 6 o'clock in the even I walked up to Mr. Tho. Fuller's in order to spend the evening there, where my wife and I supped (in company with Mr. Will. Piper and his wife and brother, Mr. John Vine Jr. and his wife, and Mr. French) on a buttock of beef boiled, a hind quarter of venison roasted, two raisin suet puddings, turnips, potatoes, gravy sauce, pickles etc. We played at brag, and my wife and I (though contrary to custom) won 4½*d*. It being a very wet night (and indeed it was one continued day and night of rain), we stayed till near 7 o'clock in the morn when Mr. Tho. Fuller brought home my wife, and we all broke up, and that very sober. My wife and self gave Mr. Fuller's servant 6*d*. each.

Sun. 8 Jan. . . . Tho. Davy at our house in the latter part of the even to whom I read the last of *The Complaint* and part of Sherlock on death. I now having read *The Complaint* through, think it an extreme good book, the author having treated many parts of religion in a very noble and spiritual manner wherein I think every deist, free-thinker, as also every irreligious person may read himself a fool. For what is wit or wisdom (without religion) but foolishness?

[2] Christopher Coates, the steward at Halland, was the child's grandfather.
[3] A nail was a measure of length for cloth equal to 2¼ inches or a sixteenth part of a yard.

Mon. 9 Jan. . . . Mr. Elless and I agreed upon the rent of the chamber over the school, which I now use as a warehouse, *viz.*, I am to pay 20s. for the rent of it from the time I first took it until next Lady Day, N.S.[4] (being near 2 years), ten shillings of which sum we agreed to outset for the tables, forms etc. which he had of me when I resigned the school to him. The remaining ten is to be passed to the credit of his account when it is become due. Mr. Elless and James Marchant and myself and wife sat down to whist about 7 o'clock and played all night till 7 in the morning. My wife and I won 11d. apiece. We spent the night very pleasant and, I think I may say, with innocent mirth, there being no oaths nor imprecations sounding from side to side, as too often is the case at cards.

Weds. 11 Jan. This day I gave a man 6d., who came about a-begging for the prisoners in Horsham gaol, three of which are clergymen; two of them in for acting contrary to the laws of man (but, in my opinion, not to the laws of God), that is, for marrying contrary to the Marriage Act;[5] the other is in for stealing of some linen (but I hope he is innocent) . . .

Tues. 17 Jan . . . Gave Tho. German at Mr. Porter's 1 hat, value 7s. 6d. it being in full for 1 year's winding up the church clock, at Easter next.

Thurs. 19 Jan. . . . In the even finished reading of Horneck's *Great Law of Consideration*, which I think a very good subject, and I am thoroughly persuaded that the only motive the author had in writing it was the salvation of men's souls. But in my own private opinion it is not written so well as many pieces of divinity which I have read, there being too great a redundancy of words to express one and the same thing . . .

Fri. 20 Jan. . . . A very sharp day; a considerable quantity of snow fell today. In the day read part of *The Universal Magazine* for December, and in the evening read a pamphlet entitled *Primitive Christianity propounded or an Essay To revive the Ancient Mode or manner of Preaching the Gospel*. This is a pamphlet which I imagine to be wrote by a Baptist preacher[6] in favour of preaching without notes. I must in

---

[4] See above, p. 7, note 6.

[5] Turner here refers to Hardwicke's Marriage Act of 1753 (26 George II, c. 33). This Act put an end to many notorious abuses by enacting that no marriage was valid unless solemnized by an Anglican clergyman after the banns had been called for three successive Sundays in the parish church. The royal family, Jews and Quakers were dispensed from its provisions: but dissenters and Roman Catholics were not. It was perhaps these for whom Turner had some sympathy.

[6] Turner was right. Daniel Dobel (1700–82) whose book was published in 1755 was

my own private opinion say that I can see no harm consequent on our method of reading, as the author is pleased to call it. But this I must acknowledge, that the idle lazy way of preaching, which many of our clergy are got into, seeming rather to make self-interest the motive for the exercising their profession than the eternal happiness and salvation of men's souls. To which if we add the intolerable degree of pride and covetousness predominant in too many of our clergy, we need not wonder at our degeneracy from the strict piety with which our fore-fathers worshipped God in the first ages of Christianity. I would not by this digression be thought to derogate from the Established Church of England. No! but only I think the precept and practice in most (that is, in too many of us) is an opposite to each other.

Mon. 23 Jan. . . . In the even about 7.20 I went down to James Marchant's, there being a concert of three violins and a German flute. I came home about 11.10 . . .

Tues. 24 Jan. . . . This day gave a man 6*d*. which came to the door under pretence of a broken chandler and whose misfortune appeared to proceed from being surety for a friend . . .

Weds. 25 Jan. . . . Mr. French this day made me a present of a fine goose and gave my wife and I an invitation to come and see him and Mrs. French tomorrow . . .

Thurs. 26 Jan. . . . We dined on the remains of yesterday's dinner with the addition of a boiled duck and some turnips. About 3.20 Tho. Fuller Jr. went with my wife down to Whyly, she riding on one of Joseph Fuller's horses, and about 6.30 I walked down to Mr. French's where we stayed and supped (in company with Joseph Fuller, his wife, two daughters and son Joseph) on three boiled ducks, two rabbits roasted, part of a cold loin of roasted pork, sausages, hog's puddings and pear pie. We played at cards; my wife and I won 7½*d*. We gave their servant 6*d*. each; that is, 12*d*. between us. We came home between twelve and one o'clock, and I may say quite sober, considering the house we was at, though undoubtedly the worst for drinking, having I believe contracted a slight impediment in my speech, occasioned by the fumes of the liquor operating too furiously on my brain.

Fri. 27 Jan. At home all day. We dined on the remains of Wednesday and yesterday's dinners with the addition of a cheap kind of soup, the receipt for making of which I took out of *The Universal Magazine* for December as recommended (by James Stonhouse MD at

a shoemaker who became a Baptist minister. W. T. Whitley, *A Baptist bibliography*, I (1916), pp. 174, 215.

Northampton) to all poor families as a very cheap and nourishing food.[7] The following is the receipt, *viz.*,

> 'Take half a pound of beef, mutton, or pork, cut into small pieces; half a pint of peas, three sliced turnips, and three potatoes cut very small, an onion or two, or a few leeks; put to them three quarts and a pint of water; let it boil gently on a very slow fire about two hours and an half, then thicken it with a quarter of a pound of ground rice, and half a quarter of a pound of oatmeal (or a quarter of a pound of oatmeal and no rice); boil it for a quarter of an hour after the thickening is put in stirring it all the time; then season it with salt, ground pepper or pounded ginger to your taste.'

This in my opinion is a very good, palatable, cheap, nourishing diet . . .

Sat. 28 Jan. . . . In the forenoon Mr. Jer. French and one Mr. Fletcher of Eastbourne called on me to go with them to take an account of the goods of Tho. Fowler, who, going to London about 6 weeks ago, has not since been heard of, so that it is conjectured he must have lost his life by some accident of one kind or another, he being a very old man—I believe above 70. We accordingly broke open his chests, boxes etc., but found nothing of any value but common household furniture, which we took an account of . . . I came home about 1.20. We dined on the remains of yesterday's dinner. In the afternoon Robt. Diggens called at our house, and I walked over to Framfield with him to get a grey, yard-wide gown for Mrs. Porter, but my mother had none. I stayed with my mother about an hour and came home about 5.20. In the even went down to Mrs. Porter's and acquainted her I could not get her gown before Monday, who treated me with all the affability, courtesy, and good humour imaginable. Oh, what pleasure would it be to serve them were they always in such a temper. It would even induce one to almost forget to take a just profit. In the even read part of the *New Whole Duty of Man*.

Sun. 29 Jan. . . . We had a proclamation read today for the holding a general fast throughout England, the Dominion of Wales and Town of Berwick-upon-Tweed on Friday the 17th February next, in order for our humiliation and confession of sins to Almighty God, to implore his blessing and protection on our fleets and armies. We dined on a

[7] *The universal magazine of knowledge and pleasure*, XXI (Dec. 1757), pp. 268–71. This is one of six recipes given by James Stonhouse under the title 'Expedients for alleviating the distress occasioned by the present dearness of corn (as published in the *Northampton Mercury* of November 28, 1757) intended principally for those, who endeavour to conceal their wants; and for such benevolent persons as would give away, or recommend, a cheap provision'.

giblet and a plain suet pudding, Jerusalem artichokes and turnips . . .

Mon. 30 Jan. My wife at church in the morn (this being the martyrdom of King Charles the 1st) . . . My brother Moses came over in the afternoon and brought me a present of a bantam cock and pullet . . . Very busy all day . . . After I had done writing my London letters, being near 8.40, I walked up to Mr. Piper's where I supped (in company with Mr. French, Tho. Fuller, his wife, son, and daughter, my wife, and Mr. Piper's family) on a sparerib roasted, a leg of pork boiled, a currant pond pudding, apple-sauce and parsnips. We played at brag till 12 o'clock; my wife and I won 3½*d.* We came home about 12.30, Mr. Piper's servant bringing my wife home. My wife gave the maid 6*d.* and I gave the man 6*d.* But sure of all the houses I was ever at, I never saw a more stingy, I may say mean-spirited, old man than Mr. Piper. And for his wife, she deserves more to be pitied than blamed, being a good-natured, poor, harmless, ignorant, inoffensive creature . . .

Thurs. 2 Feb. . . . This afternoon Mr. Burges brought me a warrant from George Courthope Esq., to bring Mary Hubbard, the late servant to Thomas Osborne, before him or some other justice in order for her to swear the father of the child of which she has been delivered of about a month. About 6.30 I walked up to Mr. Joseph Fuller's, where my wife and I supped (in company with Mr. French and his wife, Mr. Calverley, Molly, John, and Sam. French, Jos. Durrant and his wife, and all Mr. Fuller's own family) on a piece of cold roast beef, a cold giblet pasty, some cold roast goose, some cold neat's tongue, cold apple pasty and bread and cheese. We played at brag; my wife and I won 3½*d.* We stayed and spent the even with a great deal of mirth till between 1 and 2 o'clock. Tho. Fuller brought my wife home upon his back. I cannot say that I came home sober, though I was far from being bad company, for I think we spent this even with a great deal of pleasure.

Fri. 3 Feb. In the morn my brother Moses came over to stand the shop for me. And about 7.50 myself, Jn. Watford and Robt. Hook (he being headborough) went to take up the late servant of Mr. Tho. Osborne in order for her to swear the child she was delivered of the 4th ult. When we came to the house of Tho. Lewer where this girl lay in, we found her just gone, having received intelligence of our intentions. Master Hook and I went down to Rippington's, calling at Page's and Martin's as we went and taking Richd. Page with us. We left John Watford at Lewer's. After we had stayed at Rippington's about 2 hours to quench that insatiable thirst I last night contracted (and instead of allaying it, rather had increased the intenseness of the heat), John Watford came and informed us he had heard she was at Paris's, where we immediately went and took her and walked back to Thomas Lewer's, where John Watford took her up behind him and carried her to Uckfield, Master Hook and myself walking. We arrived at Uckfield about 3.50 where we was presently met with by Tho. Osborne, son of the supposed father of this child, who promised if the child was not sworn, he would give the parish an indemnifying bond with his father on Sunday next, which I agreed to upon condition that he would deposit a sum of money in my hands sufficient to defray all expenses we had already been at with her, or should be, till the bond was signed, which he almost as soon complied with as I proposed. We then went up to Mr. Courthope's, Mr. Hook to be sworn into his office of headborough, and I to inquire about pressing.[8] Mr. Courthope swore Master Hook and informed me that if we had any men between the ages of 17 and 45, 5 feet 4 inches without shoes, that worked at their own hands or had no visible employment, we must bring them into the sitting at Maresfield on Monday next, or we would be liable to be fined. We also asked his advice about our affair, who said the best way would be to swear her, though I cannot see for what except the helping of his servant to two shillings. But, however, in compliance with his opinion we had her sworn; she swore the father of her child to be Tho. Osborne Sr.[9] I accordingly took out a warrant to take him up. We dined at Mr. Halland's at *The Maiden Head* on a shoulder of mutton roasted (my family at home dining on a piece of beef boiled, and turnips). Master Hook and I came home as far as his brother Jenner's when it was very dark, rained, and I pretty much in liquor; so I thought it most prudent to go no farther, but stay all night, which I did and sent Master Hook home. But oh, fie upon my weak and feeble resolutions and also upon my bad conduct! But sure it is impossible for me to leave this almost worst of vices. Let me once more try and now begin again. My brains are so weak that almost the sight of [a] glass of anything except water destroys my reason.

Sat. 4 Feb. Awakened very early by the alarming dictates of a guilty

[8] Impressing men to serve in the army or the navy. See above, p. 37, note 13.

[9] The parish officers of East Hoathly had been trying to get Mary Hubbard to name the father of her illegitimate child since their suspicions were first aroused in July 1757. In fact, by law (as in the case of Elizabeth Elless, see above, p. 50, note 31), this was the earliest date that she could have been taken before a justice and sworn to tell the truth about the child (born exactly one month before) even though Thomas Osborne had promised to give a bond to 'save the parish harmless' from any expense in connection with the child's upbringing.

conscience. Oh, insuperable burden! I came home about 8.20. Spent this journey on the parish account by signing the warrant etc. 9s. 0d. . . . In the even Mr. Burges, Master Hook and myself went and searched Jones's house, Prall's, Watford's and Bridgman's barns for vagrants, but found none . . . Received of Mr. Burges in cash £10 in order to pay for Tho. Daw's iron. I accordingly sent it in a parcel of magazines . . . to my brother for him to carry it to the forge-man at Buxted[10] for us.

Sun. 5 Feb. In the morn myself and two boys at church . . . After Mr. Porter had taken his text I went and searched John Jones's, but found no one there but his own family, and then I went back to church . . . After churchtime Mr. Tho. Osborne and his son Tho. came and assured us they would come a-Thursday next and jointly enter into a bond to save harmless this parish, and also to pay the expense we have already been at. In the even Mr. Burges, Master Hook and myself went and searched Mr. Burges's and Potter's barns, as also William Eldridge's house, but found nobody . . .

Mon. 6 Feb. . . . T. Davy at our house. In the day read part of Burn's *Justice*[11] . . .

Tues. 7 Feb. . . . At home all day. This being Shrove Tuesday, I was indifferently busy . . . Gave some girls which came a-singing 2d.

Ash Weds. 8 Feb. . . . In the afternoon wrote a bond of indemnity from Osbornes to this parish for the bastard child of Mary Hubbard.[12] In the even Tho. Daw signed the mortgage of his house etc. in Waldron to Mr. Will. Piper for £20 and also a bond to perform covenants, both of which were dated 20 December 1757 . . . After the mortgage was signed, Mr. Burges and I gave Tho. Daw the note of hand for £20 which he gave us 12 November last. I lent Mr. Burges in cash £7 . . . which money he gave Tho. Daw which, with the 20s. Daw allowed towards making writings etc., makes £8 which was agreed by a public vestry 24 October last to be given to Daw and also ½ ton of iron which he has had and 1 chaldron of coals which is yet more to come[13] . . .

Thurs. 9 Feb. In the forenoon Mr. Tho. Osborne of this parish and his son Tho. Osborne, blacksmith of Uckfield, came and signed a bond, the obligation of which was £40, and the condition was for them

[10] Christopher Gibbs.

[11] The Rev. Richard Burn's *The Justice of the peace and parish officer* was first published in 1755. It was the standard work on the duties of parish officers, and, for a conscientious overseer such as Turner, must have been essential reading.

[12] This bond, which is still amongst the parish records of East Hoathly, in E.S.R.O., PAR 378, is reproduced in Plate 5.

[13] See above, p. 84, note 9.

jointly or severally, their, or either of their heirs etc. to indemnify and save harmless this parish from all charges whatsoever that shall or may at any time hereafter be on account of the female bastard child of Mary Hubbard, of which she was delivered the 4th ult. in this parish, and which she declared Thomas Osborne the elder to be the father of in her examination taken in writing upon oath before George Courthope Esq. the 3rd of this instant. The said bond was witnessed by Robt. Hook. Tho. Osborne the younger also gave me his note of hand, payable to me or bearer on demand for value received £1 1s. 8½d. which was also witnessed by Robt. Hook. The money was for expenses we have already been at on the girl's account, *viz.*,

For carrying her to Uckfield the 28th Oct., 1757   4s. 2½d.

Do. the 3rd instant   9s. 0d.

To John Watford for himself and horse   2s. 6d.

Do. to making the bond   6s. 0d.

Sat. 11 Feb. . . . Received Budgen's account of the hurricane in Sussex which I gave my cousin John Bennett 12d. to buy for me the 1st ult. . . .

Tues. 14 Feb. . . . In the even read part of Leadbetter's *General Gauger*.

Weds. 15 Feb. . . . Mrs. French, Joseph Fuller and his wife, Dame Durrant, Mr. Calverley and Sam. French drank tea with us and stayed and spent the even with us. We played at brag (my wife and I won 1d.) in company with Mr. French and Jos. Durrant. They all supped with us on a piece of boiled salt fish, 2 rabbits in a pasty cold, a cold baked rice pudding, an apple and gooseberry pie, egg sauce and parsnips. They stayed and spent the even with us till near one o'clock, and then we parted all sober, and I hope, after spending the even agreeable, for I am sure it was so to me.

Thurs. 16 Feb. . . . In the afternoon posted my day book, and in the even Mr. Elless at our house a-learning the use of the sliding rule[14] . . .

Fri. 17 Feb. This being the day appointed by proclamation for a general fast and humiliation before Almighty God for obtaining pardon of our sins and for averting those heavy judgments which our manifold provocations have most justly deserved, and imploring His blessing and assistance on the arms of his Majesty by sea and land, for restoring and perpetuating peace, safety and prosperity to himself and to this kingdom, myself, two boys and servant were at church in the

---

[14] The frontispiece (Plate A) to Charles Leadbetter, *The royal gauger* . . . (4th ed., 1755) which Turner had been reading two days earlier illustrates the use of Everard's sliding rule. See Plate 8.

morn; the text in the latter part of *Amos* 4.10: 'Your young men have I slain with the sword, and have taken away your horses; and I have made the stink of your camps to come up unto your nostrils: yet have ye not returned unto me, saith the Lord.' Myself, two boys and servant at church in the afternoon; we had only prayers. After even service we dined on the remains of yesterday's dinner with the addition of a plain suet pudding and some turnips. In the day read part of the *New Whole Duty of Man*. And in the even Tho. Davy at our house, to whom I read part of Sherlock on death. This fast-day to all outward appearance has (in this parish) been observed with a great deal of decorum and, I hope, true piety, the church in the morning being more thronged than I have seen it lately. Oh, may religion once more rear up her head in this wicked and impious nation and triumph over vice and immorality! Then may we once more hope for success from our fleets and armies when our commanders shall be inspired with the love of God and his most holy religion. Then (and not till then) will all private interest and connection of friends give way and become subordinate to the love of their king and country. What then might not the sons of Britain expect but that the forces of the proud Gaul, so fond of universal monarchy, would give way as they did once to an Edward and an Henry. Then might we also hope to see justice impartially delivered from the bench, and rapine and violence banished from among the sons of Albion, and the holy gospel of Christ preached with that energy and ardour as would become the profession of the preacher.

Sat. 18 Feb. . . . A remarkable cold day, but no frost. I begin to find trade once more to grow very dull and that it is almost next to impossible to get in any money due on book.

Tues. 21 Feb. . . . Tho. Davy at our house in the even in order for me to instruct him in gauging and the use of the sliding rule.[15]

Weds. 22 Feb. About 1.10 Mr. French sent his servant with a horse for my wife, who accordingly went with him and dined at Mr. French's. Myself and family at home dined on the remains of Wednesday's supper and a dish of cheap soup. Tho. Davy dined with us in order to taste our soup. About 6.40 I walked down to Whyly, where we played at brag the first part of the even; myself and wife won 1*s*. 2*d*. About 10.20 we went to supper on 4 boiled chickens, 4 boiled ducks, some minced veal, sausages, cold roast goose, cold chicken pasty, cold ham, damson [and] gooseberry tarts, marmalade, and raspberry puffs. Our company was Mr. and Mrs. Porter, Mr. and Mrs.

[15] See above, p. 136, note 14.

Coates, Mrs. Atkins, Mrs. Hicks, Mr. Piper and his wife, Joseph Fuller and his wife, Tho. Fuller and his wife, Dame Durrant, myself and wife and Mr. French's family. After supper our behaviour was far from that of serious, harmless, mirth for it was downright obstreperous mirth mixed with a great deal of folly and stupidity. Our diversion was dancing (or jumping about) without a violin or any music, singing of foolish and bawdy healths and more such-like stupidity, and drinking all the time as fast as could be well poured down; and the parson of the parish was one amongst the mixed multitude all the time, so doubtless in point of sound divinity it was all harmless. But if conscience dictates right from wrong, as doubtless it sometimes does, mine is one that we may say is soon offended. For I must say I am always very uneasy at such behaviour, thinking it is not like the behaviour of the primitive Christians, which I imagine was most in conformity to our Saviour's gospel. Nor would I on the other hand be thought to be either a cynic or a stoic, but let social improving discourse pass around the company. But, however, about 3.30, finding myself to have as much liquor as would do me good, I slipped away unobserved, leaving my wife to make my excuse; for sure it was rude, but still ill-manners are preferable to drunkenness (though I was far from being sober). However, I came home, thank God, very safe and well without ever tumbling or any other misfortune, and Mr. French's servant brought my wife home about 5.10 . . .

Thurs. 23 Feb. This morn about 6 o'clock, just as my wife was gladly got to bed and had laid herself down to rest, we was awakened by Mrs. Porter, who pretended she wanted some cream of tartar. But as soon as my wife got out of bed, she vowed she should come down, which she complied with and found she, Mr. Porter, Mr. Fuller and his wife with a lighted candle, part of a bottle of wine and a glass. Then the next thing in course must be to have me downstairs, which I being apprized of, fastened my door. But, however, upstairs they came and threatened as also attempted to break open my door, which I found they would do; so I therefore ordered the boys to open it. But as soon as ever it was open, they poured into my room, and as modesty forbid me to get out of my bed in the presence of women, so I refrained. But their immodesty permitted them to draw me out of bed (as the common phrase is) tipsy turvy. But, however, at the intercession of Mr. Porter they permitted me to put on my breeches (though it was no more than to cast a veil over what undoubtedly they had before that time discovered); as also, instead of my clothes, they gave me time to put on my wife's petticoat. In this manner they made me dance with them without shoes or stockings until they had emptied their bottle of wine and also a bottle of my beer. They then contented themselves with sitting down to breakfast on a dish of coffee etc. They then obliged my wife to accompany them to Joseph Durrant's, where they again breakfasted on tea etc. They then all adjourned to Mr. Fuller's, where they again breakfasted on tea, and there they also stayed and dined; and about 3.30 they all found their ways to their respective homes, beginning by that time to be a little serious, and in my opinion ashamed of their stupid enterprise, or drunken perambulation. Now let anyone but call in reason to his assistance and seriously reflect on what I have before recited, and they must I think join with me in thinking that the precepts delivered from the pulpit on Sundays by Mr. Porter, though delivered with the greatest ardour, must lose a great deal of their efficacy by such examples. Myself and family at home dined on the remains of yesterday's dinner. Mr. Jordan called on me but did not stay. Mr. Elless and Joseph Fuller in the evening called in to ask me how I did after my fatigue and stayed and smoked a pipe with me. And so this ends the silliest frolic as I think I ever knew, and one that must cast an odium on Mr. and Mrs. P. and Mrs. F. so long as it shall be remembered.

Fri. 24 Feb. . . . Sadly indisposed.

Sat. 25 Feb. . . . At home all day except going down to Mr. Porter's with 2 salt fish, whom, when I went into the parlour, I found a-drinking coffee, though he had not generosity or goodness enough to ask me to drink a dish with him. So one may see that the most profusest from home (that is, freest at other people's houses) are the most abstemious at home, nay even as is now the case, to a degree of mean-spiritedness, or if there can be a worse name found it deserves that.

Thurs. 2 Mar. In the morn about 7.10 I set out on Mr. Piper's horse for Uckfield. I called at John Browne's and breakfasted and also took him along with me. After we had been at Uckfield about 15 minutes, Master Hook and Hen. Osborne came there to me. We then proceeded to Mr. Courthope's in order for Browne and Osborne to swear their parishes; and upon the examination of Hen. Osborne it appeared that he belonged to the parish of St. Mary Westout in the borough of Lewes in the county of Sussex, and Mr. Courthope accordingly granted me an order to remove him and Hannah his wife to the said parish.[16] And upon examination of John Browne it appeared

---

[16] Henry Osborne and his wife obtained a settlement certificate (see below, p. 143). Hannah died and was buried in East Hoathly on 4 June 1759—an occasion for one of Turner's moral reflections. See below, p. 184.

he belonged to the Parish of Brede in this county, but he at the same time affirmed that he had hired £10 a year in this parish, which he should enter upon at Lady Day next; so Mr. Courthope thought it would be needless to have an order . . . In the even Mr. Burges and I was down to Mr. Porter's, when he drawed up a petition to be signed by the inhabitants of this parish to forfeit £20 each if they employed John Browne any more unless he got a certificate . . .

Fri. 3 Mar. In the morn John Browne came up to my house and I persuaded him to permit Mr. Porter to write to his parish for a certificate . . . In the even I went down to Jones's, where we had a vestry to consult on moving Osborne tomorrow and many other things of small note . . . Oh, what pleasure it is to think we have compromised the affair between John Browne and the parish.[17]

Sun. 5 Mar. In the morn myself, Philip and servant at church; the text in *Proverbs* 18.21: 'Death and life are in the power of the tongue', from which words we had as good a sermon as I ever heard Mr. Porter preach, it being against swearing . . .

Mon. 6 Mar. . . . After dinner my brother went to Uckfield to hire 2 horses for us to ride to Seaford upon tomorrow, there being a sale of goods saved out of a vessel stranded on that coast . . .

Tues. 7 Mar. In the morn about 5 o'clock my brother and I set out on our intended journey. We arrived at Seaford about 8.20 where, after viewing the goods (which consisted of about 26 quarters of peas, 18 quarters groats,[18] 5230 lbs of Smyrna raisins, and 20 bags of hops—all very much damaged with sea-water) in company with Mr. Geo. Beard, we then walked down to the sea-side. The sale begun about 11.20 when the peas was sold from 15s. to 22s. per quarter, and the groats nearly the same, the raisins from about 14s. to 18s. per cwt. But they having lost much of their goodness, neither Mr. Beard or myself bought any. The sale ended about 1 o'clock . . . After I came home, my wife and I went down to Jos. Fuller's where we drank tea. We stayed and played at brag with the company hereafter mentioned.

---

[17] In this case the parishioners, apparently under pressure from the vicar, were attempting to prevent John Browne from gaining a settlement in East Hoathly. Since, however, he was on the point of renting a tenement in East Hoathly at £10 *per annum* which would confer settlement upon him, the local justice of the peace had deemed it unnecessary for him to provide himself with a certificate from his last legal place of settlement. John Browne had married Elizabeth Hutson the year before, and the child with which she had been pregnant had died within three months of its birth. He gained settlement in East Hoathly, and was chosen as electioner to the overseer in March 1761. Within a week, however, he had died and was buried, aged 30, on 2 April 1761.

[18] Crushed grain, chiefly oats.

My wife and I won 18*d*. We stayed and supped there on two boiled chickens, a roasted shoulder of mutton, part of a cold ham, cold tongue, a cold veal pasty, tarts etc. in company with Mr. and Mrs. Porter, Mr. Coates, Mr. and Mrs. French, Mr. Calverley, Tho. Fuller and his wife, Dame Durrant, Master Fuller's family and Mrs. Atkins. After supper my wife being very ill, she went home, as would I very gladly, making several vigorous attempts, but was still opposed by Mr. Porter, so that at last I was obliged to sit myself down contentedly and make myself a beast for fashion sake, or else be stigmatized with the name of bad company. There we continued drinking like horses (as the vulgar phrase is) and singing till many of us was very drunk, and then we went to dancing and pulling off wigs, caps and hats. And there we continued in this frantic manner (behaving more like mad people than they that profess the name of Christians) till 9 o'clock when I deserted them and was twice pursued, but at last got clear off with first being well-rolled in the dirt. I came home far from being sober, though I must charge all this upon our reverend clergyman, whose behaviour I am sorry to see, for I shall always think it is contrary as well to the Christian religion as my own conscience. They then continued their perambulation from house to house till 12 o'clock when they got home and with imprudence and impudence declared themselves neither sick nor sorry. Now whether this is consistent to the wise saying of Solomon, let anyone judge: 'Wine is a mocker, strong drink is raging: and he that is deceived thereby is not wise.' Gave Molly Fuller 12*d*.

Weds. 8 Mar.   Abed all day . . .

Thurs. 9 Mar.   . . . At home all day. Very uneasy for yesterday morn's frolic, though I am still of the same mind that it was contrary to my mind and that it was quite by force.

Fri. 10 Mar.   . . . About 3.50 my wife went down to Mr. Porter's previous to an invitation given us yesterday by Mr. Porter and about 7.20 I went down. We played at brag the 1st part of the even; my wife and self won 4*s*. 4½*d*. We stayed and supped at Mr. Porter's on a shoulder [of] mutton roasted, a cold veal pasty, some fried veal, a cold ham, tarts, etc. in company with Mr. Gibbs and his wife, Mr. Piper and his wife, Tho. Fuller and his wife and Mrs. Virgoe. After supper the old sport went on, such as dancing, pulling off of hats, wigs, caps and shoes etc., with a variety of such-like frantic tricks, but no swearing or ill words, by which reason Mr. Porter calls it innocent mirth, though I in opinion differ much therefrom, for I think it abounds too much with libertinism to be called innocent. Poor Mr. Piper had a very great fall, but received little hurt. We stayed and breakfasted at Mr. Porter's and came home about 8.30, and I think not sober. My wife and I gave the servants 2s. 6d.

Sat. 11 Mar. ... At home all day, very piteous ... Read part of *The London Magazine* for February. Merry nights produce heavy days. Oh, what is more sharp than the bitter and tormenting stings of a conscience which is conscious to itself of having acted wrong, not only against the rules of decorum (which getting drunk certainly is), but also against the most holy laws of God. Oh! how little do we, and especially myself, deserve the name of Christians! For it is too too plain, if I judge right, that it must consist in nought but the name when our actions differ so far from it. Oh, the very thought of it is intolerable—To think that through fear of anger and being stigmatized with the name of being singular and a poor conceited particular wretch, one must be guilty of enough to forever plunge ourselves into that lake which ever burneth, and into that fire which shall never be quenched. For without holiness no man shall see the Lord.

Sun. 12 Mar. In the morn my brother Will. came to see me. My servant and nephew only at church in the morn. Just before dinner Mr. Tho. Scrase of Lewes came in, who also stayed and dined with us on a knuckle of veal and bacon, a beef pudding and a hard pudding and turnips. My wife, maid and nephew at church in the afternoon. Mr. Scrase cut off my brother's hair. They both stayed and drank tea and then both went away. A day quite misspent and lost, though I believe worse employed than in doing nothing by reason I was not at church.

Tues. 14 Mar. ... Mrs. French, Mrs. Coates, Mr. and Mrs. Porter, Tho. Fuller and his wife, Mr. Piper and his wife and Mr. Calverley drank tea with us. They, together with Mr. Coates, Mr. French and Joseph Fuller stayed and spent the evening with us and played at brag. My wife and I lost 3s. 6d. They all stayed and supped with us on some salt fish, a dish of Scotch collops[19] with force meat balls, a piece of cold roast beef, some potted beef, a cold baked rice pudding, bullace and gooseberry tarts, celery, watercresses, egg sauce, cold ham and parsnips. They all stayed, except Mrs. Coates, until near 6 o'clock, and many of them not sober. The old frantic sports went on as usual. But now I hope all revelling for this season is over, and may I never more be discomposed with too much drink or the noise of an obstreperous multitude, but may I once more calm my troubled mind and soothe my disturbed conscience with future goodness. Oh, may all the transitory, fleeting, foolish, pleasures of this life be no more in my thoughts, but

---

[19] Normally sliced veal and ham.

let me lay hold on that durable and permanent happiness that fadeth not away, but remaineth eternal in the heavens. Oh, may I have the unspeakable pleasure to hear that comfortable sentence pronounced of: 'Come, ye blessed of my Father, inherit the kingdom prepared for you from the beginning.'

Weds. 15 Mar. At home all day and better than I have once been after such a revel . . .

Weds. 22 Mar. . . . This day received a certificate from Henry Osborne, dated the 4th instant, whereby the churchwardens and overseers of the parish of St. Peter and St. Mary Westout do allow him and Hannah his wife to be inhabitants legally settled in their said parish . . .

Thurs. 23 Mar. . . . A very melancholy time occasioned by the dearness of corn, though not proceeding from a real scarcity, but from the iniquitous practice of engrossers, forestalling etc.[20] My trade is but very small, and what I shall do for an honest livelihood I cannot think. I am and hope ever shall be content to put up with two meals a day, and both of them I am also willing should be of pudding; that is, I am not desirous of eating meat above once or at the most twice a week. My common drink is only water, and which I make my choice when I am not involved in company that I can have it. As I am mortal, so have I my faults and failings common with other mortals. I believe by a too eager thirst after knowledge I have oftentimes, to gratify that insatiable humour, been at too great an expense in buying books and spending rather too much time in reading, for it seems to be the only diversion that I have any appetite for. Reading and study (might I be allowed the phrase) would in a manner be both meat and drink to me, were my circumstances but independent.

Mon. 27 Mar. . . . After dinner I went down to the parish meeting at Jones's, where I made up the accounts between Mr. Joseph Burges the present overseer and the parish, and there remains due to the parish £11 7s. 6d. There was the most unanimity at this vestry that I ever did see at any one before, there being not the least discord imaginable; nor that I observed above 2 oaths sworn during the whole time of the vestry, which was from between 12 and 1 o'clock to ten. We stayed till about 10.10 and then came home, and all sober. We spent 3d. apiece after the money allowed by the parish was spent . . . The officers nominated for the year ensuing were Jos. Burges, churchwarden, Jos. Fuller, electioner; Tho. Fuller, overseer, Jn. Cayley, electioner . . .

[20] Those who buy commodities cheaply but do not release them on to the market until the price has risen to a high level.

Tues. 28 Mar. ... This day my wife paid Mary Martin our servant in cash and goods 40*s.* in full for one year's wages and she accordingly went away ...

Wcds. 29 Mar. In the morn went up to the dwelling-house of the late James Hutson,[21] there being a sale; and when I came there I found he was absconded. Oh, ungrateful news, considering he owes me £10! ... After dinner Mr. Burges and I walked over to Uckfield, I in order to consult an attorney, and he to consult Mr. Courthope concerning Mr. Hutson's affair. When we came there, we found no attorney at home, and Mr. Courthope informed Mr. Burges that he could not tell what to do in his affair; that is, how far he could distrain for poor tax ...

Thurs. 30 Mar. After breakfast Mr. Burges and I set off on foot for Lewes in order to know more concerning our affair with Mr. Hutson. When we arrived at Mr. Spence's he was not at home, being at Lewes. When we came to Lewes, I found Mr. Rideout was not at home, he being at the assize at East Grinstead, so that I could no way know which way to proceed in the affair. We found Mr. Bridger and Mr. Spence both at Lewes, who granted Mr. Burges a summons for Mr. Hutson to appear before them a-Saturday next, but if he could not find him to serve the summons upon, he must leave it in some conspicuous part of his late dwelling-house; and provided he should not appear at the time by them appointed, they would then grant him their warrant to distrain and follow for 30 days any goods that were in his possession or on his premises at the time of his absenting himself, let them since have been bought, or any other wise in the possession of what person soever ...

Sun. 2 Apr. ... In the even Master Hook and myself went and searched John Jones's and Prall's in order to see if there were any disorderly fellows, that we might have them into the sitting tomorrow in order to send them to sea etc. We found none that we thought proper to send. We spent 1½*d.* at Jones's and came home about 9.20.

Mon. 3 Apr. After breakfast Mr. Jos. Burges came and desired I would accompany him to Maresfield (the sitting being there today) ... About 11.10 we set out on foot ... We dined at *The Chequer* in Maresfield (at the 2nd sitting) in company with my father Slater and near or quite 30 more ... Mr. Burges verified his book[22] and got his

---

[21] Turner meant 'the late dwelling-house of James Hutson' not that Hutson was dead.

[22] Burges took his accounts as overseer of the poor of East Hoathly for the preceding year to have them passed and signed by the justices sitting at Quarter Sessions.

summons, and we came away about 4.10 (he paying my expenses) . . . In the even went down to Jones's, where there was a public meeting. We came home about 11.05, and very sober . . . Our parish affairs in my opinion seem to move on in a better manner than formerly, there being now unanimity in almost all the vestries we have, when heretofore it was all noise and discord . . .

Weds. 5 Apr. . . . At home all day and very busy . . . In the even Mary Heath came to our house in order to brew for us tomorrow. She supped with us and lodged at our house. In the even read part of Collins's *Peerage of England.* A very cold sharp day, and a very hard frost last night. Oh, what can be a greater pleasure than to be employed in an honest calling all day, and in the even to unbend and relax one's thoughts by endeavouring to improve the mental and more noble part of man!

Thurs. 6 Apr. . . . Mary Heath a-brewing for us all day . . . Paid Mary Heath 9*d.* for brewing. Very busy all day. Some very sharp storms of snow and hail . . .

Sat. 8 Apr. . . . Gave James Bull 2*s.* 2*d.*, he having a petition for to ask the charity of his friends and neighbours, he being lately fined £5 for only letting a person have part of 2 oz. tea which he had just bought[23] . . .

Sun. 9 Apr. . . . I received of Mr. Porter 12*d.* in order to carry some gingerbread etc. with me tomorrow to distribute among the boys when we walk the bounds of the parish against Chiddingly . . .

Mon. 10 Apr. About 8.20 Mr. Porter, Mr. French, Tho. Fuller, Mr. Elless [and] myself, together with near 30 boys, set out in order to walk the bounds of this parish against that of Chiddingly. We went down to Whitesmith where, after staying some time, we were met according to appointment by the Rev. Mr. Herring, vicar of Chiddingly, Mr. Tho. Funnell, Will. Jenner and Mr. Tho. Hicks the butcher. We begun our walk where our parish, Laughton and Chiddingly join. We proceeded on our walk until we came to what is called the Etchingly Wood where each parish took a separate way, and neither of them had any farther evidence than that of hearsay (there being no person in either parish that ever walked it till about 4 years since); so that of consequence it must remain upon dispute though the whole quantity of land which is disputed for does not appear to me to be above 10 or 12 acres. I think by the arguments I heard of both parties makes it plain that it must

[23] Bull had apparently been caught by the excise officers and convicted for retailing tea without a licence contrary to 15 Charles II, c. 11. For the complex regulations concerning the sale of tea, coffee and chocolate at this time, see *Burn, sub* excise.

belong to this parish and that at this time (and I believe for 30 years past) is [not] taxed to Chiddingly, nor to this parish till about 4 years ago . . . Trade I hope is at present somewhat brisker than it has been, but oh, money comes in but dull! I have now I think retrenched my expenses as much as it is possible for me to do, so that with the blessing of the Almighty I hope I shall once again get the better of my misfortunes.

Thurs. 13 Apr. . . . In the even Tho. Cornwell made my wife a present of a pike 16 inches long from eye to fork, weighing about 2½ lbs. I gave him 6*d*.

Sat. 15 Apr. . . . We dined on a pike baked in Durrant's oven, a light pudding, some calves' liver, and lights boiled and minced . . . This day I received a letter from Mr. Sterry wherein he informs me that the temporary bridge, lately erected of wood for the benefit of passage whilst London Bridge was taken down in order to be repaired, took fire on Tuesday night last about 11 o'clock and was entirely consumed and burnt down, even to the water edge; and it is supposed to be set on fire by some malicious and evil-disposed person. Now could it be thought that in a Christian land, and more especial, among a people that profess the protestant religion in its primitive purity, that there could be a wretch among us so wicked as to perpetrate so black and horrid a crime? But oh, what a convincing proof is this of the predominancy of vice and wickedness in this irreligious age when any crime is not so much deemed a vice in the perpetrator when the cause of it proceeds from something that was in appearance a nuisance to his private interest!

Mon. 17 Apr. . . . At home all day and very busy. In the day read part of the 1st volume of *The Peerage of England*. Oh, what an unspeakable pleasure it is to be busied in one's trade and at a leisure hour to unbend one's mind by reading!

Fri. 21 Apr. . . . Now for some time past I have had a very busy time, which has rendered me happy in that respect, having, I hope, some prospect of the world's once more smiling on me. But, oh! how is my pleasure palled by the scurrilous treatment I have this morn received from my wife, and to the best of my knowledge without any the least provocation. Oh, how unhappy is that life that is continually perplexed with domestic disquietudes and matrimonial discord! How fleeting does all worldly happiness appear to me now. I, who was last night all serenity and calmness in my breast, am now almost depressed with trouble (though to my unspeakable comfort I can say, not with passion). What I can do I know not, for it is impossible for my trade to be carried on to advantage amidst such trouble. For I must act in all respects to the will of my wife, or otherwise have I such a clamour. I should never care how hard I worked, nor even how coarse soever my fare, could I live but happy with that person whom I have chosen for a partner during life. And what is still the bitterest misfortune of all is that I am desirous of leading such life here upon earth that I may through the mercies and merits of our blessed Saviour ever live a life of eternity in heaven, which such disturbances can no ways be consistent with. No, they quite discompose and distract my thoughts and almost drive me to despair. For almost those five years past, nay, even from the very day of marriage have I had such distracting tumults at every short interval of time. Not that I would be thought to charge them all upon my wife, for doubtless, as I am mortal, I am also liable to human frailties and have often, too too often, acted very indiscreet. I have oftentimes thought time and patience would put an end to all the discords subsisting between us as they have arisen chiefly from an unhappy temper, but now I despond of their ever ceasing till death, or at least till a separation, the thought of which strikes me to the heart and makes me almost fit for Bedlam. Oh, what disturbances has my unhappy marriage made amongst my friends! I am become a stranger unto my brethren and an alien unto my mother's children . . .

Mon. 24 Apr. . . . In the even went down to the vestry where there was Mr. Porter, Mr. Piper, Joseph Durrant, Tho. Fuller, Mr. French and myself. We spent 6½d. each and came home sober about 10.20 after a great deal of wrangling . . .

Weds. 26 Apr. About 9.20 Halland gardener called on me in order to go to our club feast at Mayfield,[24] and we accordingly both set out together, where we arrived about 11.20. I called and stayed a while with Mrs. Day and dined at *The Star* on 2 pieces of roast beef, 1 piece of boiled, a fillet of veal roasted and plum puddings, in company with near 40 more. We stayed till about 6.20 when I came home very safe, but very much in liquor about 8.20. Oh, how does the thought of it torment my poor distracted conscience. Sure I am possessed with some demon that I must be so stupid! But what can I do? I cannot undo what I have too often done; no, I cannot. But this I will do: never to be guilty of the like again . . .

Thurs. 27 Apr. . . . At home all day. Read part of *The Peerage of England.* Oh, the torment of a guilty conscience! But how do I hope I

---

[24] See above, p. 99, note 25.

shall yet, through the grace of Almighty God, have strength to resist any such temptations for the future! How am I convinced of the great frailty of human beings! Well might Our Saviour say the spirit indeed was willing but the flesh was weak. Therefore how circumspect should we all be if ever we hope for salvation.

Sun. 30 Apr. Just before churchtime Mr. Stone called and bleeded me, upon which account I did not go to church (it being very cold) . . .

Mon. 1 May. . . . In the even went down to Jones's to the vestry where there were Mr. Porter, Mr. French, Jos. and Tho. Fuller, Joseph Burges, Will. Piper, Richd. Page, Ed. Foord, John Cayley, Joseph Durrant and myself, where we stayed till about 10.20, but of all the out-of-the-way quarrelsome people I ever saw, I think no one ever came up to Mr. French. For there is no one in company must in any way the least so ever contradict or thwart what he proposes, though as for himself he seldom if ever fails to oppose that which anyone else shall happen to start, or give as his opinion. Then he has the greatest skill imaginable in foretelling and judging right of things when they are past. For if there is ever anything turns out to the disadvantage of the parish, why then it is always his opinion that it was wrong (though perhaps at the time of its being done, one of the eagerest for it), and on the other hand, if anything turns out to the advantage of the parish, why then he always knew it to be right (though perhaps the only person that should ever have opposed it). And what still renders him the more disagreeable company is that there is almost an impossibility of any person to speak a word beside himself; so the noise of his clamour with the hoarse and grating sound of his huge big oaths almost deafens the ears of any of his audience. But then the best remedy to bring him into good humour and change his perverseness into obscenity and raillery is to give him 2 or 3 drams of old English gin. But, however, after all to be serious upon the affair it is a most melancholy thing that a man who professes the Christian religion and is constant at divine worship and is in many respects a good neighbour should behave in the manner he does. For undoubtedly it is very contrary to that religion which he professes himself a member of.

Tues. 2 May. . . . In the even Tho. Fuller came in and informed me that the late servant of Mr. Hutson's was come to the house of Will. Eldridge big with child, and asked me to go down to Mr. Porter's with him, which I did . . .

Weds. 3 May. In the morning I went up to Tho. Fuller's and from thence he and I went to Will. Eldridge's, where we talked with Mary Evenden, who owned her being with child, but declared she would not swear the father of it as yet (though I think according to all circumstances nobody need think of any other person for the father but Mr. Hutson). I came home about 9.20 . . .

Sat. 13 May. . . . At home all day and very busy. In the even finished reading Wake's *Catechism*, which I think a very good book and proper for all families, there being good instructions in it and also something which is prodigious moving. It is wrote in a lively, brisk manner and not as if the author wrote more out of form than for the good of people's souls, and at the same time it is in a very plain, familiar style, suitable I think to the meanest capacities that can read. And so far as I can judge there is everything contained in it necessary to a man's salvation.

Sun. 21 May. . . . This day my mother informed me that she had lately been told that Master Darby owed so much money that it was expected he would soon be sent to gaol (and by someone in Lewes). Oh, melancholy news, for I believe he owes me near £20.

Mon. 22 May. In the morn as soon as I had breakfasted I set out for Lewes in order for to commit the management of the debt due from Master Darby to me into the hands of Mr. Rideout, but when I came to Lewes Mr. Rideout was not at home; and, fearing a delay in the affair might prove of a dangerous consequence (I mean as to my getting of the debt), I therefore committed the same into the care of Mr. Burtenshaw, attorney at law, who proposes to send for a writ this day and to arrest him a-Saturday next. The debt I swore to be due to me is £17 though I am pretty confident it is more than £18. Oh, what a confusion and tumult there is in my breast about this affair! To think what a terrible thing it is to arrest a person, for by this means he may be entirely torn to pieces, who might otherwise recover himself and pay everyone their own. But then on the other hand let me consider some of this debt hath been standing above 4 years, and the greatest part of it above three years. I have tried very hard to get it these two years and cannot get one farthing. They have almost quite forsaken my shop, buying nothing of me that amounts to any value, but every time they want anything of value, they go to Lewes. And I have just reason to suspect they must be deep in debt at other places, for undoubtedly no people of £200 a year go gayer than Mrs. Darby and her two daughters; so that I greatly fear they, instead of getting out of debt, go farther in. And as I at this time am so oppressed for want of money that I know not which way to turn [it] makes me more eager to get it. For I really and sincerely have no other motive in doing this but to secure my just due, and I think there is no probability of ever getting it but by doing this (and I really am afraid I shall never have it at all) . . . I came home thank God very safe and sober about 7 o'clock . . .

Weds. 24 May. At home all day . . . In the afternoon my wife went up to the common to see a cricket match, the parish of Laughton and ours playing, but there was not time enough to play it out.[25] In the even read the 6th book of Milton's *Paradise Lost.* This day has been a most remarkable hot day. Oh, very uneasy concerning Master Darby's affair, though I think there is no just room for the world nor him to complain of ill usage.

Sat. 27 May. . . . I have been in a continual anxiety of mind all this day by expecting to hear every minute of my success in Master Darby's affair.

Sun. 28 May. . . . Tho. Durrant, Tho. Davy and Mr. Elless at our house in the even a-drinking of syllabub, Tho. Durrant finding milk, and we cider etc.

Mon. 29 May. In the morn sent Mr. Sam. Durrant in Lewes (by the post John Streeter) in cash £26 in order for him to send me a bill for the same sum in lieu thereof, *viz.,*

	£	s.	d.
5 thirty-six shilling pieces	9	0	0
1 twenty-seven *do*.	1	.7	0
8 guineas	8	8	0
6 half-guineas	3	3	0
1 four and sixpence		4	6
11 half crowns	1	7	6
43 shillings	2	3	0
14 sixpences[26]		7	0
	26	0	0

At work in my garden part of the forenoon . . .

Tues. 30 May. . . . In the even went down into the fields to see Mr. Elless, Thomas Fuller and Tho. Durrant run a foot-race of 24 rods, which was won by T. Durrant . . . At work in my garden part of the day and also a-sawing some wood. No news yet about Master Darby's affair. Lent Joseph Durrant my horse to ride to Mayfield Fair upon.

Weds. 31 May. In the morn Mr. Charles Cooper the sheriff's bailiff called upon me and informed me that he had not yet arrested Master Darby, but was then a-going to do it . . .

[25] See above, p. 9, note 9.
[26] For the availability of coin see Appendix C. The thirty-six shilling piece was the double-pistole or doubloon, the twenty-seven shilling piece the moidore, and the four-and-sixpenny piece the Spanish dollar or piece of eight.

Sun. 4 June. In the morn about 5.40 my wife and I set out for Lewes on our roan mare, where we arrived about 7.20. We breakfasted at Mr. Tho. Scrase's and dined at Mrs. Roase's on a shoulder of veal stuffed, a pigeon pudding and green salad. We stayed and drank tea at Mrs. Roase's . . . We called on Mr. Bennett and Mr. Madgwick, but did not stop at either house. We also went to see the Castle Mount, which I think a most beautiful sight, it being so well adorned with a great variety of shrubs and flowers, and so exceeding high that you have a command of the prospect of all the circumjacent country round. We came home, thank God, very safe, sober and well about 8.30 . . .

Mon. 5 June. In the morn Mr. French and Mr. Piper came before I was up and called me in order to make the land tax books, for in reality they must be carried into the sitting today; so accordingly I made them, though I do not remember that I was either thanked or offered to be paid for the paper. But as to Mr. French's part, he has so often conferred favours upon me that I lay under great obligation to him, but as for Mr. Piper, I know of none he ever conferred on me though sometimes solicited thereto; so that I think I may without partiality say he is a mean-spirited old wretch. But, however, let him go as he is, for he will, I doubt, always remain poor in the midst of riches . . . In the even went down to a public vestry at Jones's, where our company was Jos. and Tho. Fuller, Mr. French, Jos. Durrant, Will. Piper, Richd. Page and myself. There was nothing of any moment agreed to but that Peter Adams should be released from a bond which he gave the parish as security for the child born of the body of Anne Caine, of which he is the putative father, upon condition that he shall pay to the churchwarden or overseer on Monday the 3rd of July next the sum of four pounds, and also at the same time to pay to them, or either of them, a sum sufficient after the rate of 18*d.* per week to defray the expenses of keeping the said child until it shall arrive at the full age of 7 years. Came home about 9.45 when I found my wife very ill. I also received a letter from my father Slater that my sister-in-law Ann Slater was very dangerously ill and given over by the doctors. Spent at Jones's 1½*d.*

Tues. 6 June. In the morn sent Tho. Davy to Hartfield to know how my wife's sister did. Also sent for Dr. Stone to visit my wife. Molly Hook came today as a servant during our pleasure. We dined on some boiled plaice . . . In the even Mr. Stone paid my wife a visit and declared his opinion of her illness that it was a rheumatic disorder with the gravel in her kidneys. Tho. Davy came back in the even and brought us news that my sister was very bad, and I suppose he had a great deal of my mother's nonsense, which she is very full of, having a great volubility of tongue for invective, and especially if I are the subject, though what the good woman wants with me I know not, unless it be that I have offended her by being too careful of her daughter, who, poor creature, has enjoyed but little pleasure of her life in the marriage state, being almost continually (to our great misfortune) afflicted with illness. But who is the person that should repine at what the sovereign Disposer of all events pleases to afflict us poor mortals with here below?

Weds. 7 June. . . . Mr. Stone paid my wife another visit today and let her blood . . . This day I was the spectator, and not an idle spectator only, but a gainful partaker of that which I think no man ever was—that is, of Master Piper's generosity, for he this day made my wife a present of 2 fine carp. We dined on the carp boiled and a light pudding . . .

Thurs. 8 June. . . . In the even Master Darby and Charles Cooper came to my house, he having just arrested Master Darby. I agreed to take a bill of sale of Master Darby's goods. They both lodged at our house all night. My wife very ill all day . . . A very melancholy time with me; my wife very ill, and I am prodigious uneasy about Master Darby's affair for fear I should have been guilty of any harsh or inhuman usage. Oh, that I lived in solitude and had not the occasion to act in trade. But still I hope and think I have done nought but what is consistent with self-preservation and the laws of equity.

Fri. 9 June. . . . Dr. Stone paid my wife another visit, she being very ill.

Sun. 11 June. . . . My wife continues ill though much better. Mr. Stone paid her another visit today . . .

Mon. 12 June. . . . A journeyman carpenter at work for me today a-jobbing, who dined with me on the remains of yesterday's dinner. This day we had a very fine shower, which appears to be very seasonable, there having been almost a continual drought for near 10 weeks past . . . Paid the carpenter 2s. At home all day, and my wife very ill. A melancholy time; trade very dull and my wife very ill. But this is the hand of God; therefore I hope to bear it patiently, knowing that our manifold wickednesses deserve many and great afflictions. Oh, may those misfortunes have that effect upon us which they should have upon all mankind and that they may rouse us from a supine and lethargic state of life unto a state of holiness and that we may have a just sense of our past follies and say with holy David that it is good for us that we have been afflicted; that we may learn Thy statutes. The gardener at Halland made my wife a present of some garden beans.

Thurs. 13 June. . . . Mr. Mason of Eastbourne[27] called on me (previous to a letter sent him some time since) in order to treat with Mr. Elless and myself about instructing us in the art of land-measuring. Now it is very probable the busy world may be greatly concerned to know what I can want to learn to measure land for, and I myself can give no other reason than as my trade is very bad and misfortunes attend me, and I have at present no other prospect than that of an approaching state of poverty, therefore I humbly hope this may one time or other be of some service to me. And should it so happen that it is not, I hope it will never be of any disservice but only the small expense of learning the same. My wife continues very ill . . .

Weds. 14 June. In the morn Mr. Mason called on me again; I gave him 2*s*. 6*d*. for his journey of coming. And whenever he has a farm to measure and map, or if we can get one, he will then instruct us. If it is a farm that he has to measure, then we are to give him 10*s*. 6*d*. per week each; but if we can get one, then he is to instruct us *gratis*, that is for only the expenses of measuring the said land. We dined on the remains of Sunday's and yesterday's dinners. Mr. Stone paid my wife another visit today. This day a match of cricket was played on our common between this parish and Waldron when our parish was victors, they having 8 wickets to go down.[28] At home all day. In the afternoon at work in my garden. In the even read the 12th and last book of Milton's *Paradise Lost*, which I have now read twice through and in my opinion it exceeds anything I ever read for sublimity of language and beauty of similes; and I think the depravity of human nature entailed upon us by our first parent is finely drawn.

Thurs. 15 June. . . . My wife continues very ill. This day the boys of Chiddingly and our parish played at cricket on our common, when the boys of our parish came off victors, having 5 wickets to go down.[29]

Mon. 19 June. . . . At home all day, but very little to do. In the afternoon posted my day book whereby I find I have entered in my book since I posted last, which was the 25th ult., between £11 and £12. Oh, melancholy work! The long credit that I am obliged to give must greatly hurt my trade. This day I saw in the *Lewes Journal*, which was an extract from *The Gazette*, that our troops under the command of the Duke of Marlborough had landed at St. Malo in the province of Brittany (in France) and had burnt and otherwise destroyed 137

---

[27] Christopher Mason, mathematician, a leading Sussex surveyor and mapmaker. For examples of his maps, see the works cited above, p. 109, note 38.
[28] See above, p. 9, note 9.   [29] See above, p. 9, note 9.

vessels of all denominations; that is, one man-of-war of 50 guns, about 30 privateers from 30 to 14 guns, a great number of merchants, and some small craft. And that after destroying the above vessels, he reembarked his men without any loss (in the reembarkation) in order to proceed according to his instructions. It is conjectured that they are to proceed to attack Brest or Rochefort. This success of our army must doubtless greatly weaken and distress the French (who I believe are already in a very poor state), but yet should there be nothing more done, I do not imagine this to be a loss to the French nation adequate to the charge which our nation are at in fitting out and equipping such a fleet as went upon this expedition,[30] though still I think it is acting the more humane part than burning and destroying of a town and thereby probably destroy, ruin and take away the lives of thousands of poor innocent wretches that perhaps never did or thought of any hurt to the British nation . . .

Tues. 20 June.   This is my birthday and the day on which I enter into the thirtieth year of my age. How many ere they have arrived at this age have been cut off and taken out of this world, probably in the midst of their sins! Therefore as it hath pleased the Almighty Disposer of all events to give me my health and life, how careful should I be that I live not in vain, that as I daily increase in age, so may I also improve and increase in all virtue and godliness of life, for if we only look back and reflect upon the time that is past, we shall find him that lives to the greatest age will have room to say with holy psalmist that our days are passed as it were a tale that is told. Therefore my sincere wish is that I may ever endeavour to lay hold on the present minute, that when my exit shall be, I may evermore live a life of happiness and bliss . . . After dinner my brother came over, and I rode behind him to Eason's Green (where there was a cricket match a-playing between Framfield and Isfield) . . . Framfield beat their antagonists at one innings (though there were two played)[31] . . .

Weds. 21 June.   About 10.10 my brother came over upon the mare and immediately my wife set out for Hartfield (Daniel Wicker carrying her) to see her sister, who is like to live but a very little time . . .

Thurs. 22 June.   . . . A very lonely time, having nobody all the day but a poor, wild girl to take care of the household affairs. At home all day and thank God pretty busy . . .

---

[30] Turner's opinion reflects the feeling common at the time that this hit-and-run raid on the shipping at St. Malo, led by the third Duke of Marlborough, was not pressed home to its greatest advantage.

[31] See above, p. 9, note 9. *The Cricket Quarterly*, loc. cit., omits this match.

Fri. 23 June. . . . In the even Tho. Davy at our house. I gave him one pair hose value 3s. 8d. for the many journeys he has been for me and many other jobs and kind offices he hath done for me, for which he would never take no money.

Sat. 24 June. . . . Prodigious uneasy to think my wife did not come home according to her appointment, which was last night, neither for my brother to come over as he promised. It quite astonishes me to see how I am, as it were, deserted by all my friends, though it is no other than what I have seen approaching these 5 years past (and better). But, however, what to attribute the coldness and indifference with which I am treated by my friends and relations I am at a loss to guess. Sometimes I think I must be a prodigy that all my relations in general seem to be so indifferent to me, but when I come to take a more nearer view, I can find amongst the greatest part of their behaviour something of self-interest intermixed with it, to which if we add that easy temper of mine (that is, easy to be imposed upon) I believe it will easily solve the appearing oddity; so that I shall find I have nought to trust to except the Divine Providence and my own industry . . . About 1.20 my brother Sam. Slater brought home my wife and dined with us . . .

Mon. 26 June. . . . In the forenoon I was sent for down to Mr. Porter's, where I got myself pretty well huffed for only my looks, which I think verifies the old proverb that a man's looks will hang him; though if mine was deserving of blame, it was innocently. The case was this: some time ago Mr. Porter desired I would send for a Suffolk cheese for him, which I accordingly did; and after he had had it some time, he sent for me and desired I would take half of it again and he would allow me, for he did no ways desire I should be a loser by it. Now I readily took it without any the least reluctance, no ways thinking, as he asked me to take but half, that I must offer to take the whole (which I believe was his quarrel against me). So that if my looks deserved blame, it was contrary to my inclination and therefore of consequence innocently; though I believe if I had understood Mr. Porter's language, that is the tongue to speak a different language from what the intentions of the heart are, and had offered to take the whole, all had been well. But however after he had vented his passion, he gave me orders for a coat . . .

Thurs. 29 June. . . . This day we had a rejoicing by ringing the bells etc. for a victory gained over the French by Prince Ferdinand of Brunswick near the lower Rhine, the particulars of which are not yet arrived,[32] and Mr. Coates this day gave me an invitation to come down

[32] This was the victory at Crefeld on 23 June.

tomorrow night to see him in order to rejoice there on this occasion. I think this is not the proper way of rejoicing on such occasions, for I doubt there is little thought of returning thanks to Him that gives success in war. But I think it is more probable to be guilty of what we know not to be right by getting in liquor and being guilty of indecencies thereby.

Fri. 30 June. . . . I think I have a very great dread upon my spirits about tonight's entertainment, for as I drink anything strong so seldom, I am thoroughly sensible a very little will make me drunk. Oh, a melancholy thing to deprive oneself of reason and even to render ourselves beasts! But what can I do in this affair? If I stay at home, I shall be stigmatized with the name of being a poor, proud, ill-natured, wretch, and perhaps disoblige Mr. Coates. And if I go, I must drink just as they please, or otherwise I shall be called a poor singular fellow; so I must be guilty of an indecency to please the multitude . . . About 5.20 I went down to Halland where, after casting up a large account of wood, faggots etc. for Mr. Coates, I entered the list of drinkers. Our company were Mr. Coates, the Rev. Mr. Porter, the Rev. Mr. Fletcher, Mr. Robt. Turner, Mr. Will. Shoesmith, Mr. James Shoesmith, Mr. Sam. Gibbs, Mr. John Goldsmith, Mr. Robt. Saxby, Mr. Jer. French, Mr. Jos. Fuller, Mr. Tho. Fuller, Mr. Will. Piper, Mr. Jos. Burges, Mr. Jn. Clinch, Mr. Calverley, Mr. Fran. Elless, Mr. Richard Bridgman, Joseph Durrant and myself. We supped with Mr. Coates on two pieces of cold roast beef, a cold coast of lamb, a cold chicken pasty and green salad. We drank health and success to his Majesty and the royal family, the King of Prussia, Prince Ferdinand of Brunswick, Lord Anson, His Grace the Duke of Newcastle and his Duchess, Lord Abergavenny, Admiral Boscawen, Mr. Pelham of Stanmer, the Earl of Ancram, the Earl of Ashburnham, Lord Gage, Marshal Keith and several more loyal healths.[33] But about 10.10,

---

[33] Besides the royal family, Frederick II 'the Great' of Prussia (1750–86) and Prince Ferdinand of Brunswick (1721–92), those toasted form an interesting mixture of the locally powerful and the internationally successful. The Duke and Duchess of Newcastle were, of course, the local magnates; Mr. Thomas Pelham (1728–1805) was the Duke's cousin who became Baron Pelham of Stanmer on the Duke's death in 1768 and Earl of Chichester in 1801; George Nevill, Lord (later Earl) Abergavenny was from 1757 until 1761 Lord Lieutenant of Sussex; his predecessor as Lord Lieutenant from 1754 until 1757 was John, Earl of Ashburnham (1724–1812); William, Viscount Gage (1718–91) was also a local member of the peerage who subsequently became Baron Gage of Firle. The remainder were, like Frederick the Great and Prince Ferdinand, successful on the field of battle: James Francis Edward Keith (1696–1758) had been made a Field Marshal by Frederick and was shortly to be killed at Hochkirch on 14 October 1758; William Henry Kerr, Earl of Ancram and later Marquess of Lothian (d. 1775), was Colonel of the

finding myself to begin to be in liquor, and finding it impossible to sit there without drinking of bumpers as fast as could well be poured out, I deserted and came safe home, but, to my shame do I mention it, very much in liquor; though I have this to extenuate my crime, that it was with great reluctance I entered the list, and nothing but the fear of disobliging could ever have prevailed on me to have gone. And I think I made all the efforts I could to evade drinking, though all proved ineffectual. Before I came away I think I may say there was not one sober person in company, though some more so than others, for I was the fifth that deserted. Now let us seriously reflect upon this transaction and look upon things in their proper light. I doubt not but we shall find it a very improper way of rejoicing, for instead of rejoicing in spirit and giving thanks and praises to Him that has given our armies success in battle, we have, as it must appear to any considerate person, been endeavouring to draw down vengeance and misfortunes from heaven upon our armies. For if there is a God (as undoubtedly there is), and one who delights in that which is good, then the crime of drunkenness and the many oaths and execrations that often resounded from all sides of the table could never be pleasing in his sight. Oh, the depravity of human nature! When will poor mortal man learn to be wise and think justly of his latter need? Sure the degeneracy of this present age must almost startle anyone that would but make a stop and consider duly the eternal torments which are pronounced against wickedness and impiety . . .

Sat. 1 July.   Terrible bad with the headache . . . Very bad all day, though no more than I deserve . . .

Sun. 2 July.   . . . There was a brief read for to repair the groins[34] and fortifications of the town of Brighthelmstone[35] in this county against the encroachments made by the sea on that coast, which, if not timely prevented, will in all probability eat in and destroy the said town, several houses etc. having in a few years been swallowed up by the sea. The sum necessary for repairing and amending the present groins and for building and establishing new ones is £2250 and upwards. I collected on the said brief 23*d*.; and 2*d*. I am to put in for Joseph Fuller, 2*d*. for Mr. Jos. Burges, 1*d*. for Mr. Piper, and 2*d*. I intend to give myself makes collected in the whole on this brief the sum of 2*s*. 6*d*.

11th Dragoons; George, Lord Anson (1697–1762) was Admiral of the Fleet and First Lord of the Admiralty; and Edward Boscawen (1711–61) was the Commander-in-Chief of the fleet at the siege of Louisburg.

[34] Frameworks of timber or masonry built out into the sea to prevent erosion and encroachment. For briefs see below, p. 158, note 37.   [35] Brighton.

. . . Sadly disordered all day, not having recovered [from] Friday night's debauch . . .

Mon. 3 July. . . . In the even went down to Jones's, there being a public vestry. We received of Peter Adams in cash £8 15s. 6d. and we gave him up his bond (and a proper receipt upon it) which he gave this parish to pay them the 18d. per week towards the maintenance of a female bastard child born of the body of Ann Caine, of which child he was the putative father. The said bond was dated the 4th April, 1752; so he is now forever quit of the charge of the said child. The money we received was as under:

To money which Mr. Burges had paid before Easter for keeping it	£1 7s. 0d.
To keeping it from Easter till 7 years of age	£3 8s. 6d.
To cash given the parish to take it	£4 0s. 0d.

Weds. 5 July. . . . About 3 o'clock I walked down in the park where there was a match of cricket played between our parish and the parish of Framfield when our parish was beat by about 57 runs. There was two innings played out of both sides though Framfield fetched as many their first innings as this parish did at both, and in my judgment Framfield are the best gamesters in each part.[36] I lost 1½d. and came home about 8.40 . . .

Thurs. 6 July. . . . This day Molly Hook went away; we gave her 2s. 6d. for the time the girl has been with us . . .

Sat. 8 July. . . . About 10.30 my brother-in-law Sam. Slater came to inform us that my wife's sister died yesterday about 4 o'clock . . .

Sun. 9 July. . . . I for Lewes . . . I paid Mr. Ed. Verral in cash 17s. 6d. which, with 4d. in bad copper which he allowed, makes 17s. 10d. and is in full for the sum total collected in this present year upon all the briefs read in our church[37] . . .

Weds. 12 July. In the morn my brother came over and breakfasted

[36] See above, p. 9, note 9.

[37] Briefs were royal mandates for collections towards some deserving object. The brief, addressed to the minister and churchwardens, was read from the pulpit. At the close of the service a 'collection upon the brief' was made at the church door, and occasionally on a house-to-house basis round the parish (as, for example, at East Hoathly on 5 June 1757 and 4 and 18 March 1765). The amount collected was then endorsed on the brief which, together with the money, was returned to a duly authorised agent who, himself, paid it over to the professional 'undertaker' for the brief. Here Turner was handing over a series all at once. The system which was very expensive to maintain and liable to much abuse was, to all intents and purposes, abolished in 1828. See *Tate*, pp. 120–25, and W. A. Bewes, *Church briefs* (1896). For details of the kind of sums received from individuals on a brief for a local cause see below, p. 316.

with us, and after breakfast my wife and I set out for Hartfield where
we arrived about 12.20. We dined at my father Slater's on some bacon
and beans and peas (my brother at home dining on the remains of
yesterday's dinner). My wife's sister Ann Slater was buried about 8.10.
She was aged 18 years. A very melancholy funeral, for I think it must
affect anyone to see a poor young creature cut off just in the bloom of
youth. We stayed at my father Slater's all night, as did my brother at
our house . . .

Sat. 15 July.   At home all day . . . a very wet afternoon, and I think
this is the 16th day which we have had successively whereon it hath
rained. A most prodigious melancholy time; very little to do, for I think
that luxury increases so fast in this part of the nation that people have
very little or no money to spare to buy what is really necessary. For the
too frequent use of spirituous liquors and the exorbitant practice of tea-
drinking have in such a manner corrupted the morals of people of
almost all ranks that they have rendered industry a stranger to the
people in those parts. Therefore poverty must of consequence be
always a concomitant of idleness, as is now too truly manifest[38] . . .

Mon. 17 July.   In the morning wrote out the window tax books for
Mr. Jos. Burges and Jos. Fuller . . .

Tues. 18 July.   . . . After dinner I rode over to Framfield . . . Came
home about 5.30. In the afternoon John Watford a-cutting my
grapevine. After I came home, Mr. Fra. Elless played at cricket against
Tho. Durrant and myself for one pound of gingerbread; he beat us
7 runs at one innings.[39] This is the 19th day on which it has rained
successively . . .

Weds. 19 July.   This day about 1.30 we had a very remarkable shower
of rain, for I conject for about 6 or 7 minutes I saw it rain extreme hard at
about 40 rods distance, though at the same time we had none, and the
boys that was at play before the door would often cry out that they
played both in fine and wet weather. But at last the shower came over,
and a very heavy one it was. We heard one clap of thunder, which I
think was a pretty way off, by the interval of time which passed
between the flash of lightning and the clap of thunder, but it was very
loud . . . This is the 20th day on which it has rained successively[40] . . .

---

[38] Turner here and elsewhere (see for example below, p. 280) voiced a common
complaint about tea as a luxury leading to moral decline. See N. McKendrick, J. Brewer
and J. H. Plumb, *The birth of a consumer society: the commercialization of eighteenth-century
England* (1982), pp. 28–9.
[39] See above, p. 9, note 9. *The Cricket Quarterly*, loc. cit., omits this match.
[40] The volume of the original diary covering the period between 23 July and

Fri. 21 July. In the forenoon Mr. George Tomlin's son, a hop-factor in Southwark, called on me. We walked down to Mr. French's where we stayed about half an hour; he informs me hops in Kent are like to be very indifferent . . . Thank God pretty busy all day, but a melancholy time for money, it being almost impossible to get in any debts, and where I owe anything, people press hard for their money. Oh, my easy temper! My wife extremely lame with her leg again.

Sat. 22 July. . . . This day has been a very busy one, having done I believe 8 pounds' worth of business . . . at home all day. This is the 23rd day on which we have had rain successively . . .

Tues. 25 July. Oh what a misfortune it is upon me my wife's being lame again, but let me not repine, since it is the Divine will. This is the twenty-ninth day on which we have had rain successively.

Weds. 2 Aug. I completed the reading of Gay's *Fables*, which I think contains a very good lesson of morality; and I think the language very healthy, being very natural.

Thurs. 3 Aug. In the even the Duke of Newcastle came to Halland, as did Lord Gage, Sir Francis Poole, Mr. Shelley, Colonel Pelham, Mr. Pelham and several more, and stayed all night. What seems very surprising to me in the Duke of Newcastle is that he countenances so many Frenchmen, there being ten of his servants, cooks etc. which was down here of that nation.

Sat. 5 Aug. Mr. Blake's rider called on me and he and I rode together to Lewes, when I think I see the finest horse-race that ever I see run on that down or any other. There was four horses started for the purse, £50. There was a numerous, but I think not a brilliant, company. I came home in company with Mr. Francis Elless, about ten; but, to my shame do I say it, very much in liquor.

Sun. 6 Aug. Pretty bad all day, with the stings of a guilty and tormenting conscience.

Mon. 14 Aug. At home all day and thank GOD extremely busy. Was every day to be productive of as much business as today, I should in no wise envy the rich and great their continual rounds of ease and pleasure. No, it would add fresh vigour to my drooping spirits and give an agreeable elasticity to my ardent desire of carrying on my trade with vigour; then would I exert my utmost power in buying in my goods, that I might run them out with a quick return.

Sat. 19 Aug. I entertained my sister Sally and my brother's wife

---

24 August is missing. The entries for the missing period are taken from *Blencowe and Lower*, pp. 199–200.

with the sight of the modern microcosm,[41] which I think is a very pretty curious sight, for we see the whole solar system move by clockwork, in the same manner they do in the heavens.

Weds. 23 Aug. About 4 p.m. I walked down to Halland with several more of my neighbours, in order for a rejoicing for the taking of Cape Breton etc.,[42] where there was a bonfire of six hundred of faggots, the cannon fired and two barrels of beer given to the populace, and a very good supper provided for the principal tradesmen of this and the neighbouring parishes, as there had been a dinner for the gentlemen of Lewes and the neighbouring parishes. After supper we drank a great many loyal healths, and I came home in a manner quite sober. There was I believe near one hundred people entertained at Halland this day, besides the populace, and so far as I see everything was carried on with decency and regularity; though I must think the most proper way of rejoicing is by having a general thanksgiving, that the whole nation may give thanks to Him that gives success to our armies, both by sea and land; and I think to show our outward joy it might be more properly done by distributing something to the poor.

Sun. 27 Aug. In the morn my wife and self both at church . . . We had a form of thanksgiving read today on the account of the taking of the harbour of Louisburg and Island of Cape Breton etc. . . .

Tues. 29 Aug. . . . About 10.20 I set out for Lewes where I dined with Mr. Tho. Scrase on a shoulder of mutton, French beans and a butter pudding cake . . . Received of Mr. Henry Burtenshaw a bond and a bill of sale from Tho. Darby as a security for his debt. The bond was for £22 and the bill of sale as a collateral security for the same. My bill on Darby was £18, and the charges . . . £4 1s. 0d. . . . I met at Lewes Mr. Gadsden, where we stayed tippling so long that I was constrained to stay at *The White Horse* all night . . .

Weds. 30 Aug. Came home in the morn about 6.20 . . . Oh, how do I

[41] Because this is one of the surviving extracts from the now-vanished volume of the original diary it is not possible to say where Turner saw the microcosm. It may be that he possessed a print of one, but it is more likely that he saw the elaborate automaton called 'The Microcosm' made or designed by Henry Bridges in about 1734. It was widely exhibited throughout England between *c.*1740 and *c.*1774 and was certainly in London in January 1758 and January 1759. It could well have been in Sussex, perhaps at Lewes, in the summer of 1758. I am grateful for this information to John R. Millburn, co-author (with Henry C. King) of *Geared to the stars: the evolution of planetariums, orreries, and astronomical clocks* (Bristol, 1978).

[42] Following the celebration of successes in the European theatre of war in June (see above, p. 156), there were further rejoicings in August for success in the North American theatre where a concentrated effort was being made to drive the French from Canada. Louisburg and Cape Breton Island at the mouth of the St. Lawrence were captured by General Amherst and Admiral Boscawen on 26 July.

lament my misconduct! Sure I must be one of the worst of fellows, so often as I have been overtaken in liquor, and that I still must remain so silly as I know even the smell of liquor almost makes me drunk that I should have no more resolution. What shall I do? I am even as it were drove to distraction. It is true my trouble is very great in regard to family misunderstandings, that when I are a-drinking my thoughts are elevated to such a degree that I have no guard upon my actions. But still this is making that which is already too bad worse, and at the same time it is what I so much detest and abhor. Oh, how unhappy am I that I have no more resolution! How much greater is the glory to have a just command of one's actions than all the knowledge in the world besides. Oh, how frail is mortal beings! How much do we stand in need of the divine assistance of grace, and how do we by our manifold transgressions grieve the Holy Spirit! Oh, the torments of a wounded conscience! How can I expect mercy that have so often broke my resolutions of becoming a better Christian, and especially when I think of the goodness of God to me that I have never lost my life in such a condition? Let me ever and incessantly implore the divine grace to guide my weak and frail endeavours that I may overcome sin and that wicked one which too, ah! too plainly dwells in me. Oh, much should my meditations be taken up in meditating on and bewailing my sinful course of life. I hope I may yet with the assistance of divine grace one day conquer my unruly passions, for without that I am sensible I can do no good thing. Dame Durrant dined with us on a hare roasted . . . My wife a-picking of hops for Joseph Fuller in the afternoon, and I drank tea at Mrs. Weller's. Oh, the insuperable burden of a wounded conscience!

Thurs. 31 Aug. . . . After dinner rode to Fielder's kiln to look upon some tiles. At home all day except going there, and tormented even to despair by my unpardonable folly. Read part of Salmon *On Marriage.* Very busy all day.

Fri. 1 Sept. . . . Joseph Fuller Jr. smoked a pipe with me in the even. How am I tormented with the reflections of my past actions! Oh, sure it is meet to be said if I done amiss, I will do so no more, which God grant I may not. Then shall I once more know peace and quietness in the inward parts.

Sat. 2 Sept. At home all day. In the afternoon posted my day book. We dined on a hand of pork boiled and some carrots. Oh, my troubled mind feels but little comfort. Dame Watford drank tea at our house in the afternoon.

Fri. 8 Sept. . . . In the even sold Mr. Porter's pole-pullers' neckcloth,[43] which was a silent company this time in comparison of what we have had some years past.

Mon. 11 Sept. James Marchant, John Babcock and Tho. Davy at work for me part of the forenoon in helping me remove earthenware from out of the garret to the school-house to prevent its being broken by the workmen who are a-going to new heal this house[44] . . .

Tues. 12 Sept. . . . At home all day. In the even finished reading Salmon *On Marriage*, which I think to be a very indifferent thing, for the author appears to me to be a very bad logician.

Weds. 13 Sept. The carpenters and masons at work upon the house all day . . . At home all day. Drank tea at Mrs. Weller's. In the even Tho. Davy at our house; he and I played at cribbage. I won one penny. He sat up to watch our house till between 3 and 4 o'clock.

Thurs. 14 Sept. . . . The workmen at work upon the house all day. Received by Richd. Page from Mr. Plumer 13 sacks salt. At home all day. James Marchant supped with us, and he and I sat up to watch the house and played at cribbage (though we played for diversion, staking no money) till past 4 o'clock.

Fri. 15 Sept. The workmen a-repairing our house all day. We dined on a piece of bacon, a duck and some carrots boiled . . . In the even read part of Josephus's *Jewish Antiquities*. Tho. Davy sat up again the first part of the night for us to watch our house.

Sat. 16 Sept. . . . The workmen a-repairing the house again today. At home all day, but not busy. A good deal indisposed with a cold. In the evening read part of the *Jewish Antiquities*.

Weds. 27 Sept. In the morn my brother and self set out for Eastbourne where we arrived about 7.30. We breakfasted at Mr. Sam. Beckett's, where we also dined on a shoulder of lamb roasted with onion sauce and potatoes (my family at home dining on a sheep's head, lights etc. boiled; Master Watford, being at work for us, dined with them). We came home about 10.20, but not sober, and I may say by the providence of God my life preserved. For being very drunk, my horse took the wrong way and run into a traverse with me and beat me off, but, thanks be to God, received no damage. Oh, what a poor creature I am after so many weak endeavours to prevent getting in liquor, that I must still remain so silly! What can I say, or what shall I do? I still think and even strive to prevent my weakness in this point, but how frail are the endeavours of poor mortals. Oh, how does sin abound in the flesh! Therefore how much and how earnest should our

[43] See above, p. 64, note 48.
[44] i.e. to replace the roof-tiles.

prayers be to be endued with the Holy Spirit of God! Only let me reflect how often I have (when I have been in liquor) been protected by the providence of Almighty God and rescued as it were from the jaws of death and received no hurt. And how many instances do we almost daily see of people's receiving hurt when in liquor; nay, even death itself has often, too, too often been their unhappy lot. Therefore how loudly does the protecting providence of Almighty God call on me to break off my sins by repentance and to become a new creature! For can I think myself more righteous than many poor unhappy creatures who have often fallen as victims to their intemperance and debaucheries? No, I am not. I am a sinner and a poor frail child of the dust. I even loathe and abhor myself. My very best actions are sinful. Therefore how can I look up to the Almighty for mercy or hope for his blessing on my temporal concerns so long as I abound in sin and wickedness. Let me therefore pray for the Holy Spirit of Christ that I may ask such things as be proper for me and that I may never more be guilty of such crimes—crimes which must inevitably bring me to destruction both of soul and body. And if my own conscience condemn me thus, what can I expect? For God is greater than my conscience and knoweth all things . . .

Sun. 1 Oct. . . . Not at church all day, for I am so ashamed that even my very shadow as it were alarms and terrifies me. For I think everyone who looks me in the face thinks of my past misconduct.

Mon. 2 Oct. John Watford Sr. at work for me all day. Tho. Davy and John Babcock also a-helping me gather my apples the greatest part of the day . . . Paid Mr. Blackwell, paper-maker of Hawkhurst, in cash £2 14s. 0d. in full for 12 ream of paper received of him today.

Fri. 6 Oct. At home all day and very busy a-setting my goods in order after the confusion we have been in by the house being repaired . . . This day entertained my family at 3d. expense with the sight of a raccoon. At home all day . . .

Sat. 7 Oct. . . . In the forenoon my father Slater called on us in his way from Lewes; he dined with us on the remains of Thursday's dinner, and after dinner he went home, carrying my wife with him. Oh, the arbitrary temper of a wife that must be master! Not all the entreaties and expostulations could persuade my wife to postpone this journey, though no other reason could occasion this journey than the fantastical, odd, capricious humour of her mother when undoubtedly (if ever she is a help to me) I now have the greatest occasion for it to put my accounts and other affairs in order (after so great a confusion) as they must naturally require it. But the humour of a parent must be first consulted and then the interest of a husband, as humour and fancy leads. Not but I think parental authority and filial duty should have a just deference paid to it when it is not founded on a basis opposite to that of a more near (and should be) a more dear relation. Oh, how happy must that man be whose more than happy lot it is to whom an agreeable company for life falls, one in whom he sees and enjoys all that this world can give. He can open the most inmost recesses of his soul to her and receive mutual and pleasing comfort to soothe the anxious and tumultuous thoughts that must many times arise in the breast of any man in trade, occasioned by the many losses and disappointments that must naturally occur in business, one whose heart and interest is as his own, and not only so by marriage, but inclination only. Ah, happy, thrice happy must that man be, and on the contrary (for I speak by woeful experience) how miserable must they be where there is nothing else but matrimonial discord and domestic disquietude! They drag on life, as it were with a galling and wearisome chain, and are only a burden to each other. They not only frustrate each other's temporal interest but, it is to be doubted, their eternal also. How does this thought wreck my tumultuous breast! It even chills the purple current in my veins and almost bids nature stand still. A thought so pungent must I think pierce a heart that is as hard as the nether millstone. Oh, how are those delusive hopes and prospects of happiness before marriage turned into briars and thorns and seem as if they never existed. But as happiness is debarred me in this affair, I sincerely wish it to all those that shall ever tie the Gordian knot . . .

Sun. 8 Oct. . . . After dinner I and Tho. Davy walked over to Framfield where I stayed at my mother's till past 6 o'clock. Tho. Davy spent the evening with me, to whom I read 2 of Tillotson's sermons. Oh, the intolerable and tormenting pang of a breast filled with horror like mine! If I look back, I can see the time when I was really happy, but if I take a view of some few years that are now just past, as also the present time, together with the prospect of future time (so far as we poor mortals can judge), how unhappy and miserable is the prospect! Thus by one imprudent step I have ruined, in all probability, my happiness for ever. Oh, woman, ungrateful woman! thou that was the last and most completest of the creation and designed by Almighty God for a comfort and companion to mankind, to smoothe and make even the rough and uneven paths of this life art often, ah! too, too often the very bane and destroyer of our felicity. Thou not only takest away our happiness, but givest us in lieu thereof trouble and vexation of spirit.

Mon. 9 Oct. ... Last night we had a very remarkable high wind, which continued all night ... In the afternoon my brother came over, and I received of him from my mother in cash £2 0s. 6d. in full for a parcel of spices he had today. I then gave him cash £2 6s. 6d. in order for him to pay for some sacks etc. at Lewes tomorrow for me ...

Weds. 11 Oct. ... About 6 o'clock my brother Sam. Slater brought my wife home; he stayed all night. Tho. Cornwell made me a present of some eels and a bream.

Thurs. 12 Oct. In the forenoon my brother and Jn. Babcock assisted me in packing up my rags, which we did in 4 bags containing 11 cwt. 2 qr. 17 lbs. They both dined with me on the remains of Sunday's dinner with the addition of some fried eels, a currant rice pudding, some turnips and carrots. My brother went away about 3.15 ...

Sat. 14 Oct. This day sent my rags by Geo. Ranger in order for him to carry them to Maidstone Fair where John Durrant of Waldron is to sell them for me[45] ...

Weds. 25 Oct. Received of Mr. Sam. Durrant of Waldron in cash £16 16s. 0d. in full for the rags which his nephew sold for me at Maidstone Fair ...

Sat. 28 Oct. ... After I had been in bed about half an hour, I was seized with a violent pang in one of my teeth which lasted about half an hour and then went off with the assistance of some lint steeped in spirits of hartshorn[46] and put in the hollow of the tooth.

Sat. 4 Nov. After breakfast Mr. Elless and myself went to Springham where we met Mr. Mason and measured a very large coppice. We came home about 6.50 quite tired, having neither eaten or drank in the time of our absence from home, unless it was about 2 oz. ginger-bread ... After I came home, I read part of *The London Magazine* for October, as also a poor empty piece of tautology called *A Serious Advice to the Public to Avoid the Danger of Inoculation*, in which he says a physician can only know and be the proper person to perform the operation, and that a surgeon can know nothing about it.

Mon. 6 Nov. ... About 1.45 I went down to Mr. Porter's, where I dined on a buttock of beef and another piece boiled, a sirloin roasted and three raisin suet puddings, in company with Mr. Porter, Mr. Piper, Jos. Burges, Wm. Jenner, Jos. Durrant, John Cayley, Ed. Hope, Richd. Hope, John Mugridge, John French, Mr. Morris, John Martin, Tho. Rice and Edward Foord. I paid Mr. Porter 8s. in full for one year's tithes and came home about 5.50 ...

[45] The rags were sold to paper mills to be beaten out for paper.
[46] The aqueous solution of ammonia.

Tues. 7 Nov. After breakfast Mr. Elless and I walked to Mr. Jos. Bonwick's at Waldron in order to meet Mr. Mason to survey a farm called Summersbrook, but he not being come there, we went and took a view of the Furnace and came home about 2.10 . . .

Weds. 8 Nov. In the morn as soon as I had breakfasted Mr. Elless and I walked to Mr. Jos. Bonwick's where Mr. Mason and we surveyed 2 pieces of land. We came home about 6.30 . . .

Thurs. 9 Nov. This day I took some Sal. Glauber,[47] my side being much out of order. Sent by John Streeter in cash 3s. 9d. for my club arrears, this being a quarterly night.[48] In the even read part of Wiseman's *Chyrurgery*.

Fri. 10 Nov. . . . In the even Mr. Mason and Mr. Elless at our house a-plotting what land we have measured. My wife stayed and supped with Mrs. Coates and came home about 10.30. In the even read part of the examination of Francis Delap Esq., late Provost Marshal General of the Island of Jamaica . . . before a committee of the Honourable House of Assembly, Nov. 10th, 1756, wherein I find him cruelly used in person, goods, fortune and reputation by Governor Knowles.[49]

Fri. 17 Nov. After breakfast Mr. Mason called on me, and we walked over to Waldron where we surveyed several pieces of land and came home about 3.40 . . . In my absence our late servant, Mary Martin, came to see us . . .

Sat. 18 Nov. . . . Mr. Mason and our late servant dined with us on the remains of Wednesday's dinner and also drank tea with us, Mr. Mason being at our house all day a-plotting some land which we have measured. At home all day . . .

Tues. 21 Nov. . . . In the forenoon Mr. Elless and I walked over to Waldron where we met Mr. Mason, and there we surveyed till about 4.20 and then came home . . . Joseph Fuller Jr. smoked a pipe or 2 with me in the even . . . Oh, most melancholy time—very little to do!

Thurs. 23 Nov. . . . In the afternoon my wife went down to Joseph Fuller's and about 7.20 I went to accompany my wife home. We stayed and played at brag in company with Master Fuller's family and Mr. French. My wife and I won 2s. 2d. and came home about 10.20. But

[47] Glauber salt, a purge.
[48] See above, p. 99, note 25.
[49] Turner, presumably, read of this affair in one of his periodicals. Francis Delap was Deputy Provost Marshal in Jamaica. Following a dispute in 1754 with the Governor of the island, Admiral Charles Knowles, Delap was suspended from office, fined £500, thrown into irons, and imprisoned without a proper trial for a year and a day. G. Metcalf, *Royal government and political conflict in Jamaica, 1729–1783*. Royal Commonwealth Society, Imperial Studies Series XXVII (1965), pp. 128–9.

sure of all the women I ever saw, Mrs. Fuller bears the bell for oddness of temper. For some time past she and many of the family have been almost incessantly inviting my wife to come and see them, but I imagine my wife went of the wrong day, for I think without express words Mrs. Fuller could not more plainer have told my wife she wanted us to be absent than she did by her indifferency and by her finding fault [that] we drank too much liquor, and many more such-like oddities . . .

Sat. 25 Nov. . . . Paid John Streeter the post in cash 11*s.* 8*d.* and by 7 yds Irish 9*s.* 4*d.* which makes together the sum of £1 1*s.* 0*d.* and is in full for 1 wig which I received by him today from Mr. Will. Brooker at Mayfield . . .

Tues. 28 Nov. . . . I received of Will. Stone a parcel of linen and cake-soap, which he found on the turnpike road between Lewes and the Broyle Gate and is the property of the Hailsham post, and which I am to send him by Streeter the Mayfield post. I was obliged to give him 2*s.* 6*d.* for finding it . . .

Thurs. 30 Nov. Received of John Streeter 3*s.* for the money I paid Wm. Stone the 28th instant for finding a parcel, and also 6*d.* for my trouble for going to Framfield for it . . .

Fri. 8 Dec. . . . After dinner I went down to Whyly in order to call Mr. French where, after staying some time, we set out for Buxted Place, he being a-going there to collect some quit-rent of Mr. Medley . . . When we came to Buxted, Mr. Medley was not at home. We went and spoke to Mr. Turner and then went to Uckfield where we smoked a pipe with Mr. Miller and came home by my mother's, where we stayed a while. From thence we came to Whyly where I stayed about half an hour and came home about 8.10, somewhat the worse for drinking; that is, I was what is commonly called pert. A very hard frost and excessive cold.

Weds. 13 Dec. . . . In the even made a will for my brother Will. wherein he has given my brother John 20*s.* to be paid in 3 months after his decease, to Bett. £5 if demanded in 2 years, but if not, to her son Philip, as also £20 more to him to be put out to interest by his executor and the interest money paid him yearly till he is 22, and then he is to have the whole. But if he dies before, it is to be divided between John, Moses, Richard and Sally. He has given Philip his best *Bible, Common Prayer Book, New Whole Duty of Man* and *The Poor Man's Help, or Young Man's Guide.* To me he has given all the rest and made me his executor, and I am to make the most I can of it and then pay his debts, funeral expenses etc. and then take £5 myself and any

6 books I like and then divide the residue between Moses, Richard and Sally.

Thurs. 14 Dec. . . . Paid John Streeter in cash 3*s.* 11*d.* for 1 pair silver clasps which he brought me from Lewes a-Monday. In the afternoon my wife went to see Mrs. Porter, and about 7.30 I went down, where we played a few games of whist, but neither won or lost. We supped at Mr. Porter's on some roasted potatoes, sausages and bread and cheese. We stayed chatting with Mrs. Porter and Mr. Porter till near 12 o'clock. A very melancholy time—but little to do, and really no getting in of any debts that I cannot think what will become of me. Oh, could I be in some place of business where I could carry on business in its proper channel! Then should I be acting in the sphere that is my choice. Not that I would desire to amass a large sum of money together. No, only that I might have the prospect removed from before my eyes of an approaching poverty.

Thurs. 21 Dec. . . . This being St. Thomas's Day, gave [33] people one penny and a draught of beer for a Christmas gift . . .

Fri. 22 Dec. . . . At home all day. In the even read part of Addison's *Evidences of the Christian Religion.* Sure never was a more melancholy time than now. What the reason is I know not, but I have so little trade and my trust is so great that I think I must be ruined. And how to extricate myself out of my difficulties I am quite at a loss. I should not care how poor soever my own living was, so I had but a prospect of not losing that little I once had.

Sat. 23 Dec. . . . My wife a good deal indisposed with the pang in her side and hath an ulcer broke out on one of her legs. Oh, heavy and great misfortunes! But let me not repine since it is the will of Almighty God.

Mon. 25 Dec. This being Christmas Day, myself and wife at church in the morn . . . My wife and I both stayed the communion. My wife gave 6*d.*, but they not asking me, I gave nothing. Oh, may the God of all mercy pour into our hearts His Holy Spirit to strengthen this our weak endeavours that we may increase in faith and good works and maintain and keep the good intentions that I hope we have this day taken up, through the merits and intercession of our blessed Redeemer Jesus Christ. The widow and James Marchant dined with us on a piece of roast beef and suet plum pudding . . . Tho. Davy at our house in the evening . . . .

Thurs. 28 Dec. . . . Gave John Streeter the post 12*d.* to his box[50] . . .

[50] Streeter did not appreciate this gift. See Turner's comments on 4 January 1759, and again on 29 December 1759.

Fri. 29 Dec.   About 10.20 I went down to Whyly, from whence Mr. French and I set out for Buxted Place, he being a-going to gather some quit-rent of Mr. Medley . . . After Mr. French had done his business with Mr. Medley, we were prodigious civilly entertained with some bread and cheese, wine and beer etc. . . . We also was showed the house all over, which undoubtedly is a very fine place, being built in the modern taste, though as yet it is not completely finished.[51] We then went to Uckfield where I called on Thomas Osborne for some money, but could get none. Then we went to Mr. Miller's where Mr. French had more quit-rent to gather and stayed and smoked a pipe and came back to my mother's. We stayed but a very little time there. My cousin Moses Bennett being at my mother's, he came home along with me. We came home about 5.20. We called in at Mr. French's but did not stay though, I must say, not sober. Oh, what a trouble is it to me that my head is so weak, for the liquor I have drank today I am sure has been but little, and I tried as much as possible to avoid drinking that little I did drink! It even makes me almost despair to think that I am well assured I have no value for liquor, neither am I fond of drinking, and yet when I am in company, I must always appear singular or else this misfortune happens to me. Mrs. Virgoe and her two children drank tea at our house. This even a meteor was seen by several people in this neighbourhood, which appeared like a ball of fire that was falling from the clouds to the earth and seemed as if it fell about Waldron, leaving a train of sparks behind it as it descended, and its bigness was at last about the size of a large ball, though at first almost like a moon, and extremely light. I imagine fear and surprise hath exaggerated many of the above circumstances.

Sat. 30 Dec.   . . . Gave Richard Fuller 6*d.* and the carrier's servant 12*d.* to their Christmas box as it is called. At home all day, but a heavy heart and a tormenting conscience though there was no consent of the will in this.

1759

Tues. 2 Jan.   . . . About 7.30 I went down to Mr. Porter's, where I supped on some veal cutlets, 3 roasted chicken, a cold ham, sausages, a cold chicken pasty and tarts, in company with Mr. and Mrs. Coates,

---

[51] For Buxted Place see above, p. 30, note 9.

Jos. Fuller, Mr. French and his wife, Mr. Calverley, Tho. Fuller and his wife, Mrs. Atkins and Mrs. Eliz. Hicks. We played at brag in the even. My wife and I won 12*d.* which we gave to the servants. We came home about 2.30 in the morn but I cannot say quite sober, that is, in regard to myself . . .

Weds. 3 Jan. . . . A melancholy headache today, but nothing more of moment except giving some boys who came a-singing 3*d.* . . .

Thurs. 4 Jan. About 1.20 my wife went up to Tho. Fuller's, where she dined . . . About 7.30 I walked up to Tho. Fuller's, where I supped on 2 roasted rabbits, a cold giblet pie, some cold goose, a neck of veal roasted and tarts, in company with Mr. and Mrs. Porter, Mr. Wm. and John Piper, Mr. Sam. Gibbs and his wife, Mr. and Mrs. French, Mr. Calverley, Mrs. Vine and Joseph Fuller. We played at brag in the even. My wife and I won 2*s.* 2½*d.* and gave Mr. Fuller's servant 6*d.* apiece and came home in very good order about 12.45, but sure a very dark night. This day John Streeter the post brought me the 12*d.* I gave him the 28th ult. and told me it was [not] enough, which I gladly received and then gave him just nothing.

Fri. 5 Jan. . . . A very wet day. Tho. Davy, Mr. Elless, James Marchant and my wife and self played at loo[1] in the evening; my wife and self lost 4½*d.* They supped with us on some content,[2] it being a wager which was lost among us.

Sat. 6 Jan. . . . Gave some men that came a-singing 3*d.*

Sun. 7 Jan. . . . In the afternoon Tho. Durrant and I walked over to Framfield where we drank tea at my mother's, and from thence we walked round by John Midmire's at Barnet Wood to ask him for some money, when I received the comfortable news of his keeping out of sight for debt. Sure I am a most unfortunate man! What will become of me I cannot think. I must certainly fail and leave off trade . . .

Tues. 9 Jan. . . . About 2.30 Mr. Jn. Collison came in, and he and I balanced our accounts . . . Mr. Collison spent the even with me, but about 7.30 word came that I must go down to Halland; so I was obliged to leave Mr. Collison alone. I supped at Mr. Coates's on some cold chicken pasty, some boiled chickens and oyster sauce, cold tongue, some hashed duck, a shoulder of mutton roasted, a cold buttock of beef and tarts, in company with Mr. and Mrs. Porter, Mr. and Mrs.

[1] Loo was a card game somewhat like whist. A player failing to take a trick or breaking the game's laws was 'looed'—required to pay a forfeit into the pool.

[2] *O.E.D.* has two eighteenth-century quotations for 'content' in this sense: that from 1700 defines it as 'a thick liquor, made up in rolls in imitation of chocolate, sold in some coffee-houses'. How this settled a wager, and what the wager was about, Turner does not reveal.

French, Mr. and Mrs. Gibbs, Tho. Fuller and his wife, Mr. Calverley, Mrs. Virgoe and Jos. Fuller. We played at brag in the even. My wife and I won 3s. 8d. We gave the servants 18d. We came home in good order about 2.15 . . .

Tues. 16 Jan. . . . About 12.50 my wife walked down to Whyly to dinner, myself and servant at home dining on a piece of beef boiled and turnips and potatoes . . . About 6.50 Joseph Durrant and I walk down to Whyly where I supped on some boiled chicken, cold turkey minced, a shoulder of mutton roasted, a cold chine, a cold ham, tarts etc., in company with Mr. Porter and his wife, Jos. Durrant and his wife, Mrs. Coates and Mrs. Atkins, Mrs. Virgoe and Mrs. Vine, Tho. Fuller and his wife, Mr. Will. Piper and his brother and Mrs. Gibbs. We played at brag in the even. My wife and I lost 3s. 7d. and gave the maid 6d. each. We came home about 1.40 in good order, though I am quite sick of this trade, for it must certainly be useless or hurtful to a tradesman. Neither do I think it consistent with religion, and I should much rather be left out of the number and should think it a greater honour to be absent than present at any of their entertainments.

Tues. 23 Jan. . . . My wife and sister went down to Mrs. Atkins's to dinner, and myself and servant dined at home on what remained of yesterday's dinner . . . About 8.10 I went down to Mrs. Atkins's where I supped on a piece of cold roast beef, a loin of veal roasted, tarts etc. in company with Mr. Porter and his wife, Mr. French and his wife, Mr. Calverley, Tho. Fuller and his wife, Joseph Fuller, and Mrs. Virgoe. We played at brag in the even. My wife and I won 3s. 2d. We gave the maid only 6d. We came home about 1.10 in very good order.

Weds. 24 Jan. We dined on some boiled bacon. In the even went up to Tho. Fuller's to borrow 12 bottles of strong beer. At home all day except as above.

Thurs. 25 Jan. . . . Mr. and Mrs. Porter, Mrs. Atkins, Mrs. Coates, Mrs. Gibbs, Miss Bett. Hicks and Mr. Calverley drank tea at our house. We played at brag in the even. My wife and I lost 1d. The foregoing people, together with Mr. French, Tho. Fuller, [and] John and Will. Piper supped with us on a piece of fresh salmon boiled, some Scotch collops, 3 boiled chicken, a cold chick pie, some cold slices of ham, sturgeon, a cold tongue, potted beef, tarts etc. They stayed till about 1.10 and all went away in very good order. At home all day.

Fri. 26 Jan. In the forenoon walked down to Whyly, but did not stay. We dined on the remains of last night's supper. Molly, Bett. and Nanny Fuller, Frances Weller, Molly and Sam. French and Lucy

Durrant drank tea at our house, and they, together with James and
Joseph Fuller [and] John French, supped at our house on the remains
of last night's supper. We played at brag in the even. My wife and I
won 19½d. They stayed till 1.30, and went away all sober and in good
order. And what is very remarkable, there was not that I could observe
one oath swore all the even. Huzza! the keeping Christmas I hope is
now over, and I think I was never more overjoyed, for besides the
expenses attending it there is something in it that is quite foreign to my
taste or inclinations, I rather choosing a recluse and steady way of
living that may allow time for reason to exercise her proper faculties
and to breathe as it were into the mind of man a serene and pleasing
happiness, which in my opinion never can be enjoyed when it is so
often disjointed and confused by such tumultuous, or at least merry,
meetings.

Thurs. 1 Feb. . . . My wife and I this day saw a chameleon and
salamander which was preserved in spirits and carried about for a
show; it cost 1*d.* each.

Mon. 5 Feb. . . . In the even went down to the vestry at Jones's,
where was Mr. French, Joseph and Tho. Fuller, Mr. Burges, John
Vine Jr, Joseph Durrant and Richd. Page. There was no business of
moment to transact, though oaths and imprecations seem[ed] to
resound from all sides of the room, and the sound of which seemed to
be harsh and grating, so that I came home about 7.20. I believe was
there to be the penalty paid (which is assigned by the legislature) by
every person that swears, and even supposing the persons to be only
common labourers that constitute our vestries, there would need no
tax to be levied to maintain our poor. For I think the fund arising from
such forfeitures would be more than sufficient to defray all the
expenses of the poor.

Sun. 11 Feb. . . . After churchtime my wife and self, with Tho.
Davy, walked down into the land belonging to Halland to see some
Turkey, or Oatland Sheep, which are there a-keeping. They are very
like our country sheep for size and make, but they have no horns, and
their coats are hair instead of wool, and their colour seems tending to a
brown, or like anything which has been singed with the fire. There
were some young lambs which were of a perfect liver-colour. They are
extremely poor, though they are fed with hay twice a day and oats once
and are in grounds where there is good grazing. Their poorness I
imagine proceeds from the different degrees of heat between our
climate and that from which they are brought.[3] I believe it is as mild a

---

[3] I am grateful to Dr. M. L. Ryder, whose book *Sheep and man* (1983) deals with this

time, considering the season of the year, as hath been known in the memory of man, everything having the appearance and carrying with it the face of April rather than of February (the bloom of trees only excepted). The meadows now are as verdant as sometimes in May; the birds chirping their melodious harmony, and the footwalks dry and pleasant . . .

Mon. 12 Feb. My wife being very ill in the morn, sent Tho. Davy to Doctor Stone's. He came back and breakfasted with us. Mr. Stone paid my wife a visit and also let my blood . . .

Weds. 14 Feb. . . . Mr. Stone paid my wife another visit today and let her blood . . .

Fri. 16 Feb. This being the day appointed by proclamation for a general fast and humiliation before Almighty God for obtaining pardon of our sins and for averting those heavy judgments which our manifold provocations have most justly deserved and imploring His blessing and assistance on the arms of His Majesty by sea and land and for restoring and perpetuating peace, safety and prosperity to His Majesty and these his kingdoms, myself and servant at church in the morn (my wife staying at home only upon account of her illness) . . . The text is *Psalm* 122.6: 'Pray for the peace of Jerusalem: they shall prosper that love thee.' Myself and servant at church in the afternoon; we had only prayers. After we came home, we dined on the remains of

---

subject, for the following comment: 'Despite several clear pieces of biological evidence in this most interesting description it is not possible to identify either the breed or the type of sheep. The reference to 'hair' suggests the 'hair' type of the tropics. The only present-day breeds in Britain approaching this description are the brown Soay of St. Kilda and the Wiltshire (which is white): but both are horned. The brown Turkish Imroz breed does not have a 'hair' coat, and the Ödemis which has a 'hair' coat is white. The 'liver-colour' lambs and 'singed' adults seem to indicate that they were born red and became tan with age. Modern breeds in this category are the Guirra of Spain and the Sologue and Rouge de Roussillon of France. The latter has a tradition of originating in North Africa. One cannot take 'Turkey' at its face value since there is also a long tradition of animals being erroneously named after a supposed source. They are much more likely to have originated in Britain, and the areas from which they could have come are Orkney/Shetland, the Hebrides, Wales (the Welsh Mountain breed has a coarse red hair known as kemp) or the south west (the Portland breed has tan fibres): but of these only the Orkney/Shetland and Welsh breeds have ewes lacking horns. The designation 'poor' (despite good feed) indicates the primitive nature of the sheep and was not due to the changed climate, as Turner supposed. Heredity is far more important than environment in this respect.'

Some five and a half years later, when Abraham Baley had taken over the administration of the Halland estate, he wrote: 'The sheep at Halland Park are so poor I can get no butcher that will give anything for them, nor indeed, if they were fat, would scarce anybody here eat the mutton; so that, to lessen the quantity a little, I have ordered a few of them to be kill'd at proper times and given to the poor.' Letter of 3 November 1764 in Abraham Baley's letter book 1763–73, E.S.R.O., S.A.S. HA 310.

yesterday's dinner with the addition of some hog bones, which Dame Durrant made my wife a present of. The fast in this place hath seemingly been kept with great strictness and, I hope, with a sincere and unaffected piety, our church in the morning being crowded with a numerous audience. God grant we may every one of us reform our wicked ways and that every Briton may inspire his neighbour with a sense of religion and the fear of God by his own good example. For I think no nation had ever greater occasion to adore the Almighty Disposer of all events than Albion, whose forces meet with success in almost all quarters of the world, and where plenty once more rears her pleasing aspect. The pestilential distemper is now no more among our herds; and there now seems to reign a spirit of unity in our national councils, [and] a king sitting upon the British throne whose whole intention seems to be that of making the happiness of every individual of his subjects the same as his own. Oh, let Britain think on these blessings and adore the giver of them! Let us all with sincerity and pure devotion in one voice continue to supplicate the blessing of the Almighty on this our happy isle!

Tues. 20 Feb. . . . A very sharp cold day, but no frost; the dust flies as much as in May sometimes.

Weds. 28 Feb. alias Ash Wednesday. . . . This day that shadow of a man Mr. Will. Piper came to inform me that his boy was to be made a Christian of tomorrow and that he expected I should stand godfather for it, I having in some measure before promised him—that is, so far as if he could get nobody else. But however the poor old wretch told my wife that he would [not] have she should come up tomorrow night, for he did not invite any to dinner with him but them only that stand sponsors. For if he were to invite his neighbours to dinner with him (as in gratitude he ought), they would not come half, nor he could not nor would not have the plague and trouble of getting a dinner for so many people, though I believe had not niggardliness been the only motive to prevent his asking his neighbours, there would have been no fear of their coming.

Thurs. 1 Mar. . . . About 12.20 Mr. Piper came along with his son, and Mrs. Fuller, Mrs. Burges, John Piper and myself went and stood for it and gave it the name of William, it being its father's name. We then came back to our house when John Piper earnestly begged of my wife to go up to dinner with them, protesting he thought his brother to be a strange man, and accordingly she went to Mr. Piper's, where we, together with Mr. and Mrs. Burges and Mrs. Fuller, dined on a knuckle of pork boiled and greens, a pond currant pudding and a cold duck pie designed as a pasty. We played at brag in the even. My wife and I lost 2s. 10d. We stayed and supped with the poor old wretch on a loin of veal which was 4 hours in roasting and then in part roasted, a ham of bacon boiled and greens, the cold duck pasty, a hot buttered apple pie and a hot baked rice pudding, in company with Mr. and Mrs. Porter, Tho. Fuller and his wife, Joseph Fuller and his wife, Mr. Burges and his wife, John Vine and his wife, Joseph Durrant and his wife, Mr. French and his wife, Mr. Calverley and Miss Tealing and Tho. Diplock. We came home about 2.40, and all sober. My wife and I gave the nurse and maid 3s. A very dear night's entertainment, and I am sure a very bad one, for I never spent an evening with less pleasure in my life. There was not any liquor I suppose worth drinking, though I tasted only some small beer and that came like drops of blood. To see the niggardly behaviour of the poor old man and the raillery of the company that was continually a-playing upon the old wretch made it quite irksome and, to me, very disagreeable. Undoubtedly he deserves very justly to be ridiculed, considering how handsomely he has been entertained at all his neighbours this year, and then the many ungenerous expressions the old man has been known to use concerning the expense of the christening and the too great quantity of liquor the women drank when his wife was brought abed, which confirms me in the opinion I have long since maintained that a thoroughly covetous man is at all time a most unsociable creature and even a pest to society. Not but I would have every man be frugal, but not to the degree of niggard when he can so well afford it as Mr. Piper can. But, however, let him be pitied and go on in his own way, which is to fix all his happiness in accumulating together a heap of sordid drops, and for no other use but only to say or think he is worth so much. Oh, the pernicious consequence of money! What will it not make poor frail mankind to do when he hath once made it his only aim! . . .

Sun. 4 Mar. . . . After churchtime Mr. French, Tho. Fuller and myself went down to Mr. Osborne's to get some money of him, but could get nothing but fair promises. Oh, melancholy time! There seems to be little else but fraud and deceit among too, ah, too many of mankind . . .

Weds. 7 Mar. In the forenoon I received a letter from my brother wherein he acquainted me my mother was very ill, and I then walked over to Framfield where I found my mother very ill with diabetes . . .

Tues. 13 Mar. . . . In the afternoon our servant went over to Framfield to know how my mother did and found her never likely to get the better of her illness . . .

Thurs. 15 Mar.   After breakfast walked over to Framfield to see my mother, whom I found extremely ill, and I think like to hold it but a very little time . . .

Sat. 17 Mar.   . . . In the forenoon went to Lewes on a horse lent me by Mr. Ben. Shelley. I gave Mr. Sam. Durrant in cash £70 and received of him one bill on Sir Joseph Hankey and partners . . . value £130, and for the remainder (£60) I gave him my note of hand, payable to him or order on demand with interest . . .

Mon. 19 Mar.   About 6.20 I set out for London.[4] Breakfasted at *The Bell* at Godstone; got into London about 11.30. I dined at Mr. Tho. Standing's on a piece of fresh salmon boiled and some minced veal. Drank tea with Mr. Margesson, and spent the even and supped with Mr. Sterry, where I also lodged. Paid Mr. Tho. Standing one bill on Margesson and Collison 30 days date dated today value £22 14s. 0d. No. 379 and which [is] in full. Paid Mr. John Crouch one bill on Margesson and Collison dated today 30 days date No. 380 value £20 and by cash 3s. which leaves due to Mr. Crouch £41. Balanced accounts with Mr. Joseph Hillier and gave him one bill on Margesson and Collison dated today 30 days date No. 381 value £20 18s. 0d. of which bill £10 5s. 0d. is in full on my own account and the remaining £10 13s. 0d. is in full on account of Mr. Sam. Slater. Paid Mr. Calverley and Son one bill on Messrs. Margesson and Collison dated today 30 days date No. 382 value £10. Paid Margesson and Collison today on my cash account the bill value £130 which I received of Mr. Sam. Durrant a-Saturday last as also a bank bill value £20. Paid Mr. John Wathin in cash £3 16s. 0d. in full on my own account. Paid Cruttenden and Burgess in cash 5s. 6d. in full on my own account. Paid Margesson and Collison in cash £1 9s. 6d. in full on Mr. Sam. Slater's account. Spent today:

At Godstone: horse	4d.
Ostler	1d.
Breakfast 8d. ½ pint wine	1s. 2d.
Turnpike	2d.
	1s. 9d.

Looked out goods today at Levy's, Gore's, Crowley's etc.

Tues. 20 Mar.   Balanced account with Mr. Michael Gatfield and paid him one bill on Margesson and Collison dated today 30 days date No. 383 value £16 4s. 0d. and which is in full. Paid Messrs. Barlow and Wigginton one bill on Margesson and Collison dated today 30 days

---

[4] For Turner's suppliers whom he visited on this trip to London see Appendix C.

date No. 384 value £12 18*s*. 0*d*. and which is in full. Paid Mr. James
Blake one bill on Messrs. Margesson and Collison dated today 30 days
date No. 385 value £36 14*s*. 0*d*. of which bill £25 is on account of my
mother and the remaining £11 14*s*. 0*d*. is in full on account of Mr.
Sam. Slater. Paid Messrs. Rushton and Kendall one bill on Margesson
and Collison dated today 30 days date No. 386 value £29 which bill is
in full of my account. Paid Gore and Perchard one bill on Margesson
and Collison dated today No. 387 30 days date value £6. Paid Mr.
Richd. Sharp in cash £1 17*s*. 0*d*. in full on account of Mr. Sam. Slater.
Paid Smith and Bickham in cash £3 10*s*. 6*d*. in full for Mr. Sam.
Slater. Paid Cruttenden and Burgess in cash £1 15*s*. 6*d*. in full for Mr.
Sam. Slater. Paid Gore and Perchard 5*s*. for a Gunter's 4-rod chain.
Paid Mrs. Mary Reynolds in cash £2 14*s*. 6*d*. in full on my account.
Paid Mr. Tho. Neatby in cash £1 16*s*. 0*d*. on my account. Paid Mr.
John Levy in cash £1 16*s*. 6*d*. in full on my account. Paid Daker and
Stringer in cash £1 0*s*. 8*d*. in full. Paid them afterwards in cash £2 8*s*. 0*d*.
in full for a parcel bought of them today. Sent by a servant of Mr.
Standen's to Mr. Wm. Farnworth's in cash £4 1*s*. 0*d*. and for which he
brought me a receipt from Mr. Wm. Farnworth as received in full for
Mr. Ralph Heale of Calne in Wilts on my account. Messrs. Margesson
and Collison paid Mrs. Crowley and Co. on my account in full £5 10*s*. 8*d*.
I also received on Margesson and Collison on my cash account
£5 5*s*. 0*d*. I breakfasted at Margesson and Collison's, dined at Barlow
and Wigginton's on a leg of pork roasted and apple-sauce and supped
at Mr. Neatby's and spent part of the even there. The remaining part
at Mr. Sterry's, where I lodged. Spent 2*d*.

Weds. 21 Mar.   Balanced accounts with Mr. Geo. Otway (where I
breakfasted) and paid him a bill on Margesson and Collison dated
today 30 days date No. 388 value £19 18*s*. 10½*d*. which is in full of my
account to the 1st of February last. Paid Mr. John Albiston in cash £7
which is in full on account of Mr. Sam. Slater. Received of Mr.
Norfolk, pewterer, in cash 17*s*. 0*d*. for goods sold to him today, *viz*.,

To 15⅜ lb. pewter, which was my mother's	7*s*.	8*d*.
To 18¾ lb. my own	9*s*.	4*d*.

Received of Mr. Nuns in cash 16*s*. 2*d*. for goods sold him today,
viz., ·

3 lb. brass, my mother's	1*s*.	6*d*.
10¼ lb. *do*. my own	5*s*.	1½*d*.
11¼ lb. *do*.	7*s*.	8*d*.
Copper 2¼ lb.	1*s*.	10½*d*.

I then paid him in cash 12s. 3d. for goods bought of him today, viz.,

One tea kettle	9s.	6d.
2 pound weights	2s.	0d.
1 half pound do.		6d.
1 quarter do.		3d.

Received of Gore and Perchard for horns 15d. and then I paid him in former account as also for horse-hair delivered to him today. Received of Messrs. Thomson and West in cash £4 15s. 0d. which was the money they sold a pocket of hops belonging to Joseph Fuller for. Received of Gore and Perchard for horns 15d. and then I paid him in cash £1 which was a mistake they had made yesterday, my bill being £7 instead of £6. About 2.45 I came out of London and thank God arrived safe at home about 10.45.

Spent and paid today as under:

Spent in town		9d.
Horse and ostler	4s.	0d.
Gave Mr. Sterry's servant	1s.	0d.
Do. the porter	1s.	0d.
At Godstone, my horse		4d.
Dinner		4d.
½ pint wine		6d.
Ostler		1d.
Forest Row, my horse		4d.
Gill wine		3d.
Turnpike		4d.
		8s. 11d.

The true state on my journey in regard to all my disbursements etc. . . . . £251 12s. 1d. . . . . Disbursed etc. on account of Mr. Sam. Slater . . . £40 13s. 6d. . . . . For Mrs. Turner . . . Paid £25 . . . Received £30 9s. 11d. . . . Due to Mrs. Turner £5 9s. 11d.

Thurs. 22 Mar. At home all day . . . Mrs. Vine drank tea with us. In the afternoon sent our servant to Framfield, who brought me word my mother was rather worse.

Sun. 25 Mar. . . . After dinner Tho. Davy and I walked over to Framfield, and Richard Fuller on their horse carried my wife. We found my mother very ill . . .

Sun. 1 Apr. About 6.20 my brother Moses sent a man to acquaint me with the melancholy news of my mother's death, which happened this morn about 2.30. A melancholy theme to say anything upon. No, what can we do under such misfortunes but to submit to the will of Almighty God, and from such instances of the mortality of our nature learn to supplicate the Father of mercies to give us grace that we may meet death at any time (when He shall think proper to call) without fear, by being prepared by a virtuous and pious life, through the mercies and merits of our blessed Saviour. I walked over with the the messenger and breakfasted with my brother and sister and also dined there in company with my brother Dicky. In the afternoon Tho. Durrant brought my wife over on Mr. Thornton's horse. We stayed at Framfield and drank tea and came home about 7.30 . . .

Tues. 3 Apr.   After breakfast went down to Halland with a pattern of cloth for his livery. We dined on the remains of yesterday's dinner. After dinner walked over to Framfield where I drank tea. Came home about 7.10 . . .

Weds. 4 Apr.   . . . Our late servant Mary Martin came in to assist my wife tomorrow in my absence . . .

Thurs. 5 Apr.   . . . In the morn took up a livery coat for Mr. Coates's servant and a fustian frock for Mr. Porter's son and servant. About 10.20 I walked over to Framfield and dined at my brother's on a leg of mutton roasted, a piece of beef boiled, greens, potatoes, and a pond currant pudding (my family at home dining on the remains of Sunday's dinner) in company with my brother and sister, brothers Dicky and John and William, my uncle Hill and cousin Molly, my aunt Moon, Uncle Will. Ovendean and his wife and my cousin Tho. Ovendean. About 8.30 we buried my mother. Ah, melancholy scene! She was in the 62nd year of her age. We are now left as it were without any head, quite mother and fatherless, and it seems just as if we was now a-going to turn out in a wild world without any friends. Oh, may the God of all mercy pour His holy spirit into our hearts that we may grow in grace and unite together with brotherly love and kindness and always think of our high calling through our blessed Saviour and Redeemer Jesus Christ. I stayed at Framfield all night.

Sat. 7 Apr.   . . . Our late servant went home and we made her a present of a handkerchief, value 13½*d.*, for her trouble of coming over . . . A remarkable shower of hail today between 12 and 1 o'clock and which I imagine continued 40 minutes. The hail was very large and came so fast that it lodged on the ground as it were like snow.

Good Friday, 13 Apr.   After breakfast walked over to Framfield in order to meet my brother Richard to read my mother's will, but he did not come, so that at present the content of her will is still a secret to me . . .

Mon. 16 Apr.   . . . About 3.20 I went down to Jones's to the vestry . . . We were very unanimous at our vestry; nothing but harmony and good humour seemed to be in every one of the people there . . . The officers nominated for the year ensuing were Jn. Cayley, overseer, Jn. Watford, electioner; Thos. Fuller, churchwarden, Jos. Fuller, electioner. . .

Fri. 20 Apr.   In the forenoon I walked over to Framfield where I met my brother Richard, and we according to our agreement opened my mother's will and read it and wherein she has given all her personal estate to be divided between my brothers Moses and Richard and my sister Sally. Her farm at Chiddingstone in the county of Kent she hath given between myself, my brothers Moses and Richard and my sister Sally as tenants in common, and hath charged my part with the payment of six pounds to my sister-in-law[5] Elizabeth if she be living at the time of my mother's death, but if she be not, I am to pay the same to a natural or base-born son of my said sister's, named Philip. And she has made, constituted and appointed my brother Moses executor to her will[6] . . . Oh, may the God of all mercy pour down his blessing upon my brothers and sister, as also upon myself, that we may live in peace and unity amongst ourselves and that we may have continual in our view the joyful hope of an eternity in heaven (though not by any works of our own, but only through the merits and mediation of Jesus Christ).

Sat. 21 Apr.   . . . In the morn about 5 o'clock we had a very great stroke of thunder, or, as some think, it was not but something of an explosion in the air, different from that of thunder . . .

Sun. 22 Apr.   . . . My wife and self at church in the morn where we had a sermon preached by the Rev. Mr. Tho. Hurdis, curate of Laughton (Mr. Porter being gone to Laughton to administer the sacrament), from the latter part of *Luke* 19.44: 'Because thou knewest not the time of thy visitation.' We dined on a leg of mutton roasted. In the afternoon our late servant Mary Martin came over, and she together with myself and wife and maid at church in the afternoon when Mr. Hurdis preached us another sermon from *Luke* 6.41,42 . . . In my opinion he is as fine a churchman as almost I ever heard . . .

Tues. 24 Apr.   . . . In the forenoon Mr. Stone came and made me an issue upon my back by eating it in with caustic, and then scarifying it . . .

[5] i.e. half-sister. Elizabeth was the daughter of Turner's father by his first marriage. See Appendix A.
[6] A copy of the will, made on 9 October 1754 but not proved by Moses Turner until 28 Agust 1777 (the same day that Thomas Turner proved their father's will) is in E.S.R.O., South Malling wills register D9, p. 236.

Thurs. 26 Apr. . . . After breakfast walked over to Framfield. From thence my brother and I walked over to Uckfield where I paid Mr. Cheeseman, alias Chesmer, 6s. 8d. for a copy of the court roll for my house in Framfield, he being steward of the Manor of Framfield, which copy he is to send me the 14 May next . . .

Fri. 27 Apr. In the forenoon Tho. Davy at our house a-taking up a coat and waistcoat. I paid him one shilling which is for the same sum he paid for me at Mayfield yesterday, it being our club feast[7] . . .

Sat. 28 Apr. . . . In prodigious pain all day with my back.

Mon. 30 Apr. . . . In the afternoon Mr. John Breeden of Pevensey came to our house in order to take an inventory of my mother's stock-in-trade. He together with Richd. Page and his wife drank tea at our house, and after that Mr. Breeden and I walked into a neighbouring field to see a person pick up 100 stones laid one yard distance from each other, which he performed in less than 46 minutes, but he should have done it in 45. The distance is near 5¾ miles . . .

Tues. 1 May. In the morn Mr. Breeden and I walked over to Framfield. I carried with me the leg of mutton bought a-Saturday, which I made my brother and sister a present of. After breakfast Mr. Breeden and I proceeded to take an inventory of my mother's effects. As my brother proposes keeping forward the shop, we also valued the goods to him. We dined there on a leg of mutton roasted and a green salad. Stayed at Framfield all night.

Weds. 2 May. We continued very busy appraising the goods. My brother Richard came over to us and dined at Framfield on a shoulder of mutton roasted and green salad. We stayed at Framfield all night.

Thurs. 3 May. Mr. Breeden and I dined at Framfield on the remains of Tuesday's and yesterday's dinner with a plain rice pudding. Today we completed the appraising the stock-in-trade and stayed and drank tea at Framfield and then came home about 7.10. Mr. Breeden supped and lodged at our house. In the even ill with a sore throat, swelled face and the fever. Tho. Davy supped with us on account he rode Mr. Breeden's mare over to Framfield, but we was come away before.

Fri. 4 May. In the morn Mr. Breeden breakfasted [with] us and then set out for Nutley Fair. I being ill, we concluded to postpone any further appraising till Thursday next . . .

Sat. 5 May. At home all day and very much indisposed . . .

---

[7] See above, p. 99, note 25.

Sun. 6 May. . . . At home all day and, thank God, much better.

Sun. 13 May. Myself and wife at church in the morn . . . We dined on a leg of lamb boiled, a piece of bacon, a gooseberry pudding and some greens. My wife, self and servant at church in the afternoon . . . In the afternoon my brother Moses came over and stayed and drank coffee with us and then went home. In the even and the day read two of Tillotson's sermons and part of the 2nd volume of Hervey's *Meditations*. Oh, what a most delightful time it is—the birds tuning their melodious throats and hymning their creator's praise whilst perhaps man, ah! frail and degenerate man, lies supinely stretched on a bed of luxury and ease, or else so immersed in the vain and empty pleasures of this world that he is utterly forgetful of the goodness of the Supreme Being that showers down His blessings upon him, and sheds plenteousness around his table. Also every tree and shrub bespeaks the wisdom and goodness of an all-wise creator.

Mon. 14 May. . . . My brother Moses, James Marchant and myself walked to Uckfield Fair where I bought of Mr. Per. Hart 12 chamois skins, for which I paid him 10*s*. We went in to see a man perform several curious performances on the slack wire, and which I actually think were very curious. We paid one penny each . . .

Tues. 22 May. We dined on a pigeon pudding. After dinner I rode over to Framfield. From thence I rode to Chailey to inform my brother Richard that myself and Moses intended to go up to Chiddingstone a-Tuesday next. When I came to Mr. George Beard, I found him absent, being in London, and his wife a-visiting; one child at home very ill, the other cross and sufficiently employed one servant to look after it; the other servant a-brewing; a mantua-woman there at work, and my brother as full as he could turn his hands in the shop. Oh, strange confusion! But however, I stayed and drank tea and came home about 8.15. I called at Isfield to bespeak some oatmeal.

Fri. 25 May. Paid Mr. Lawrence Thornton, our officer of excise, in cash 5*s*. 1¼*d*., which is in full for the same sum he paid the collector of excise yesterday for an inland duty on the stock-in-hand of coffee and chocolate, *viz*.,

> For 3½ lb. coffee   4*s*. 4¼*d*.
> For 1 lb. chocolate   9*d*.   . . .

Weds. 30 May. . . . My wife very ill all day. Oh, a melancholy time! What will become of me I cannot think. Very little trade and always so afflicted with illness—but let me not repine at that, as it is the will of the Supreme Being, which always knows what is better for us than we ourselves. Possibly it is good for us that we have known affliction . . .

Sat. 2 June. . . . My wife really extremely ill.

Mon. 4 June. My brother Moses came over to stand the shop for me in my absence a-serving the funeral of the wife of Henry Osborne . . . About 2.30 I set out for the funeral where I gave away the following gloves:

2*d.* chamois women's: Mrs. Kenward, Mrs. Elphick, Goldsmith's servant, Dame Lewer, Dame Martin, the servant maid.

Men's 2*d.* chamois: Mr. Tho. Osborne, Hen. Osborne Jr., Hen. Osborne Sr., Will. Osborne, Mr. Kenward, Mr. Goldsmith, Mr. Elphick, Will. Goldsmith, Edmd. Elphick, Richard Page Jr., Tho. Lewer, The Rev. Mr. Porter, Tho. Turner. Men's in all, 13 pairs; women's in all, 6 pairs. Expended in all, 19 pairs.[8]

We brought the corpse to church about 5.50 where we had a sermon suitable to the melancholy occasion from *Acts* 17.31 . . . Nothing should, I think, have a greater effect upon the mind of man to admonish him to forsake his evil ways and prepare for eternity than such instances of mortality, unless it be the unbounded goodness of Almighty God that suffers so many of us still to continue in this life that we may repent and return unto Him whom we have behaved to as rebels and apostate creatures. I say, how striking must this be upon the minds of men to think that we are not cut off for our multiplied transgressions, nor sent down quick into the grave for our daily impieties! Oh, would man but wisely consider this and in due time make his peace with God (through the merits and intercession of our Saviour Jesus Christ) and prepare himself for an eternity, how would the fear of death approaching be turned into that of hope and comfort and even joy itself! And the putting off this mortal body would be as it were only a transition from a state of trial and probation, of trouble and pain, to an eternal state of joy and happiness in heaven where we shall with angels and archangels be incessantly praising and blessing and adoring our heavenly Father and also our only Saviour and Redeemer. In the evening went down to Jones's to the vestry . . .

Weds. 6 June. . . . Mr. Thornton and I played a few games of cribbage in the even; I won one penny. At home all day. My wife very ill; Mr. Stone paid her another visit.

Thurs. 7 June. . . . About 7.30 Tho. Davy, Tho. Durrant, and Mr. Thornton and myself went to the widow Horsecraft's at Chiddingly. The account of our going was this: the good woman buying spirituous liquors of me, I have long since proposed to go and spend 6*d.* with her,

---

[8] See above, p. 7, note 5.

which was this night agreed to. I spent 8*d.* and came home about
11.20, and in a manner quite sober . . . My wife something better.

Sat. 9 June. Borrowed of James Marchant in cash 10*s.* 6*d.* Paid T.
Bentley, broom-maker, 13*s.* in full for 202 brooms received by him
today . . . After dinner Mr. Elless and I rode to Alfriston where we
stayed and smoked a pipe or 2 with Mr. Cooter. I also spent some time
with Dr. Snelling, for whose opinion I went; that is, to know if he could
help me to any salve to dress the issue upon my back with that was
more adhesive than that I had already, or if I could have a bandage to
keep on the dressings with.[9] We came home about 8.30 . . . Spent
nothing today, only 6*d.* which I paid Dr. Snelling for some salve.

Tues. 12 June. In the forenoon my brother came over and brought
Philip who is to board at Master Hook's upon the following terms:
Master Hook is to find him in meat, drink, washing and lodging for £5
per year, though conditionally that if either party want to relinquish it,
they may at any time. And the boy accordingly went to Master Hook's
today . . . In the even went down to Mr. Elless's and smoked a pipe or 2
in company with Jos. and Tho. Fuller, Mr. Burges, Robt. Hook and
Mr. French. I came home about 10.10. The reason of our paying this
visit was a-taking of leave of Mr. Elless, he being a-going to live at
Alfriston . . . Took physic today.

Tues. 19 June. In the morn about 5.10 I set out with an intent to go to
Maidstone. I got to the Wells[10] about 7.50 and breakfasted there with
my brother. After breakfast I set forward and overtook the team that
carried my rags very soon, and went with them to Hadlow Common
where I sold my rags to the widow Burtenshaw[11] as under, *viz.*,

3 bags rags, 6 cwt. 1 qr. 9 lbs., at 30*s.*	£9	9*s.* 11½*d.*
Carriage of *do.*		9*s.* 6*d.*
	£9	19*s.* 5½*d.*

which money I received of her and drank tea with her and bought of
her 2 reams of writing paper, for which I paid her in cash £1. Finding
the smallpox to be greatly at Maidstone, I concluded not to go, but
gave Mr. John Gosling in cash £29 and also my orders and then came
back and supped and lodged at my brother's in Tunbridge Wells
where I arrived at 8.10 . . .

Weds. 20 June. . . . This day is that on which I enter into the 31st
year of my age, and, Oh, may I, as I increase in years, also increase in

---

[9] The issue had been opened by Richard Stone on 24 April.
[10] Tunbridge Wells.
[11] The Burtenshaws or Buttanshaws were paper makers at West Peckham, Kent.
See A. H. Shorter, *Paper mills and paper makers in England, 1495–1800* (1957), p. 197.

goodness! And then and then only is it that I shall know pleasure in this life and also may hope to end my days with peace and tranquility of mind as becomes a Christian.

Mon. 25 June. . . . My wife very ill. Paid John Streeter 12*d.* for 1 oz. of elixir of vitriol[12] he bought for me today.

Thurs. 28 June. . . . In the afternoon walked with Mr. Thornton to Eason's Green (called at Mr. French's and looked at his wool) to see a cricket match played between the parish of Framfield with 3 men out of Uckfield and the parish of Chiddingly with 2 men out of Chalvington and one out of Ripe. Chiddingly eleven beat only one run and all the men out on both sides.[13] I came home in company with James Marchant and Thomas Davy about 8.10, having not spent anything . . .

Sun. 1 July. Myself only at church in the morn. We had a sermon preached by the Rev. Mr. Chalice, curate of Ripe (Mr. Porter preaching himself at Ripe) . . . According to my opinion Mr. Chalice is a very indifferent orator, either for a good delivery or a good discourse . . .

Thurs. 5 July. . . . About 1.10 Charles Diggens called on me and I walked with him to the Dicker to see a game at cricket played between the same two elevens that played on Eason's Green the 28th ult. which was not played out, though if it had, in all probability Framfield would have beat, they heading the other eleven 98 runs the first innings.[14] Came home about 8.30. Spent 1½*d.* though neither ate nor drank during the time I was absent from home. A prodigious hot day, but a fine serene air, there being nothing of a sultry air.

Sat. 7 July. . . . This day received by the post the disagreeable news of the French being landed at Dover. Oh, melancholy news! but yet I hope it is only a false report set about by some credulous and fearful people without any real foundation . . . My wife very ill with a cold all day and I think somewhat dangerous.

Mon. 9 July. . . . Very busy all day and at home. This day I saw in the newspaper that instead of our being invaded by the French, we have a fleet under the command of Admiral Rodney now lying before Havre de Grace and hath burned upwards of 30 flat-bottomed boats and also were a-bombarding the town and had set it on fire in two places when the express came away. So that we have a sudden transition from sorrow to joy.[15] In the afternoon posted my day book.

[12] Aromatic sulphuric acid.
[13] See above, p. 9, note 9.
[14] See above, p. 9, note 9.
[15] The French threat to invade was very real in June and July 1759 and, as the entry

Tues. 10 July. In the morn about 6.10 rode over to Framfield where I breakfasted with my brother and also dined with him in company with my brother Richd. and Mr. Geo. Beard on some part of a leg of lamb and some beans . . . We met intending to settle my mother's affairs, but as my brother had not got in all the book debts and bills, and at the same time my brother Richd. behaving very contrary, we did not settle anything . . . My wife paid Mr. Sam. Beckett 16*d.* in full for 4 lbs of lobsters received by him today.

Weds. 11 July. This day John Cayley and his servant Ann Smith were married at our church . . . In the afternoon played at cricket some time, and I lost about 1½*d.*[16] . . . In the even bottled off some cider.

Fri. 13 July. . . . This day the Rev. Mr. Porter's son was baptized.

Sun. 15 July. . . . This morning Jas. Hutson and Mary Evenden were asked the first time . . .

Tues. 17 July. . . . In the evening [my] brother came over for some coffin letters. In the even I went out to cricket some time but long enough to play out a match[17] . . .

Weds. 18 July. After breakfast rode to Mr. Dicker's to collect in a small bill, but he was [not] at home. Called on Mrs. Peckham, but to no purpose. Called on Rippington, but to no purpose. Went to Alfriston for Mr. Jn. Snelling's opinion for my wife, but he was not at home. Dined with Mr. Fran. Elless on some bread and cheese (my family at home dining on some beans and bacon). Came home by Hellingly to find Master Darby, but to no purpose. Called on Hen. Osborne, but to no use. Came home about 5.30, and I think I never rode when it was so intense hot as it was today about 10.40. Spent today 9*d.*

Tues. 24 July. . . . In the even Mr. Thornton carried my wife out for some air. Very busy . . .

Sun. 29 July. In the morn rode to Mr. Snelling's at Alfriston to consult him further concerning my wife's illness, but he was not at home; I went forward by his servant's direction and found him at Mr. Allfrey's at Westdean. We came back to his house where I breakfasted with him and paid him in cash 3*s.* 8*d.* for medicines I had of him the 21st instant and today. I came home about 11.20. In my absence our late servant Mary Martin came to see us and dined with us

for 7 July shows, rumour was rife. Rodney's attack on Le Havre which did much to raise British morale took place between July 3 and July 7. See Julian S. Corbett, *England in the Seven Years' War* (1907), II, pp. 24–5.

[16] This reference is not included in the article in *The Cricket Quarterly* referred to above, p. 9, note 9.

[17] See above p. 9, note 9.

on some bacon, a pudding cake and some French beans. My family all
at church in the afternoon; the text is *Joshua* 24.24: 'And the people
said unto Joshua, The Lord our God will we serve, and his voice will
we obey', from which words we had an excellent sermon as I think I
ever heard . . .

Tues. 31 July. . . . Down at Halland three times today. A great deal
of rain today. In the afternoon drawed off a barrel of cider. James
Marchant supped at our house, he helping me powder some sugar for
Halland.

Weds. 1 Aug. After breakfast rode over to Framfield and stayed
there while my brother went to Uckfield to get me a pound of green
tea, for which I paid him 9s. 3d. . . . Down at Halland 3 times today . . .
What quantities of [people] begin to come [down] to Halland, and only
to prepare and make ready a provision for luxury and intemperance
against Sunday next, when perhaps hundreds of poor creatures are
lamenting for want of sustenance; and here shall be nothing but waste
and riot.

Sun. 5 Aug.[18] I spent most part of today in going to and from
Halland, there being a public day, where there was to dine with his
Grace the Duke of Newcastle, the Earls of Ashburnham and
Northampton, Lord Viscount Gage, the Lord Abergavenny and the
two judges of assize, and a great number of gentlemen, there being, I
think, upwards of forty coaches, chariots etc. I came home about seven,
not thoroughly sober. I think it is a scene that loudly calls for the
detestation of all serious and considerating people, to see the sabbath
profaned and turned into a day of luxury and debauchery, there being
no less than ten cooks, four of which are French, and perhaps fifty
more, as busy as if it had been a rejoicing day. There was such huzzaing
that made the very foundations (almost) of the house to shake, and all
this by the order and the approbation of almost the next man to the
King. Oh, what countenance does such behaviour in a person of his
Grace's rank give to levity, drunkenness and all sorts of immorality!

Sat. 8 Sept. . . . Paid John Jenner the hatter in cash and goods
£1 7s. 0d. in full for 12 hats received by him today, *viz.*,

6 men's hats at 2s.	12s. 0d.
6 *do*. 2s. 6d.	15s. 0d.

Paid Mr. John Gosling in cash and goods 4s. 4½d. for goods received
by him today, *viz.*,

---

[18] The volume of the original diary covering the period between 2 August and
2 September is missing. The entry for 5 August is taken from *Blencowe and Lower*,
pp. 203–4.

3 aprons	1s. 7½d.
3 skins	1s. 7½d.
6 throat hasps[19]	1s. 1½d.

. . . In the afternoon Mr. Thornton carried my wife to Eason's Green in order to see a cricket match.[20] At home all day and thank God very busy.

Fri. 14 Sept. Master Hook made my wife a present of a carp. We dined on a carp boiled, a batter pudding, a piece of beef boiled and some carrots. In the afternoon, Mr. Thornton carried my wife out for air . . .

Sat. 15 Sept. . . . Tho. Cornwell made my wife a present of a few eels . . .

Sun. 16 Sept. No service at our church in the forenoon, Mr. Porter preaching at Laughton. We dined on some eels parboiled and broiled, a lamb's head and pluck boiled, potatoes and turnips . . .

Sat. 29 Sept. In the morn Mr. Elless called on me and breakfasted with me, and after breakfast we rode to Waldron Thorns where we met Mr. Mason and surveyed a farm of Dr. Duke's of Battle. We came home about 4.20 . . . Spent (that is, gave a boy) today 3d. for drawing the chain.

Mon. 1 Oct. . . . Mr. John Vine smoked a pipe or 2 with me. James Marchant and Thomas Davy supped with us, who had been gathering of walnuts for us all the afternoon.

Thurs. 4 Oct. In the forenoon walked down in the park to look at an old pollard from whence a swarm of bees had been taken . . . In the even went down to Jones's to make up the following trifling affair, *viz.*, some time in the summer Master Bull of Whitesmith and a little boy of Francis Rich's, being together in Halland Park, found a swarm of bees which they agreed to divide between them, and accordingly sometime after, they sent a person to Mr. Gibbs the keeper to ask his consent to take the bees at the proper time for taking them. The answer he brought again was that Mr. Gibbs gave them his free consent so to do (though the fellow had never seen Mr. Gibbs, and only told them lies all the time) but however they, knowing no other but what they had the keeper's good will and free consent for taking them ventured (innocently enough) to proceed to action and about 3 weeks ago took the bees. Somebody next day told the keeper of it and he before night committed it into the hands of an attorney, and now it was agreed to be

[19] Either a fastening for the neck of a garment, or perhaps the same as a throat-latch: a strap passing beneath a horse's throat to keep the bridle in position.

[20] See above, p. 9, note 9.

left to me (in behalf of Rich and Bull) and Mr. John Goldsmith (in behalf of the keeper) to decide, when it was agreed between us for them to pay 2s. 6d. each, the value of the honey and wax, spend one shilling each and pay the lawyer's letter, which they did, and then I came away. But the keeper saying he did not want the money, I do imagine he spent most of the 5s. upon them, but sure it must appear cruel in the keeper to use his power in so arbitrary a manner, for he owned that he was thoroughly persuaded the men were honest and that they would by no means have taken the bees if they had not been assured in their own minds that they had his permission for so doing, but yet as it was in some measure contrary to law, he was determined to show his power that no one for the future should dare transgress, but what they must expect the law. Or at least if they disobliged him and the law open against them, they must expect to know something of the charge of law. But however I think that if showing of power tend only to oppress the honest and industrious poor (as it did in the aforegoing cause) happy is the man that hath least of it.

Fri. 5 Oct. . . . I hear today that Mr. Gibbs, instead of giving the men any of the money he received of them yesterday on the bee affair, again stayed and spent the greatest part of it upon them, which I think was but a small recompense, and specially as it was attended with a worse consequence, for one of the poor creatures, not being used to drinking, being very much in liquor, happened to drop some expression at which the keeper took umbrage, and I suppose positively declared he would [be] his professed enemy . . .

Mon. 15 Oct. In the forenoon my cousin Charles Hill came to see me, and dined with us on a shoulder of mutton roasted, a light and bread pudding and potatoes . . . Thank God very busy most part of the day. My cousin Hill, having been a surgeon aboard a man of war, as also a letter of marque[21] to Jamaica, entertained us in the even with some of the incidents he met with in his voyages, but all in a modest manner, very unlike most modern travellers.

Sat. 20 Oct. . . . Paid John Gosling Jr. as under, *viz.*,

To 2 cwt. 1 qr. 26 lb. of rags which he sold for me at Maidstone Fair at 31s. 6d. per cwt.	£3 18s.	3d.
In cash today		6d.
	£3 18s.	9d.

[21] A ship having the sovereign's licence to be armed and to capture the merchant shipping of a hostile nation. The actions of such ships were recognized by international law as legal and therefore not acts of piracy. Such ships were called privateers or corsairs.

Per contra Cr.

To 2 cwt. 3 qr. 2½ lb. Cheshire cheese
 received by him today at 27s.   £3 14s. 10½d.

Carriage of rags   3s. 9d.
                £3 18s. 7½d.

I then paid him in cash 12s. 10½d. which was for a mistake made the
22nd June last. I then received of him 1½ cwt. of sugar at 50s. instead
of 1¼ cwt., which with the carriage is 12s. 10½d. . . . In the even read
the extraordinary *Gazette* for Wednesday, which gives an account of
our army in America, under the command of General Wolfe, beating
the French army under General Montcalm (near the city of Quebec)
wherein both the generals were killed, as also two more of the French
generals, and the English general Monckton, who took the command
after General Wolfe was killed, was shot through the body, but is like
to do very well; as also the surrender of the city of Quebec, with the
articles of capitulation.[22] Oh, what pleasure is it to every true Briton to
see with what success it pleases Almighty God to bless His Majesty's
arms with, they having success at this time in Europe, Asia, Africa and
America, and I think in this affair our generals, officers and common
men have behaved with uncommon courage and resolution, having
many and great difficulties to encounter before they could bring the
city to surrender.

Mon. 22 Oct. . . . At home all day and thank God pretty busy. My
wife very ill. In the even Mr. Thornton and I played at cribbage; I won
6d. Oh, how melancholy time I have, my wife being so continual ill, but
as it is the will of Almighty God let me always submit thereto with
patience and resignation.

Tues. 23 Oct. . . . My wife and I had an invitation to Halland to
drink tea and sup there with a great many more, there being a rejoicing
on account of the taking of Quebec (but as my wife was very ill neither
of us went). In the even there was a bonfire of 9 hundred of faggots and
there was about a hogshead of beer given to the populace, and several
discharge of cannon. In the day down at Halland twice. Thomas Davy
spent the even with me. I think this is not proper rejoicing on so great a
victory, for this is only occasioning of people to get drunk and run
themselves farther into sin and in reality of no service to any individual,
whereas I think as God giveth the victory, to Him should thanks and
praises be returned, and at the same time let our outward joy be
expressed by giving to the laborious and industrious poor what is too

[22] The expedition under General Wolfe and Admiral Saunders up the St. Lawrence
took Quebec from Montcalm on 18 September.

often consumed in vanity and luxurious folly and I doubt too often to the increase of vice and irreligion. I only mention this as my opinion (for I would leave every man to judge for himself) nor would I willingly condemn or blame any one for dissenting from me in any point.

Sun. 28 Oct. . . . Both at Framfield and at Hoathly we had a thanksgiving prayer read for the success with which it has pleased Almighty God to bless His Majesty's arms both by land and sea, and in my opinion it was extremely well composed. In my absence my wife was taken so extremely ill with a fit of the colic that they did not expect her to live until I came home, and accordingly sent for me by John Babcock, but as I stayed no longer at my brother's than just drinking of tea he missed me almost at home. I came home about 6.10 and found my wife really extremely ill. Paid John Babcock 6*d*. for going to Framfield. Tho. Cornwell, Tho. Durrant and Mr. Thornton supped at our house, on account they had been a-catching my horse.

Mon. 29 Oct. In the morn Tho. Davy went for Dr. Stone, who came along with him. Tho. Davy and Dame Durrant dined with us on the remains of yesterday's dinner. My wife so very ill that I was obliged to send Tho. Davy again for Mr. Stone, who accordingly came and paid her another visit. In the afternoon my brother came over, and he and Dame Gower and her daughter drank tea at our house. In the even my father Slater came to see us, who stayed all night. My brother also stayed with me till 8.30. My wife has really been so ill today that I did not see any probability of a recovery, though thank God she is much better in the even. Her illness was a violent colic, the gravel and an obstruction of the catamenia[23] . . .

Tues. 30 Oct. . . . Mr. Beckett made my wife a present of some fine whiting and a flounder. Mr. Stone paid my wife another visit . . .

Sat. 3 Nov. . . . In the afternoon my brother came over and drank tea with us. I received of him in cash 13*s*. 3*d*., *viz.*, to 6 slats 3*s*. 6*d*., to 4 testaments 5*s*. 2*d*., to 1 hive 4*s*. 3*d*., to 1 dozen buttons 4*d*. At home all day. My wife continues very ill, though rather in a mending way. Master Durrant smoked a pipe with me in the even.

Mon. 5 Nov. . . . In the even Charles Diggens, David Brooker, Joseph Fuller and Tho. Durrant sat for a while with us. In the even, nay, we sat a-drinking of cider so long that we were almost all somewhat in liquor.

Thurs. 8 Nov. . . . In the even went down to Mr. Porter's and paid him in cash 2*s*., and 6*s*. he owed me for 12 load of dung, makes together the

---

[23] The menstrual discharge.

sum of 8s. which is in full for one year's tithe due at St. Michael last . . .

Tues. 13 Nov. . . . Charles Diggens drank tea at our house. Today wrote a petition for Master Diggens to his Grace the Duke of Newcastle, to solicit his Grace to promote his son to be a supervisor . . .

Thurs. 15 Nov. . . . After dinner set out for Alfriston in company with James Marchant, Tho. Durrant and Thomas Davy (they on foot and myself on horseback). The intention of our journey was purely to see Mr. Elless (and by his desire). We got to Alfriston about 4.30. We supped with Mr. Elless at his lodgings on a fine piece of beef roasted, a currant pond pudding, a currant suet pudding and a butter pudding cake. We, together with Mr. Elless and the person at whose house he lodges, played at brag in the even, and notwithstanding we played as low a game as possible it was my unhappy lot to lose 3s. I think almost to give over ever playing at cards again, for I think it quite inconsistent with that which is right, for if we reflect how much more service this 3s. would have done had it been given to some necessitous but industrious poor than to be fooled away in this manner, therefore of consequence if there could have been more good done with what I lost I was not a-doing right when I was a-losing it. We spent the even and night till past 3 o'clock and, excepting my loss, extreme agreeable, for we had plenty of good liquor and a hearty welcome and no swearing or quarrelling, but all seemed prodigiously delighted with each other's company, and at the same time went to bed sober. We all lodged at Mr. Elless's lodgings . . .

Fri. 16 Nov. After breakfast we all went and called on Mr. Snelling, who took a walk with us to Seaford in order to see a five-gun battery erecting there. We stayed at Seaford about an hour and a half and came back to Mr. Snelling's and drank a bottle of beer, and then went and dined with Mr. Elless on a leg of mutton boiled, turnips and a butter pudding cake. We came home (thank God, very safe) about 4.30 and I think sober . . .

Mon. 19 Nov. . . . Paid Alex. Whitfield 2s. 4d. in full for Philip's schooling to this day . . .

Weds. 21 Nov. After breakfast I set out for Alfriston, there being a sale of a shopkeeper's goods who has lately failed in trade. I dined with Mr. Elless at his lodging . . . There being no goods at the sale that I wanted which I could buy worth my money, I came home about 5.15. My brother came over in my absence and stayed all night. Spent today 7d., viz., my horse 4d., myself 3d.

Fri. 23 Nov. . . . We dined on a sheep's head and pluck boiled, some broiled herrings, turnips and potatoes. In the afternoon Tho.

Davy carried my wife down to Mr. Sam. Gibbs's, he having a son baptized today, and my wife and I was both invited to the christening. And about 7.40 I went (in company with Joseph Durrant). My wife and I supped at Mr. Gibbs's on a turkey roasted, 3 roast teal, a roast chine and 4 boiled chickens, a cold venison pasty, gooseberry and bullace pie and a cold ham (in company with Mr. and Mrs. Coates, Mr. and Mrs. Porter, Jos. Fuller and his wife, Tho. Fuller and his wife, Mr. French and his wife, Mr. Calverley, Joseph Durrant and his wife, Mr. Hurdis and Mr. Joseph Gibbs, Mr. Fran. Gibbs and his wife, Tho. Page and his wife, Mr. Goldsmith and Mr. Whitfield). We came home about 3.50, pretty indifferent, though not sober. My wife played at brag in the evening; she lost 6½d. We gave the nurse and servant 12d. . . .

Sun. 25 Nov.   My wife, self and servant at church in the morn . . . Charles Vine and Ann Fuller asked their last time. We had a proclamation read for a day of general thanksgiving to Almighty God for his manifold mercies vouchsafed to this our happy isle (on the 29th of this instant) . . .

Tues. 27 Nov.   . . . Paid James Marchant 2s. for making 8 pair of fearnought spatterdashes.[24]

Thurs. 29 Nov.   This being the day appointed by His Majesty for a general thanksgiving to God for vouchsafing such signal successes to His Majesty's arms both by sea and land, particularly by the defeat of the French army in Canada and the taking of Quebec; and for most seasonably granting to us at this time an uncommonly plentiful harvest, my wife, self and servant at church in the morn. The text in *Psalm* 50.23: 'Whoso offereth praise glorifieth me: and to him that ordereth his conversation aright will I shew the salvation of God', from which words we had in my opinion a very excellent sermon . . .

Mon. 3 Dec.   John Durrant Jr. bleeded my wife in the morn, for performing of which I gave him 6d. . . .

Sat. 8 Dec.   . . . About 5.30 walked down to Halland (previous to my invitation yesterday) there being a rejoicing on account that Admiral Hawke hath dispersed a fleet which was preparing to invade this nation. The engagement happened near Belle Isle.[25] The advantage gained by our fleet was but small, only burning two French ships, sunk

---

[24] Spatterdashes were long protective gaiters, in this case made of fearnought, a kind of stout woollen cloth.

[25] This was the naval battle off Quiberon Bay, Brittany, which took place on 20–22 November. This defeat for the French Brest fleet together with that by Admiral Boscawen over the Toulon fleet off Lagos Bay, Portugal, on 19 August, put paid to French plans to invade England and cut their communications with their forces in America and the West Indies.

two and took one, and at the same time we had the misfortune to lose two which ran ashore in the night, and, as they could not be got off, the Admiral destroyed them. This engagement is looked on as a great advantage, on account it has entirely dispersed the fleet and at the same time wholly disconcerted their schemes so that in all probability their thoughts of invading these nations must be laid by for some time. There was a fire of 8 hundred of faggots, a discharge of cannon and considerable quantity of beer given away among the populace, and we had a supper of a cold sirloin of roast beef and bread and cheese. Our company was Mr. Porter, Joseph and Tho. Fuller, Mr. Jer. French, Mr. Calverley, Mr. Burges, Joseph Durrant, Mr. Whitfield, John Browne, Sam. Gibbs, Mr. Saxby, Mr. Goldsmith, Ed. and Wm. Shoesmith Jr. and Richd. Bridgman. We drank a great many loyal healths (and even too many). I came home about 11.15 after staying in Mr. Porter's wood near an hour and a half, the liquor operating so much in the head that it rendered my legs useless. Oh, how sensible I am of the goodness of the divine providence that I am preserved from harm; and, that I may always have a just sense thereof, let me ever be meditating of the goodness of Almighty God to me, a poor sinful creature.

Sun. 9 Dec. At home all day. I took some physic . . . Our servant at church in the afternoon, but neither my wife nor myself all day. Oh, how does the stings of a guilty conscience torment me! Well might Solomon say: a wounded spirit who can bear? Oh, let this be one more admonition for me never to do so again!

Sun. 16 Dec. John Durrant Jr. bleeded my wife in the morn, for performing of which I gave him 6*d.* Myself only at church in the morn . . . We had a thanksgiving prayer read for the late victory obtained at sea over the French fleet by Admiral Hawke near Belle Isle . . .

Weds. 19 Dec. . . . This day Mrs. Atkins in a manner huffed me pretty much because I would not cut her a Cheshire cheese at the same price I sold them whole. Now I affixed a small profit upon my Cheshire cheese, even only about a farthing a pound, by reason I would not cut them, so that if I have disobliged her in this I can no ways reflect on my own ill conduct, though undoubtedly I should be loath to disoblige any person, but still, if they will be angry with me when I am strictly speaking doing them justice, I can no ways prevent it. As I told her, a-Saturday next I should have some that I would cut her any quantity of, nay, even 4 pounds, but those I have now by me being very large and old, in cutting one I must lose a considerable quantity.

Fri. 21 Dec. About 3.10 we arose to perform our task, *viz.*, some of

the ancestors of the Pelham family have ordered that on this day
(forever) there should be given to every poor man or woman that shall
come to demand it 4*d*. and every child 2*d*. and also to each a draught of
beer and a very good piece of bread. My business was to take down
their names while Mr. Coates paid them, and I believe there were
between 7 and 8 hundred people relieved, of all ages and sexes, and
near £9 distributed, besides a sack of wheat made into good bread and
near a hogshead and a half of very good beer. I came home about
11.20, having received my groat amongst the rest . . . Relieved [30]
poor persons today with 1*d*. each and a draught of beer . . .

Sat. 22 Dec. . . . Received from my father Slater by the carrier a
present of a fine turkey . . .

Sun. 23 Dec. . . . Took physic today. Mrs. Virgoe, her son and
daughter and Tho. Davy dined with us on a turkey roasted, a piece of
beef boiled, a raisin suet pudding and turnips . . .

Tues. 25 Dec. My wife and self at church in the morning . . . My
wife and I both stayed the communion; we gave 6*d*. each. The widow
Marchant, James Marchant and Eliz. Mepham dined with us on a
buttock of beef, a raisin suet pudding, turnips and potatoes. They
stayed and drank tea with us.

Weds. 26 Dec. . . . In the even I went in to Joseph Durrant's to
smoke a pipe with Mr. Thornton, where was Joseph Fuller Jr. and Mr.
Elless. We stayed until near 3.20 when I came home sober. Mr. Elless
came and lodged at our house. I think I never spent an even with less
pleasure in my life, most of the company being in liquor, and very
indifferent company, some before they was in liquor and some
afterwards, though if I speak truly I think universal good nature did not
abound . . . Paid Jos. Fuller 4*s*. for highway tax.

Sat. 29 Dec. . . . Gave John Streeter the post 2*s*. for his Christmas
box; that is, 12*d*. which I gave him last year and which he returned
again, thinking it not enough, and 12*d*. for this year.[26] Gave 12*d*. to the
carrier's servant for his Christmas box . . .

[26] See above, pp. 169, 171.

1760

Weds. 2 Jan. . . . At home all day, and but little to do. Joseph Fuller Jr. and Mr. Thornton smoked a pipe with me in the even. Oh, how pleasant has this Christmas been kept as yet, no revelling nor tumultuous meetings where there too often is little else but light and trifling discourse, no ways calculated for improvement; and it's well if it's not intermixed with some obscene talk and too often with vile and execrable oaths. Not that I am anyways an enemy to innocent mirth, but what I protest against is that which is not so.

Tues. 8 Jan. . . . Received of John Watford half a hog weighing 17 stone 6 lb. at 2*s.* the stone[1] . . .

Weds. 9 Jan. . . . Received of John Long in cash £2 8*s.* 0*d.* for Mr. Alex Whitfield, it being the money he sold the furniture of the school to him for . . .

Thurs. 10 Jan. We dined on some hog's feet, ear etc. boiled, with potatoes and a piece of beef boiled . . . at home all day. A very sharp cold day.

Fri. 11 Jan. Mr. Fran. Elless dined with us on a hog's haslet roasted and applesauce . . .

Sat. 12 Jan. . . . We dined on the remains of yesterday's dinner. Very ill all day with a violent purging and a prodigious sickness. A most remarkable cold day, it freezing even by the fireside, as also in our bedchamber. This day hired of Mr. John Long the school-chamber at 10*s.* a year.

Tues. 15 Jan. . . . My wife bad with the rheumatism in her face. Gave a man 6*d.* who went about with a petition to make up his loss sustained by fire at Wartling in this county, the damage being estimated at about £55.

Weds. 16 Jan. My wife went down to Mr. Porter's to dinner . . . Paid Tho. Cornwell 16*d.* for two rabbits bought of him today. About 7.35 I went down to Mr. Porter's where I supped on some roasted chicken, a cold goose pie, a cold chicken pasty, a cold ham, a piece of cold roast beef, some cold roast goose, tarts, etc. in company with Mr. and Mrs. Porter, Mr. Calverley, Mrs. Atkins, Mr. French and his wife, Joseph Fuller and his wife, Joseph Durrant and his wife, and Tho. Fuller. We played at cards in the even; my wife and I lost 16½*d.* We

---

[1] See above, p. 1, note 2.

gave the servant 12*d*. Came home about 2.10, very sober, and so was the whole company, in a manner.

Thurs. 17 Jan.   Took physic today . . .

Fri. 18 Jan.   . . . My wife went to Joseph Fuller's to dinner . . . About 8.05 I went up to Joseph Fuller's where we played at brag till suppertime; my wife and I won 16*d*. I supped there on some roasted chicken, a shoulder of mutton roasted, some cold neat's tongue, cold ham, part of a cold roast goose, part of a cold chicken pasty, tarts and puffs etc. in company with Mr. and Mrs. Porter, Mr. and Mrs. French, Mr. Calverley, Mrs. Atkins, Mr. Piper, Dame Durrant, and Tho. Fuller. We came home about 1.35 and I think quite sober. My wife and I gave their servant 12*d*.

Sat. 19 Jan.   . . . We dined on some pork bones roasted in the oven and applesauce. At home all day and pretty busy . . .

Mon. 21 Jan.   . . . Paid Mrs. French by a book debt and goods £1 10*s*. 0*d*. in full for 60 lbs of butter she put up for me (that is, I got it put up for Mr. Tho. Scrase at Lewes). In the forenoon Mr. Scrase came to see me and dined with us on a rabbit stewed, some sausages and the remains of yesterday's dinner. I received of him in cash £2 9*s*. 4*d*. which is in full on his account except the butter I had put up for him this year.

Thomas Scrase Dr.

To 26 yards Irish at 2*s*. 2*d*.		£2 16*s*. 4*d*.
To 60 lbs. butter ⎫ the year 1758		£2 5*s*. 6*d*.
To 31 *do*. ⎭		£5 1*s*. 10*d*.

Per contra Cr.

To 2 wigs for my brother William		£1 11*s*. 6*d*.
To one *do*. myself		£1 1*s*. 0*d*.
Received in cash today		£2 9*s*. 4*d*.
		£5 1*s*. 10*d*.

so that there remains due to me only for the butter that I got put up for him this year, which was 104 lbs. Mr. Scrase stayed and spent the afternoon with me and then went home . . .

Thurs. 24 Jan.   . . . Richard Fuller carried my wife down to Mr. French's to dinner . . . About 7.40 Master Durrant and I walked down to Mr. French's where I played at brag till suppertime. My wife and I lost 3*s*. 7*d*. I supped there on some roast ducks and roast chicken, part of a cold boiled buttock of beef, cold ham, a cold chicken pasty, tarts, puffs etc., in company with Mr. and Mrs. Porter, Jos. Durrant and his wife, Joseph and Tho. Fuller, Mrs. Virgoe and Mrs. Atkins, William and John Piper. We came home about 3.20 and thank God very sober, as was all the company (except Dame Durrant). But I must own it grieves me to lose so much money, and especially as I think it wrong, nor would I ever play, was it not upon the account of being sneered at by the company; not that I think it a crime to play at cards, no farther than the consequences accruing therefrom; as for tradesmen to lose anything considerable, it is a-lessening the stock-in-trade and perhaps injuring the creditors. My wife and I gave Mr. French's maid 6*d*. each.

Mon. 28 Jan. . . . In the even I heard that Edmund Elphick was absconded, who is about £5 in my debt, and myself and Master Hook walked down to talk with John Browne about it, and we found the report to be true, for my advantage.

Tues. 29 Jan. . . . My wife went down to Halland to dinner . . . About 7.30 I went down to Halland, where I supped on a cold roast turkey, a breast of veal roasted and hot, part of a cold chicken pasty, some ducks roasted and hot, cold neat's tongue, tarts etc. in company with Mr. and Mrs. Porter, Mr. and Mrs. French, Mr. Calverley, Mrs. Atkins, Tho. Fuller, and Joseph Fuller and his wife. We played at brag in the even and according to custom my wife and I lost 2*s*. 10½*d*. We came home about 2.20, sober. We gave among the servants 12*d*. O cruel fortune!

Thurs. 31 Jan. At home all day, and but very little to do . . . gave the post 2 brass or bad shillings for things, as under:

For 1 pack cards	1*s*. 0*d*.
1 letter and in cash	9*d*.
	1*s*. 9*d*.

Posted part of my day book, and in the even Charles Diggens came over and he, my wife, myself and Mr. Thornton played at whist till 3.40. My wife and I not playing together, we lost nothing . . .

Fri. 1 Feb. . . . In the even my wife went down to Mrs. Atkins's to drink tea, and about 8.05 I went down, where we stayed and supped on some roast chicken, a cold ham, a hot boiled green tongue, a boiled leg of mutton, fried chaps, tarts, cold fine baked puddings etc., in company with Mr. Coates, Mr. and Mrs. Porter, Joseph Fuller and his wife, Mr. and Mrs. French, Mr. Calverley and Tho. Fuller. We played at brag in the even and according to custom my wife and I lost 2*s*. 2½*d*. We came home about 2.20, very sober. My wife and I gave Mrs. Atkins's servant 6*d*. each. How tired am I of those more-than-midnight revels; how inconsistent is it with the duty of a tradesman, for how is it possible for him to perform or pursue his business with vigour, industry and pleasure when the body must be disordered by the loss of sleep and perhaps the brain too by the too great a quantity of liquor which is often drunk at those times. And then, can a tradesman gaming have any palliation? No! It is impossible, though it's true we game more for to pass away time than for thirst of gain, but what a way is it of spending that which is so valuable to mankind? Well may our great poet Mr. Young say 'When time turns a torment, then man turns a fool.'[2] And suppose a game of cards innocent in itself, yet the consequences cannot be so if what is commonly called fortune should run against any one at play, that he lose more than his income will allow of. But supposing the person to lose can afford it without any ways in the least incommoding or straitening his circumstances; I say, suppose this case, it cannot be innocent because that sum which anyone can afford to lose he can at the same time afford to dispose of in any other way. Then that sum given away in charity must have been a better way in disposing of it. Therefore, if there is a better way to dispose of the money lost, losing it is not right, and then how oft are the passions moved in such a manner by a bad run of play, that the more warm of us many times increase at that quality so much as to use oaths and execrations not fit to be heard among Christians.

Tues. 5 Feb. . . . Mr. and Mrs. Porter, Mr. Coates, Mr. and Mrs. French, Joseph Fuller and his wife, Mr. Calverley, and John and Will. Piper drank tea and supped with us on a turkey roasted, some salt fish, cold dried tongue, potted pigeons, potted beef, cold chicken pasty, cheese cakes, rice pudding, grape tarts, puffs etc. We played at brag in the even; my wife lost 6*d.* They stayed till 2.20, and not a person in company sober, and I am sure to my own shame I was as bad as anyone. Oh, I that have been so fortunate as to keep sober everywhere else to be so foolish or at least so unfortunate as to be drunk at home, but the company seeming to be wondrously pleased with their entertainment, which exhilarated my spirits that I was transported beyond the natural bounds of my temper, and by that means was left destitute of reflection and caution. Oh, may the God of all goodness give me grace never to be guilty of the same again.

Fri. 8 Feb. . . . Molly French drank coffee with my wife, and she, John and Sam. French and Mr. Long supped with us on the remains of Tuesday's supper. We played at brag in the even; my wife and I lost 11½*d.* and now I think I have taken up a resolution not to play any more at cards if I can possibly help it.

Thurs. 14 Feb. In the forenoon rode over to Framfield . . . I balanced an account with my brother and sister as under, *viz.,*

[2] Edward Young, *Night Thoughts*, night II, line 166: 'Then time turns torment, when man turns a fool'.

Myself Dr.

    To cash received of James Knight the 22nd
of October in full for half a year's rent due
the 10th October when we had deducted £2
for land tax   £11 2s. 6d.

Per contra Cr.

    To mending, schooling, clothing,
boarding etc. of Philip Turner to the 1st April   £ 5 16s. 6d.

                                   £ 5 9s. 0d.

the quarter part of which is £1 7s. 3d. for each of my brothers and myself, and I gave my brothers Moses and Richard and my sister Sally a
note of hand for £1 6s. 4½d. on demand, which was by mistake 10½d.
too little in each note . . .

   Tues. 19 Feb. About 5.40 I set out for Rye in order to take a shop
which I heard was to be let at Winchelsea (and a very sharp frosty
morning I had to ride in). I called at Battle and baited and spent 8d.,
that is 2d. for oats and 6d. for half a pint of wine. I then proceeded on my
journey, and sure I never rode in a more unpleasant day in my life, it
snowing and raining all the way and the wind most remarkably high, that
before I got to Rye I had scarce a dry thread about me. And when I got
just to Rye the tide was out, and I, a stranger to the roads, was afraid to
venture through; so that I turned back and rode to a public house
about 2 miles back, where I hired a guide and went in and dried
myself. While he was getting ready I spent:

To my horse	4d.
To beer for the person that took care of my horse	
and others that assisted in drying my things	4d.
To brandy for my guide	5d.
Beer myself	2d.
Gave the people for their trouble and wood	3d.
	1s. 6d.

I then proceeded on my journey and got safe to Rye about 2.50, though
once wet through again. I gave my guide 12d. I put up at *The George*,
where I dined on some beefsteaks and oyster sauce and spent the even
there in company with Mr. Fender, Mr. Blake's servant . . . I stayed at
The George all night.

   Weds. 20 Feb. I got up pretty early and, finding the shop I went after
was let, I stayed and breakfasted and came away immediately. I spent:

My horse and ostler	1s. 8d.
The chambermaid	6d.
Myself	5d.
	2s. 7d. . . .

Thurs. 21 Feb. In the morn my brother came over and breakfasted with us, and then he and I set out for Lewes where we met my brother Richd. previous to appointment, and we went together to Mr. Sam. Durrant's for him to settle the affairs of my late mother between my brothers and sisters, which he did not thoroughly complete, but kept the accounts for a further examination . . .

Fri. 22 Feb. Paid John Cayley 2s. for the rent of our 2 seats in the church. Paid Tho. Burfield 7s. 6d. for 1 dozen beehives received by him today. We dined on a chicken roasted, a piece of boiled pork and turnips. At home all day. My brother stayed and breakfasted with us, and then went home. Very little to do all day; a most melancholy time—my poor heart is almost sunk with trouble! Dame Durrant made my wife a present of a fat goose.

Mon. 3 Mar. At home all day . . . In the even wrote my London letters. My wife went down to Mr. Porter's, their youngest child being very ill.

Tues. 4 Mar. . . . In the even went down to Mr. Porter's, to take instructions for the funeral of their little boy, who died this morning about 8.20, aged about 9 months.[3] At home all day, but very little to do.

Fri. 7 Mar. Mr. Stone bleeded [me] in the forenoon and I received of him in cash £3 3s. 0d. which is in full for the composition of a debt due to me from Will. Elphick of £5 5s. 0d. and really it is a fine drawback, as some of the money was paid out of my pocket, and what makes me the more vexed about it is I have the greatest reason in the world to think that it is a very shuffling and unfair compound, but I was willing to have something, rather than lose all. But I must own and shall always look upon myself as guilty of mean spirit in coming into the composition and not seeing the affair to the utmost, for I really think the man when he absconded took with him a considerable sum of money . . .

Sat. 8 Mar. . . . In the afternoon went down to Mr. Porter's, to go to church with their child, who was this day buried, as was also Anne, the daughter of Francis Turner, aged 31 years; they was both buried about 5.40. I gave away at Mr. Porter's 2 pairs men's white kid gloves. This day a melancholy affair broke out in this neighbourhood, *viz.*, Lucy

---

[3] The child had been born on 8 June 1759 and baptized Thomas on 17 June.

Mott, servant to Mr. Jer. French, last night absconded herself from her service (privately and quite unknown to anyone in the family) and from many corroborating circumstances there is great probability to think she hath committed that rash action of suicide. She went off in her worst apparel, and left behind her all her money, and had taken more than common care in laying up all her clothes and collecting it together in such a manner that it might be the more easy to find by her relations. There is also the greatest reason imaginable to think she was pregnant, for by some reasons given by her mistress it seems plain, and on the 18th ult. she wanted of my wife 2 ounces jalap.[4] But upon her saying we had not such a quantity in the house and telling her she must be either mistaken in the name or quantity, for that must be enough for 20 people to take, she seemed greatly confused and pretended it was not to take (but, however, she by fortune had none) so that I should rather think her intentions might be then rather to destroy the foetus than herself.[5]

Sun. 9 Mar. . . . Before churchtime Tho. Davy, Tho. Durrant and myself went to seek after the poor creature that is lost, but could find nothing of her . . . Tho. Davy and myself searched after the poor girl again in the afternoon, but found nothing of her . . .

Mon. 10 Mar. . . . At home all day, and but little to do. Oh, melancholy time; what will become of me I cannot think, for work hard I cannot, and I think never was trade so dull . . .

Tues. 11 Mar. . . . At home all day, but nothing at all to do. O, miserable work; what shall I do?

Fri. 14 Mar. This being the day appointed by proclamation for a general fast and humiliation before Almighty God for obtaining pardon of our sins, and for averting those heavy judgments which our manifold provocation have most justly deserved, and imploring His blessing and assistance on the arms of His Majesty by sea and land, and for restoring and perpetuating peace, safety and prosperity to himself and to his kingdom, my wife, self and servant at church in the morn . . . We dined on the remains of yesterday's dinner . . . The fast in this place has been seemingly kept with a becoming decency, and God grant that the people of this nation may have a just sense of their sins and repent and return to the Almighty.

Sun. 23 Mar. . . . A good deal out of order with my side, which I am very fearful will turn to something of a cancerous humour.

[4] A purgative drug.
[5] Lucy Mott, in fact, reappeared and, whether pregnant or not, her banns were called for the last time on 1 June, see below, p. 206.

Tues. 25 Mar. . . . In the even went down to Mr. Bridgman's, where John Barnard made an entertainment, his apprenticeship being this day ended. I came home about 10.30 very sober. I gave the young man 2s.

Sat. 5 Apr. . . . At home all day and really extremely ill with a slight fever and pang in the side . . .

Mon. 7 Apr. . . . After dinner I went down to Jones's to the vestry . . . Wm. Piper was chosen overseer, Jos. Fuller electioner; Jn. Cayley churchwarden, Jos. Durrant electioner . . . In the even wrote my London letters. We had several warm arguments at our vestry today and several volleys of execrable oaths oftentime resounded from almost all sides of the room, a most rude and shocking thing at public meetings . . .

Thurs. 10 Apr. . . . Ah melancholy time! But little to do, and what lies heavy upon my spirits is I am afraid my brother at Framfield will not do well and I am so deeply concerned that, should he do ill, I must be entirely ruined.

Thurs. 17 Apr. . . . At home all day and pretty busy. In the even read part of Derham's *Physico-Theology*. Oh, the anxiety of my mind through fear my brother at Framfield will not do well in trade—not but I believe him thoroughly honest, sober and careful to the best of his knowledge, but then I doubt his abilities in regard to trade.

Fri. 18 Apr. . . . In the even a good deal indisposed with a stitch-like pain in my side, breast and back.

Sun. 20 Apr. . . . In the forenoon Mr. Stone paid me a visit and bleeded me . . . At home all day. My side bad, and I am fearful whether I shall ever get the better of it.

Mon. 21 Apr. . . . Received of Tho. Davy in cash 5s., being in part for his and Master Hook's club arrears. Sent by John Streeter 11s. 3d., it being in full for mine, Master Hook's, and Tho. Davy's club arrears due on quarterly night, which is the 24th instant.[6]

Tues. 22 Apr. In the morn rode to Master Chipper's at Laughton to speak to him for to serve the nails for Counsellor Russell's house which is a-going to be repaired . . .

Fri. 25 Apr. . . . Paid John Jenner, hatter at Hailsham, in cash and goods £1 1s. 9d. in full for the hats received by him today: 3 men's felts at 3s., 3 do. at 2s. 6d., 3 boy's do. at 21d. . . .

Sun. 4 May. . . . In the even my wife went down to see Mr. Porter's two youngest children, who are both very ill with fevers and

[6] See above, p. 99, note 25.

inflammations of the lungs. Dr. Stone paid me another visit. Thank God something better.

Tues. 6 May. ... At home all day and but very little to do. I am very doubtful that my disorder in my side etc. is of a strumous kind,[7] and what gives me the greater room to think so is that the left parotid is very much swelled.[8]

Weds. 14 May. After breakfast I rode to Heathfield where there was a sale of shop goods, one Stephen Smith and his son having failed in trade. I bought several odd things for which I paid Mr. Robert Olive, the salesman, £3 5s. 3d. From thence I went to my brother's at Framfield, where I drank tea (or rather dined) ... After tea my brother and I went to Uckfield Fair where we stayed about 1 hour and half. I came back with my brother by Framfield but did not stay. I came home about 7.50 ... A very fine pleasant day, and I think there was a large fair.

Sun. 18 May. Myself and servant at church in the morn ... We had a brief[9] read for damages by fire to John Bodmin at Chalk in Kent (to be collected from house to house), whose damage by fire amount[ed] to upwards of £1231 ... James Trill and Lucy Mott was asked at our church the first time ...

Mon. 19 May. ... This day was played in the park a cricket match between an eleven whose names were John in this parish and an eleven of any other name, which I suppose was won by the latter with ease[10] ...

Tues. 20 May. ... My brother came over in the forenoon but did not stay ... Gave Mr. Porter, who came a-collecting for the brief read a-Sunday, 12d. towards the relief of the poor unhappy sufferer. Lent Thomas German in cash £2 2s. 0d. on his watch, which I took in pawn ...

Sat. 24 May. ... At home all day and thank God very busy, though very ill all day and I am very fearful my liver is cirrhous[11] ... Oh how uncertain is the health of poor mortal man!

Weds. 28 May. ... At home all day, but very little to do in the shop. My side a great deal better, and may the God of all goodness pour into my heart the graces of His Holy Spirit, that I may continually adore and praise Him for this, His inestimable blessing to me, a poor sinful creature, through the merits and mediation of my blessed Saviour.

---

[7] Affected by scrofulous swellings.
[8] Turner evidently had swollen glands, the parotid being a gland in front of the ear.
[9] See above, p. 158, note 37.
[10] See above, p. 9, note 9.   [11] Jaundiced.

Sun. 1 June. . . . No service at our church . . . Mr. Porter being ill. Mr. Long and I walked to Chiddingly, where we had a sermon preached by the Revd. Mr. Herring . . . James Trill and Lucy Mott were out-asked.

Mon. 9 June. Our late servant Mary Martin came to see us in the forenoon and dined with us on the remains of yesterday's dinner with the addition of a veal pudding and green salad. In the afternoon I walked up to the common to see a game of cricket played between our parish and that of Chiddingly, which was won with great ease by the latter. Chiddingly the 1st innings got 86 runs, and our parish 66, Chiddingly 2nd innings got 117, and our parish 50, so that our parish was beat 87 runs.[12] In the evening wrote my London letters . . .

Thurs. 12 June. . . . In the even went down to Jones's, in company with Joseph Fuller, Mr. Long, Mr. Thornton and James Marchant, to take part of a forfeit which Joseph Fuller paid to be off a bet which he had made with Mr. Thornton. The bet was for a guinea that Mr. Saxby had either a horse or mare that would go (one pace) from the park gate at the leigh of Laughton to *The Cat* in Cats Street[13] within the hour (being between 9 and 10 miles and very uneven ground), but Mr. Thornton laid he had not. We stayed and spent 7½d. apiece besides Joseph Fuller's forfeit and I came home about 10.20 sober. But little to do all day . . .

Fri. 13 June. We dined on a piece of bacon boiled, a light pudding and spinach . . . In the even played at cricket some little time.[14] At home all day, and but very little to do. In the forenoon John Baker assisted me in packing up my rags to go to Maidstone Fair. We packed them up in 3 bags containing 6 cwt. 2 qr. 18 lb.

Sat. 14 June. . . . Paid 3d. for 1½ lbs of crabs bought today. At home all day and thank God pretty busy. In the even went to cricket a little while with several more out before my door[15] . . .

Mon. 16 June. . . . Delivered to John Gosling Jr. 4 bags of rags, weighing 6 cwt. 3 qr. 14 lb. net, which he is to sell for me at Maidstone Fair the 20th instant. At home all day and not very busy.

Tues. 17 June. Mr. Sam. Beckett made my wife a present of 3 fine mackerel . . . In the afternoon kept the school for Mr. Long, he going to a cricket match at Chiddingly[16] . . .

---

[12] See above p. 9, note 9.   [13] Cade Street, Heathfield.
[14] See above p. 9, note 9.
[15] This reference is not included in the article in *The Cricket Quarterly* referred to above, p. 9, note 9.
[16] See above, p. 9, note 9.

Thurs. 19 June. In the forenoon my brother came over and stayed and dined with us on a broiled mackerel, a piece of pork and a piece of bacon boiled, a raisin suet pudding and greens . . . Joseph Fuller Jr. smoked a pipe with me in the even. Pretty busy all day, but sure a most melancholy time for money, the greatest part of trade being trust, and doubtless in so many small articles we forget a great many, which makes it so much the worse trading.

Fri. 20 June. . . . This day hath been my birthday, and that on which I enter into the 32nd year of my age. And may the God of all mercy and goodness pour into my heart the grace of His Holy Spirit, that as I grow in years so may I increase in goodness and daily be renewed in the inner man, and so number my days that I may apply my heart unto wisdom, through the merits and intercession of my blessed Saviour and only Redeemer, Jesus Christ.

Thurs. 26 June. . . . In the afternoon made part of the land tax books, Mr. Piper and myself being assessors for the present year.[17]

Sat. 28 June. In the forenoon rode to Bayley's Lane to the crock kiln to buy some earthenware, but there being not enough burnt, I agreed to postpone it until Tuesday next. We dined on a beef pudding and cabbage leaves. In the afternoon rode up to Mr. Vine's and looked upon his wool, which he agreed I should have at the market price . . . In the even made part of the window tax books. Pretty busy in the day.

Mon. 30 June. Received of Mr. John Gosling in cash £5 3s. 7½d. which is in full for the rags I delivered to his son the 16th instant, *viz.*,

To 6 cwt. 3 qr. 14 lb. rags sold at 30s.	£10	6s.	3d.
Carriage allowed		10s.	0d.
To 5 quires paper		2s.	6d.
	£10	18s.	9d.

Per contra Cr.

To 2 ream of writing paper bought for me	£1	0s.	0d.
To 4 lumps of sugar bought for me, weight			
1 cwt. 0 qr. 15 lb. at 73s.	£4	2s.	9¾d.

Received by carriage:

	cwt.	qr.	lb.			
viz., sugar	1	0	15			
rags	6	3	14			
paper	0	0	26			
	8	0	27 at 18d. per cwt.	12s.	4¼d.	
Do. in cash today				£5	3s.	7d.
. . .				£10	18s.	9d.

[17] See below, p. 216, note 3.

Thurs. 3 July ... In the afternoon my cousin Molly Hill came to see us and after tea my wife and cousin walked to Whitesmith Fair. In the even I played at cricket some little time before the door[18] ...

Sun. 6 July. ... In the afternoon went and sat with Mr. Long some time, who has the measles ...

Mon. 7 July. ... This day finished making the land and window tax books, which I sent by Mr. Box to the sitting at Maresfield to have signed by the commissioners, and which I received back again by him again in the evening ...

Weds. 9 July. ... In the afternoon my wife walked to Whitesmith to see a mountebank perform wonders, who has a stage built there and comes once a week to cozen a parcel of poor deluded creatures out of their money, he selling packets which are to cure people of more distempers than they ever had in their lives for one shilling each, by which means he takes sometimes £8 or £9 of a day[19] ...

Thurs. 10 July. ... My wife with Dame Cornwell most part of the day, she being very ill with the measles, as is some one or other in almost every house in the parish ...

Tues. 15 July. ... After dinner I and Mr. Thornton set out for Lewes (I wanting to see Mr. Friend, to know the price of wool), but went by Alfriston, there being a game of cricket to be played there between Sam. Jenner of the same place and Storry Adams of Arlington, for two guineas each; when Adams went in first and was bowled out the first ball, then Jenner went in and [was] bowled out the first ball. Then Adams went in the second time and was again bowled out in 2 balls, and Jenner went in his second time and the first ball he had he struck and got one run, which decided the game, and I believe in less than 10 minutes. They then played again for half a guinea each; when Adams went in first and got 17 runs, then Jenner went in and got five runs. And then Adams went in his 2nd innings and got 2 more runs, and Jenner went in his last innings and was bowled out the 2nd ball, so that Adams won this game by 14 runs. I having the good fortune to meet Mr. Friend at Alfriston did not go any farther, but came home about 7.10. Spent 7*d*. today. Paid Dr. Snelling 12*d*. in full. My wife paid Mary Heath 18*d*. for two days' washing. In the morning my brother came over and brought my horse. What can betray greater stupidity in mankind than to game for such large sums?[20]

Fri. 18 July. ... In the even rode with Joseph Fuller Jr. to Waldron to look at some wool he had bought, but did not then buy it, though I

[18] See above, p. 9, note 9.
[19] See above, p. 11, note 14.   [20] See above p. 9, note 9.

bought it after I came home at 8½d. per lb. A remarkable hot day, and really but little to do in the shop. Oh, most melancholy time; what shall I do to work—I cannot!

Thurs. 24 July. . . . Paid Joseph Fuller in cash £3 16s. 10d. in full for 74 fleeces of sheep's wool received by him today weighing 3 tod 12½ lb. at 22s. 8d. . . . Played at cricket a little time in the even[21] . . .

Sun. 27 July. . . . In the even and the day read 6 of Bishop Sherlock's sermons, which I think extremely good, there being sound reasoning in them, and seem wrote with an ardent spirit of piety, being mostly levelled against the deists. .

Mon. 28 July. . . . In the even Mr. Long and I walked to Uckfield, where I called on Mr. Hart to buy some chamois skins, but he had none. We spent some time with Mr. Elless and some with Wm. Turley, and came home very sober about 9.30. Very busy all day, though I believe I barely took 20s. ready, notwithstanding I delivered above £7 worth of goods. In the evening wrote my London letters.

Weds. 30 July. In the morn rode to the keeper's to ask for the funeral of his mother, but was too late to succeed . . .

Sat. 9 Aug. . . . In the afternoon Mr. Long and I set out for Eason's Green in order to meet Mr. Elless to play a game of cricket, but meeting him on this side the Green and there coming a shower we came back, as did my brother who came along by accident[22] . . .

Weds. 13 Aug. . . . In the evening Mr. Long, Mr. Thornton and myself walked to Honey's Green. I went in order to look upon Master Dapp's wool, but they pretended they should not sell any this year, though I have had the promise of it some time, and they have even taken up goods upon it these 6 months . . .

Fri. 15 Aug. In the afternoon my wife went down to Halland to pay Mrs. Coates a visit . . . In the even after I had shut in shop walked down towards Halland to meet my wife, but not seeing of her I went in and stayed some time, in company with Mr. and Mrs. Porter, with whom we came home in company about 9.30. This day His Majesty's purse of £100 was run for on Lewes Downs, when only Lord Portmore's brown horse Bosphorus started for the same. There was also a bye-match run for a considerable sum between the Duke of Richmond's grey horse Muli-Ishmel and Sir Matthew Featherstonehaugh's grey mare Sally, which was won by the former with great difficulty, he not beating by

---

[21] See above, p. 9, note 9.
[22] See above, p. 9, note 9.

above half a length, and at the same time the knowing ones were very much taken in.[23]

Sat. 16 Aug. . . . After dinner set out for Lewes Races, when the following horses started for the Subscription Purse of £50: Mr. Bray's grey horse Valerius; Mr. Wildman's chestnut horse Mickelham; and John Wadman Esq's bay horse Ranter, which was won by the former, he getting the 2 first heats. The first was obtained with some difficulty, but the second very easy . . .

Thurs. 21 Aug. . . . In the even went out a little space of time with Mr. Long to instruct him in land surveying. At home all day, except as above, and really very little to do.

Mon. 25 Aug. . . . Very little to do all day. In the day read part of Bracken's *Pocket Farrier*, which I look upon as a very complete thing of its kind.

Thurs. 28 Aug. . . . Paid 4*d.* for 10 whiting bought today. In the even I had the accident, as I was a-going to turn my horse to grass in the orchard, to tread on a broken glass bottle, which flew up with some force and cut me pretty much across the inner process of the tibia.

Sat. 30 Aug. . . . Thank God my ankle pretty middling . . . At home all day and thank God pretty busy. Oh what pleasure it is to have some trade! How does it enliven one's spirits!

Sat. 6 Sept. . . . In the afternoon packed up my wool, with Tho. Durrant's assistance, *viz.*, 5 cloths containing 594 fleeces weighing 41 tod 20 lb. . . . Thomas Durrant supped with us.

Mon. 8 Sept. John Watford came to work for me in the morn; we cleared all my pear trees and then carried them to Thomas Bean's at Whitesmith where we made them out into perry, making near 60 gallons. I paid T. Bean 4*s.* 6*d.* for the use of his mill and press, and his own and man's assistance. I spent 6*d.* for beer to drink with our bread and cheese, which we carried with us . . . We came home without perry about 6.20 . . . Very busy all day.

Tues. 9 Sept. . . . This day sent Mr. Tho. Friend (by John Divol) mine and my brother's wool . . .

Weds. 10 Sept. In the morn so soon as I had breakfasted I set out for Goudhurst in order to buy some flannels. I called at Ticehurst and baited myself and horse and from thence I went to Goudhurst, but did no business, the man having only one piece made . . . My wife in my absence paid our servant Hannah Marchant in cash 4*s.* which, with a book debt of 36s., makes the sum of 40s. and is in full for 16 months' servitude ending Tuesday the 16th of this instant September.

Tues. 16 Sept. ... Received of Tho. German a note of hand dated today, payable to me or bearer on demand, for the 2 guineas I lent him the 20th of May last, and for which he then left me his watch as a security, and which I now delivered to him again in the lieu of the above note. At home all day. In the afternoon posted part of my day book and really a melancholy time. Very little to do except trust.

Weds. 17 Sept. ... At home all day, and really so little to do that I am quite in a panic for fear of going wrong in my trade, and as to getting in any money—there is no such thing as I can find ...

Thurs. 18 Sept. In the morn John Durrant Jr. bleeded my wife ... At home all day and but very little to do, and even what little I have is altogether trust, and my affairs are so connected with my friends that I know not how to extricate myself out of my trouble.

Tues. 23 Sept. ... In the even my brother came back from Lewes and left my horse and also left me a bill drawn by Mr. P. Warren of Warminster on Mr. James Wilkins in Basinghall Street, London, dated the 15th instant, one month's date, and payable to Mr. Tho. Friend or order, and also properly indorsed by Mr. Tho. Friend. Out of the same bill there is due to my brother £30 2s. 0d., *viz.*,

Myself Dr.

To the above bill on Wilkins	£70	0s. 0d.
To money my brother paid Mr. Madgwick		2s. 4d.
To *do.* Mr. Tho. Friend in full on my account	£ 5	15s. 4d.
	£75	17s. 8d.

Per contra Cr.

To money my brother received of Mr. Tho. Friend for the 41 tod 20 lbs of wool at 22s. per tod sent him the 9th inst.	£45	15s. 9d.
To balance	£30	2s. 0d.
	£75	17s. 9d.

Received of Tho. German the 2 guineas I lent him the 20th of last May, and for which he gave me his note of hand the 16th instant (which I now delivered up to him). About 11 o'clock we had several flashes of lightning and some claps of thunder and a very windy and rainy night.

Weds. 24 Sept. ... At home all day and very busy. In the even between 9 and 10 o'clock we had a pretty severe tempest of thunder, lightning, wind and rain. We had several loud strokes of thunder and a great deal of lightning and it remained very boisterous weather all the night.

Sat. 27 Sept. . . . In the [even] went down and sat with Mr. Porter some time, he being ill with a fluxion on the parotids and maxillary glands.[24]

Mon. 29 Sept. . . . My wife, standing sponsor for a boy of Jn. Browne's who was baptized today, she dined there . . . Charles Diggens coming over to take up a coat for Mr. Coates, stayed and smoked a pipe or two with me, and about 8.10 we both went to the christening house, where we stayed till near 12 o'clock, in company with Mr. Burges, John Watford and Fran. Gibbs. My wife and I came home about 12.30 and pretty near sober, but something the worse for drinking. My wife gave the nurse 18*d.* and the maid 12*d.* In the even as we went to Mr. Browne's we had several very severe flashes of lightning and a very loud stroke or two of thunder and it continued to rain by showers I believe the whole night and the wind extremely high. Very busy all day.

Tues. 30 Sept. . . . At home all day and pretty busy. My wife very ill all day. A very melancholy time; the weather prodigious tempestuous and has been so for some time . . .

Fri. 3 Oct. Mr. Stone paid my wife a visit . . . At home all day and, thank God, pretty busy, but my wife very ill. Oh, how melancholy a time it is, [when I], quite destitute of father or mother, am in all probability like to lose my wife, the only friend I believe I have now in this world, and the alone centre of my worldly happiness! When I indulge the serious thought, what imagery can paint the gloomy scene that seems just ready to open itself as it were for a theatre for my future troubles to be acted upon.

Sun. 5 Oct. . . . Mr. Stone paid my wife another visit and let her blood . . . my wife really very ill. A melancholy time.

Mon. 6 Oct. . . . At home all day and pretty busy, but surely my wife is extremely ill. Oh, what an agony of mind I am in with doubt and fear of my wife's illness proving mortal . . .

Tues. 7 Oct. . . . Dame Durrant and Mrs. Virgoe drank tea with my wife, who is extremely ill. In the even there was a rejoicing at Halland and a bonfire for our armies under the command of General Amherst having taken Montreal and all Canada from the French.[25] All the neighbours were regaled with a supper, wine punch and strong beer.

[24] The rector seems to have had a discharge from the facial glands.
[25] Montreal fell to General Amherst on 8 September 1760.

Today sent Thomas Durrant to Brighthelmstone[26] for Dr. Poole, who came to my wife in the even, who is prodigious ill. At home all day and thank God pretty busy.

Weds. 8 Oct. In the morn Dr. Poole (he lying at Mr. Porter's) paid my wife another visit, and I paid him for his fee 2 guineas . . . At home all day and thank God pretty busy. My wife extremely ill . . .

Thurs. 9 Oct. In the morn got Bett Mepham to come and wait on my wife during her illness. In the forenoon my father Slater came to see us and dined with us . . . At home all day and thank God my wife something better, and pretty busy all day. Dr. Poole, when he visited my wife yesterday, gave it as his opinion that her illness was the fluor albus[27] and an ulcer in the right kidney.

Weds. 15 Oct. . . . At home all day. My wife continues very ill. It is impossible for tongue or pen to express the trouble I now feel on account of my wife's illness, and the constant fears that I have she will never get the better of her illness. Oh, would the God of all blessings look down upon us in this our day of adversity and trouble and restore my wife to her former health and so pour the graces of His Holy Spirit into our hearts that we may live in love and unity and every day become better Christians!

Sat. 18 Oct. . . . Dr. Poole paid my wife another visit today and wrote her another prescription. Mr. Stone by Dr. Poole's order let her blood. At home all day and pretty busy. My wife not so well again. Oh, melancholy time! . . .

Sun. 19 Oct. . . . Myself, servant and Bett Mepham at church in the afternoon . . . We had a prayer with thanksgiving read for the success with which it has pleased Almighty God to bless His Majesty's arms in North America. In the even my wife took a vomit, who is, I hope, something better . . .

Sat. 25 Oct. . . . At home all day and, thank God, my wife something better. In the even read Gibson on lukewarmness in religion, and a sermon of his entitled *Trust in God, the best remedy against fears of all kinds*, both of which I look upon as extreme good things.

Sun. 26 Oct. . . . In the forenoon Dr. Stone came to pay my wife a visit and let my blood . . . Today we had the melancholy news of the death of His Most August Majesty George II, King and Parent of this our most happy isle. His death happened yesterday morning about 7 o'clock by a fit of an apoplexy;[28] had His Majesty lived to the 10th of

---

[26] Brighton.   [27] Leucorrhoea.

[28] George II died at Kensington of a heart attack.

November he would have been 77 years of age. He has sat upon the British throne 33 years the 21st of last June.

Thurs. 20 Nov. . . . At home all day. My wife, poor creature, most prodigious ill, and the trouble it gives me quite sinks my spirits. I have now no prospect of her recovery (that is, according to any human probability). But oh, may the God of all goodness and mercy look down upon us!

Mon. 1 Dec. After breakfast my father Slater went home upon my horse (he coming yesterday on foot). My father made my wife and myself a present of a guinea each, which I look upon as done from a principle of good nature and compassion for our misfortunes, so that I look upon it as a thing of greater value than barely the worth of the present . . .

Sat. 6 Dec. . . . At home all day; a very cold day. My wife very ill. Mr. Long made my wife a present of a woodcock and blackbird. He stayed and supped with us.

Fri. 12 Dec. . . . At home all day. My wife, poor creature, extremely ill. Oh, what a melancholy time I have; my spirits are quite sunk beneath a load of grief, and my heart overwhelmed with a sea of trouble, but let me not grieve at the divine will, but the more earnestly let me pray for grace to direct my steps and to put my trust in the Most High, who is a rock of defence to all such as call upon Him faithfully and will hear their prayer in the needful time of trouble.

Sun. 21 Dec. No service at our church in the morn, Mr. Porter preaching at Laughton. We dined on two rabbits roasted and stuffed. Our late servant only at church in the afternoon. Dr. Poole, coming to see a child of Mr. Porter's, paid my wife a visit and charged me 10*s.* 6*d.* Really a fine thing it is to be a physician who can charge just as they please and not be culpable according to any human law. At home all day. My wife, poor creature, very ill.

Mon. 22 Dec. This being the day that the poor of this parish go about asking for charity against Christmas, I relieved [30] with 1*d.* each and a draught of beer . . .

Thurs. 25 Dec. This being Christmas Day, our servant only was at church. Just at churchtime my brothers Moses and Richard came to see me and stayed with me about an hour and half. Mr. Long dined with us on a piece of beef roasted, a suet plum pudding boiled and potatoes. Mr. Long stayed and drank tea with us. In the even read one of Tillotson's sermons. At home all day. My wife, poor creature, worse again. Mr. Stone bled her today in the foot.

Fri. 26 Dec. . . . At home all day. My wife, poor creature, very ill.

This day died Mr. Edward Calverley after a long illness, aged 76 years . . .

Sat. 27 Dec. . . . In the forenoon I set out for Lewes in order to get a set of plates etc. for Mr. Calverley's coffin. I dined with Mr. John Madgwick on a beef pudding . . . I also made the following agreement with Mr. John Madgwick, *viz.*, that he is to take my brother Richard as a servant to serve in his shop for the term of 3 years from the 14th day of January next, for which service he is to give him £21, but if at the end of two years he had rather part with him than keep him the third, he is then only to pay him the sum of ten guineas, and whether he either keeps him 2 or 3 years, he is once during the time of his service to take him with him to London, and once to Maudling Hill Fair at his own expense, the hire of his horse only excepted . . .

Sun. 28 Dec. . In the morn I walked down to Whyly with a shroud, sheet etc. for Mr. Calverley. Myself and late servant at church in the morn, where we had a sermon preached by a young clergyman just come to be curate at Laughton, and I imagine this to be the first time of his preaching (Mr. Porter administering the sacrament at Laughton). The text in the former part of *John* 1.5: 'And the light shineth in darkness', from which words we had in my opinion a learned sermon and I think if the young gentleman's morals are good he will in time make a fine man . . . My wife, poor creature, very ill. Oh, melancholy situation, but still let me not repine, for doubtless many are as much afflicted.

Tues. 30 Dec. . . . After dinner I went and served part of Mr. Ed. Calverley's funeral, Mr. Lashmer also serving a part . . .

1761

Thurs. 1 Jan. In the morn set out for Chailey. Called at Framfield and breakfasted with my brother. I dined at Mr. Geo. Beard's on some hog's chap and knuckle boiled, a plain hard pudding and greens . . . This being the day that my brother Richard's apprenticeship expired, I and Mr. Beard cancelled my brother's indentures and Mr. Beard paid my brother in cash £14, agreeable to a note of hand he gave my brother when bound, and I accordingly delivered him up his note of hand. I also took up of him a note of hand which my mother gave Mr. Beard when my brother was bound, promising to pay him 24s. if he should think proper to demand it when my brother's time was expired, which

Mr. Beard did not, but generously gave me the said note up. My brother delivered to me to keep for him £12 12s. 0d. but I have given him no note or anything to show for it . . . I called at Framfield and drank tea with my brother. My wife, poor creature, very ill. Oh, most melancholy situation! Paid Mr. South 4s. 6d. in full for 3 diuretic draughts for my horse.

Fri. 2 Jan. . . . This day Dr. Poole generously called on my wife and took nothing for his fee, but told me the melancholy news that there was no prospect of my wife's recovery but by a miracle. My wife, poor creature, very ill. At home all day. This day died a daughter of Mr. Porter's, about 4 years of age.[1]

Sun. 4 Jan. . . . No service at our church in the morn, on account of the death of Miss Porter . . . In the even Tho. Davy at our house, to whom I read three of Tillotson's sermons . . .

Tues. 6 Jan. In the morn Ed. Foord, our headborough, and myself went down to Henry Osborne's with an intent to distrain his goods for window tax, but his brother paid me the same after some words, it being only 3s., without the trouble . . .

Weds. 7 Jan. In the morn went and invited the children to support Miss Porter's pall[2] . . . At home all day, and but very little to do. My wife continues very ill . . .

Thurs. 8 Jan. . . . In the afternoon went and served Miss Bett. Porter's funeral, where I gave away, with the gloves ordered which I have not served, in all 24 pair, *viz.*, 2 pair men's kid, 3 pair women's *do.*, 7 pair men's lamb, 6 pair women's lamb, 6 pair maid's *do.* . . .

Fri. 9 Jan. . . . After breakfast went to Maresfield to meet the receiver-general of the land tax, where I paid his clerk, Mr. Tho. Gerry, the sum of £62 in part for half a year's land tax due to his Majesty at St. Michael last[3] . . .

Sun. 11 Jan. . . . In the even read part of the Book of Job. My wife, poor creature, very bad. Who can describe or imagine the pang she has endured and which through God's grace she bears with patience. Oh,

---

[1] Elizabeth Porter who had been baptized on 23 July 1756.

[2] For the practice of children supporting the pall at the funeral of a child see Bertram S. Puckle, *Funeral customs, their origin and development* (1926).

[3] The procedure for the collection of the land tax (a tax based on income derived from property) had been laid down by an Act of 1693: 4 William & Mary, c. l. County commissioners were appointed from the propertied classes and they in their turn appointed assessors and collectors for each parish. The assessors and collectors were normally the same people since the assessors received no salary while the collectors did. Twice a year the moneys collected were paid to the receiver-general for the county. For the tax and its collection see W. R. Ward, *The English land tax in the eighteenth century* (1953).

5. The bastardy bond, written by Thomas Turner on 9 February 1758, by which Thomas Osborne indemnified the parish of East Hoathly against any charge which might arise from the child of Mary Hubbard: *E.S.R.O. PAR 378*. See pp. 135–6.

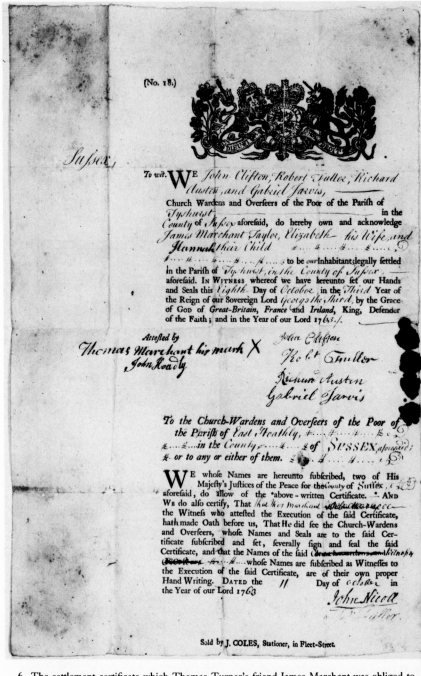

(No. 18.)

Sussex,

To wit. WE *John Clifton, Robert Fuller, Richard Austen, and Gabriel Jarvis,*

Church Wardens and Overseers of the Poor of the Parish of *Tyshurst* in the County of *Sussex* aforesaid, do hereby own and acknowledge *James Marchant Taylor, Elizabeth his Wife, and Hannah their Child* to be our Inhabitant, legally settled in the Parish of *Tyshurst, in the County of Sussex* aforesaid. IN WITNESS whereof we have hereunto set our Hands and Seals this *Eighth* Day of *October*, in the *Third* Year of the Reign of our Sovereign Lord *George the Third*, by the Grace of GOD of *Great-Britain, France* and *Ireland*, King, Defender of the Faith; and in the Year of our Lord 17*63*.

Attested by
Thomas Marchant his mark X
John Hoadly

John Clifton
Robt Fuller
Richard Austen
Gabriel Jarvis

To the Church-Wardens and Overseers of the Poor of the Parish of *East Hoathly,* in the County of **SUSSEX** aforesaid; or to any or either of them.

WE whose Names are hereunto subscribed, two of His Majesty's Justices of the Peace for the County of *Sussex* aforesaid, do allow of the above-written Certificate. AND WE do also certify, That *Thomas Marchant* the Witness who attested the Execution of the said Certificate, hath made Oath before us, That He did see the Church-Wardens and Overseers, whose Names and Seals are to the said Certificate subscribed and set, severally sign and seal the said Certificate, and that the Names of the said *Churchwardens and Overseers* whose Names are subscribed as Witnesses to the Execution of the said Certificate, are of their own proper Hand Writing. DATED the *11* Day of *October* in the Year of our Lord 17*63*

John Nicoll
T. Fuller

Sold by J. COLES, Stationer, in Fleet-Street.

6. The settlement certificate which Thomas Turner's friend James Marchant was obliged to procure from the authorities at Ticehurst: *E.S.R.O. PAR 378.* See p. 276.

if I am so unhappy [as] to lose her, what an inestimable treasure shall I lose! Alas, what shall I not lose!

Mon. 12 Jan. . . . Paid Tho. Burfield 8*s*. 9*d*. for 15 beehives received of him today . . .

Thurs. 15 Jan. . . . In the forenoon my father Slater came to see us and brought me a present of a sparerib, some sausages and hog's-puddings. He dined with us on a sparerib roasted and apple sauce. He stayed with us all night . . .

Fri. 16 Jan. After breakfast my father Slater went home . . . At home all day. But little to do. My wife, poor creature, most extreme ill; who can paint or describe my trouble? No, it is out of the power of human being to do it, and God grant my most inveterate enemies, if any I have, may never know so melancholy a situation. My soul is quite overwhelmed with grief; oh, the loss of so inestimable a treasure, even that of a sincere friend and virtuous wife!

Sat. 17 Jan. . . . In the afternoon my brother Moses came to see me and stayed and drank tea with me and stayed with me all evening. Oh, my poor wife is most prodigious bad! No, not one gleam of hope have I of her recovery. Oh, how does the thought distract my tumultuous soul. What shall I do? What will become of me?

Sun. 18 Jan. . . . In the even read part of Young's *Night Thoughts*. My wife, poor creature, most extreme ill, but oh, how charming does her fortitude appear for an example, and more so, does it yield complete comfort to her soul; and at the same time shows the purity of her conscience.

Tues. 20 Jan. . . . My wife, poor creature, very ill. Ah, a melancholy daily repetition. Who can paint or imagine my trouble, which I now labour under, and what is more, the prospect of a change severer for severe? Paid a nephew of Mr. John Wilbar's 6*s*. 3*d*. in full for brushes bought of him today . . .

Mon. 26 Jan. . . . In the even wrote my London letters. At home all day. My wife, poor creature, very ill. Oh, the tumult in my troubled breast. I must ere long lose the partner of my soul, one with whom I can converse with sincerity and freedom, one not influenced and guided by the unruly dictates of passion and sense, but whose intellectuals are directed by a more nobler motive, even that of religion, for to describe her virtues is beyond the power of my pen!

Fri. 30 Jan. . . . At home all day. My wife, poor creature, very ill . . . This 18 weeks past has she almost been in continual pain.

Sun. 1 Feb. . . . Myself at church in the morn . . . We had a proclamation read for a fast and humiliation on Friday the 13th instant for imploring the blessings of Divine Providence on fleets and armies . . .

Mon. 9 Feb. About 12.30 set out for Lewes, where I dined with Mr. Madgwick on a piece of pork boiled, a knuckle of veal and greens (my family at home dining on the remains of yesterday's dinner). Came home about 6.20. Paid Mr. Ed. Relfe 14s. in full, *viz.*, for a hand whip for myself 5s. 6d., for mending a surcingle for myself 2s. 6d., to two girths and a surcingle for Mr. Thornton, 6s. Spent upon myself, horse, ostler and turnpike, 5d. I drank tea with Mr. George Verral. My brother came over in my absence and stayed till I came home, when I lent him my horse. In the even wrote my London letters. My wife rather easier today. I wish I could say I came home thoroughly sober, though in reality I was not much the worse for drinking, for I drank but a very small quantity and wrote 8 letters after I came home, but still my conscience tells me I had in some measure impaired the use of my reason (which I believe is not right so to do). Oh, the frailty of human resolution! But yet I hope through God's grace to become more firm and settled in my resolutions and to act uniformly in all my actions. My wife very ill.

Fri. 13 Feb. . . . This being the day appointed by proclamation for a general fast and humiliation, my servant and myself at church in the morn, the text in *Jonah* 3.5: 'So the people of Nineveh believed God and proclaimed a fast, and put on sackcloth, from the greatest of them even to the least of them'. Myself and servant at church in the afternoon, where we had nothing but prayers. After we came home we dined on some boiled tripe . . .

Sun. 22 Feb. In the morn rode over to Framfield, but did not stay. Came home and breakfasted and then set out for Lewes, where I dined at Mr. Madgwick's on a neck of veal roasted, and a currant butter pudding cake boiled and greens (my family at home dining on a leg of mutton roasted, and potatoes) . . . I called on Mr. Geo. Verral and Mr. Tho. Scrase, and came home about 6.30 and, oh, could I say sober—but my frail resolution and weak brain frustrated my intentions in that particular. Oh, how does the thought of it torment my conscience; well might the wise man say: a wounded spirit who can bear? So many times as I have set a resolution not be to guilty of this crime and yet have as often broke it, sure I am of all men the most miserable . . . Drunkenness is only as an inlet to all other vices, for when reason is laid asleep, then, then, sense and passion ride triumphant, so weak is nature (or at least corrupt and fallen nature). But what I most stand aghast at is to think how miserable (nay, even past description) must my unhappy lot assuredly be should I sleep never to open my eyes again in this world (whenever I am in liquor). No, my portion must be that of the wicked person who knows not God, ah! and possibly more severe, nothing but wailing and weeping and gnashing of teeth. May I, ah! may I, always think of this momentous truth and never more 'be guilty of impairing my reason through overmuch surfeiting and drunkenness . . . Spent self and horse 17*d.*

Mon. 23 Feb. We dined on the remains of yesterday's dinner. At home all day. My conscience extreme troublesome. Oh, for one rash and inconsiderate action have I hours of pain and that of the most tormenting this world can afford! . . . My wife, I hope, something better. In the evening wrote my London letters.

Mon. 9 Mar. . . . In the morn my brother Slater came to see us and brought us a hare for a present . . . Very busy all day, though took but very little ready money. At home all day. My wife, poor creature, quite ill. In the even wrote my London letters.

Weds. 11 Mar. We dined on the hare my brother Slater brought us on Monday, roasted, with a pudding in the belly . . .

Thurs. 12 Mar. . . . At home all day. My wife most extreme ill again, poor creature. Oh, my unhappy fate should I lose her, which I think I must do in a few days! I am quite distracted with grief.

Tues. 17 Mar. . . . John Durrant made my wife a present of some small eels.

Weds. 18 Mar. Paid Mr. Howard a gingerbread baker in cash 8*s.* 9*d.* in full for gingerbread etc. bought of him today, *viz.*, 14 lb of thick, 2*s.* 3*d.*, 1 gross of sweethearts 1*s.*, thin bread 5*s.* 6*d.* We dined on the remains of yesterday's dinner with the addition of an eel pudding . . . At home all day; posted my day book. My wife, poor creature, most extreme full of pain . . .

Easter Mon. 23 Mar. . . . In the afternoon I went down to Jones's, there being a public vestry . . . The officers nominated for this ensuing year are Will. Piper churchwarden, Jos. Fuller electioner; Jer. French and Jn. Vine Jr. overseers, and Jn. Browne and Jos. Durrant electioners . . .

Fri. 27 Mar. . . . My wife somewhat easier today. In the even the Duke of Newcastle came to Halland from the election at Lewes.

Mon. 30 Mar. In the forenoon my brother Sam. Slater came to see us and brought us a present of a sparerib and some sausages and hog's-puddings, and he, together with Mary Heath who was a-washing for us all day, dined with us on the remains of yesterday's dinner. My brother went away about 3.30. Molly Fuller spent part of the afternoon with us. At home all day; Mary Heath lodged at our house all night. My wife rather better today. This day died John Browne after a few days' illness (nay, rather after a few hours' illness), aged 30 years. Oh, what a scene, or rather lesson of mortality is here! How should instances of this kind awaken the remaining part of mankind to prepare for that awful moment which we all know must come sooner or later! And from this and many other instances we may be convinced that it often happens as it were instantaneous, without any warning of the approach, and perhaps in a moment when we are not prepared for it. What poor negligent and heedless creatures too many of us are, that we cannot learn from such striking instances to live in such a manner that we may not be afraid to meet death whenever it may happen (though it should be sudden)! And why should we permit the sensitive faculties so far to get the ascendant of the spiritual as not to make it our most comfortable hopes that we may one day through God's grace enjoy a life which will have no end, where we shall not be disturbed with cares and anxieties, but enjoy the blessings prepared for them who love God? How should such thoughts spur us on to the performance of all Christian duties, always remembering He is faithful that has promised us life everlasting!

Thurs. 2 Apr. In the morning our late servant, Mary Martin, came to assist me in the shop a few days . . . In the afternoon went to the funeral of John Browne, who was buried about 6.15 . . .

Fri. 3 Apr. In the forenoon went to Lewes to buy some lump sugar for Halland. Paid Mr. Rob. Plumer in cash £4 0s. 6d. in full for 1 cwt. 0 qr. 2 lb. of lump sugar bought of him . . . Down at Halland today 4 times. Pretty busy today. My wife, poor creature, quite ill today. Oh, how am I grieved to think of her sufferings!

Sat. 4 Apr. About 4.05 Tho. Durrant and I set out for Lewes in order to get some punch bowls, but could not get any. I came home by Halland, where I breakfasted . . . To and fro at Halland all day. In the even the Duke of Newcastle came to Halland . . . Paid Mr. Faul. Bristow 5s. 3d. for 1 lb of chocolate bought of him today.

Sun. 5 Apr. . . . Not at church all day. This day there was a public day at Halland, where I believe there was 5 or 6 hundred people. At Halland almost all day. My brother went home about 3.20. Captain Lamb lodged at our house, there being not beds enough at Halland. My wife, poor creature, very ill again. How glad I am the confusion is all over at Halland.

Mon. 6 Apr. In the morn down at Halland where there was I believe nearly 5 hundred people to attend his Grace to Lewes, the election being there for the county today, but no opposition[4] . . . After tea our late servant went home . . . My wife, poor creature, extreme ill. This day we had sent us for a present from Halland a shoulder of lamb. Oh, how glad am I the hurry and confusion is over at Halland, for it quite puts me out of that regular way of life which I am so fond of, and not only so, but occasions one by a too great hurry of spirits many times to commit such actions as is not agreeable to reason and religion, for when reason is in a flustering state the passions are apt to ride triumphant.

Sat. 11 Apr. . . . My wife, poor creature, most extreme ill . . . She has now been ill these 28 weeks past yet bears it all with the greatest patience and resignation to the divine will, making it her constant and earnest prayer to God for grace to strengthen her weakness, that she may not be impatient.

Thurs. 16 Apr. . . . In the afternoon down at Mr. Porter's a-copying out an agreement which he had drawed up between himself and Mr. Joseph Burges, wherein Mr. Burges hath sold to Mr. Porter all his freehold land and a copyhold messuage and some land late Durrant's for £900, the conveyance to be signed and the money to be paid at New Michaelmas day next. Mr. Burges is to clear the said premises of all quit-rents, taxes, annuities etc. to the said time and he is not to cut any timber, trees, tillows, shaws,[5] hedges or underwoods, and is to be allowed £10 10s. 0d. for all the dung, muck, compost, mould etc. which is now made or that he shall hereafter make on the said premises, and all expenses attending the same is to be paid between them, and, as there is an annuity of £12 per annum payable yearly out of the freehold to Sarah the wife of Joseph Burges deceased, Mr. Porter agrees to pay that upon Mr. Burges's leaving £50 with him, which he thereby engages to do, and if ever Mr. Burges offers his other copyhold to sale (late Jenner's) Mr. Porter is to have the refusal of it at £200 (though Mr. Porter declared several times afterwards in mine and Joseph Fuller's presence that whenever Mr. Burges should pay off the mortgage which he is to have upon it, Mr. Burges should be at liberty to sell it to anyone he thought proper), and if any of the parties does not comply with every of the aforesaid articles he is bound to pay the other the sum of hundred pounds forfeit. After both parties had signed

---

[4] The Duke of Newcastle and the Pelham family controlled the county, and many of its boroughs, electorally for a long period during the century. See W. D. Cooper, *The parliamentary history of the County of Sussex; and of the several boroughs and Cinque Ports therein* (Lewes, 1834), and D. J. Hope-Wallace, '18th-century election methods', *S.C.M.*, 6 (1932), pp. 640–3, 698–700.

[5] Tillows were tillers or saplings; shaws were underwoods.

it, Joseph Fuller and myself, who was both in presence all the time, set our names as witnesses to the same. We stayed at Mr. Porter's till about 9.50. I am sincerely sorry I was so unfortunate as to be thought a proper person to be a witness to their agreement, for in my opinion Mr. Burges is not equally qualified with Mr. Porter to make a bargain and I think the truth is verified in this present bargain, and if I was present and witness to a thing my conscience told me not right (as mine undoubtedly did) then undoubtedly I am culpable and to blame in acting contrary to my own conscience and of consequence to the Christian religion, of which I am an unworthy member. Whoever may happen to see this when I am no more I beg he would not think I mention this to show myself more righteous or honest than the rest of mankind. No! I have no such thoughts, but am truly sensible I am a poor, wicked, and polluted creature, and perhaps my best actions may be just matter for repentance. Neither do I mention it to condemn Mr. Porter as an over-reaching and unjust person. No, I mention it purely to show how frail, how weak, our nature is, that when self interest is the concern we are always as it were partial in our own favour, even so fond are we of our dear selves as to do things which we would blush at and even condemn as blameworthy in another, and yet so blind are we like holy David as not to know we are the very persons we pass sentence on. Therefore how constant and earnest should we be in our prayers to Almighty God, that he would try us and search out the ground of our hearts and direct our ways, for who of us knows how oft he offends? . . . My wife, poor creature, most extremely ill. A dull heavy time for trade; never did I know it so dull before. I am quite as it were overwhelmed with trouble.

Fri. 17 Apr. . . . At home all day. My wife somewhat easier than yesterday, but really extremely ill. Very little to do all day. Oh, how does my poor wife's illness and her continual pain afflict me. Could it be described I think it would melt a heart as obdurate as flint into compassion and sympathetic sorrow. Gave 6*d.* to a brief[6] for fire which was collected from house to house.

Mon. 20 Apr. . . . In the afternoon wrote my London letters. In the even went up to Mr. Will. Piper's, in order to make his will, but finding it required some person of more knowledge than myself to make it according to form and the testator's meaning, I declined making it, for fear I should make some mistake that might bring trouble into the family, and [cause] the effects of the testator to be thrown away

---

[6] See above, p. 158, note 37.

amongst the lawyers, or the testator's intentions frustrated; though I believe I was rather over-timorous. Came home about 12.10 and left Mr. Piper very ill. My wife, poor creature, very ill indeed. Mr. Porter and myself gave Mr. Piper the liberty to put us in as trustees and guardians to his children, as he could not prevail on anyone else.

Tues. 21 Apr. . . . My wife somewhat easier today, though still very bad and dangerous. Went up to Master Piper's in the morning to assist him with my advice about making his will, there being an attorney come to do it.[7]

Weds. 22 Apr. . . . This morning about 7.30 died Mr. Will. Piper. We dined on some hog's chine-bones boiled and turnip greens. In the afternoon walked up to Mrs. Piper's to consult her about the burying of her husband . . .

Thurs. 23 Apr. . . . In the afternoon Mr. Porter and myself went up to Mrs. Piper's and broke [open] her late husband's will, wherein Mr. Porter, myself and John Piper are left joint trustees to the same. We sealed up his writings and came home about 6.20. Mrs. French drank tea with my wife, who is, poor creature, most extreme ill. In the even Sam. Jenner brought me Mrs. Browne's books and I put some things in them to rights and kept the books in order to write out some bills for her.

Mon. 27 Apr. . . . In the forenoon my brother Sam. Slater came to see us, as did our late servant Mary Martin, who came to stand the shop for me while I went to serve Mr. Piper's funeral. They both dined with us on the remains of yesterday's dinner. After dinner I went to the house of the late Mr. Piper where I served his funeral, giving away 39 pair gloves and 6 hatbands. We came to church about 5.50, when we had a sermon preached for him, the text in *Matthew* 25.46: 'And these shall go away into everlasting punishment: but the righteous into life eternal' . . .

Tues. 28 Apr. . . . My brother Moses called on me in the forenoon in his way to the crock-kiln at Bayley's Lane in order to take my orders for crocks etc. He also called on me in his road home and drank tea with us. There being at Jones's a person with an electrical machine, my niece and I went to see it, and though I had seen it several years ago, I think there is something in it agreeable and instructing, but yet at the same time very surprising. As to my own part I am quite at a loss to form

---

[7] The will is to be found in E.S.R.O., Lewes Archdeaconry wills register A 60, p.352. It was proved by Turner on 11 June. The witnesses were Jeremiah French, Thomas Fuller, Richard Stone and Thomas Farne who was, presumably, the attorney.

any idea of the phenomena.[8] I paid for myself and niece 6*d.* At home all day; but very little to do, and my wife, poor creature, very ill and worn to death with continual pain.

Holy Thursday, 30 Apr. . . . At home all day. My wife, poor creature, most extreme ill. Ah, how melancholy my situation! How does the prospect of losing her (who to me is the only valuable blessing this world affords to make the rugged and uneven paths of this life glide smoothly on) afflict me to see her so afflicted and can get her no relief. Oh, may she one day enjoy the happiness of that comfortable sentence of 'Come, ye blessed children of my Father, receive the kingdom prepared for you from the foundation of the world'.

Sat. 2 May. . . . At home all day and my wife, poor creature, most extreme ill, I think so ill that it would almost move even an entire stranger to shed tears at the sight of her, and to hear her continual moaning through the severity of pang.

Weds. 6 May. After breakfast I rode to Pevensey to consult my friend Jn. Breeden about a servant he recommended to me . . .

Sat. 16 May. . . . Mrs. Vine sent my wife a present of a fine tench and some eels, for which I gave the boy 6*d.* . . . At home all day and pretty busy. My wife very ill. A remarkable fine seasonable time for weather; all nature seems to strive with the greatest energy to be luxuriantly bountiful to man, nay even to poor and ungrateful man.

Sun. 17 May. . . . This day was buried at our church Francis Rich, aged 45 years, who died after a few days' illness and has left a widow and 7 children. What a moving spectacle it was (I am not able to describe it) to see an industrious and sober man, the only support of his family, followed to the grave by his widow and fatherless infants, whose tears and lamentations as it were bespoke their inward and sincere grief. Oh, would the licentious libertine take opportunities to frequent such scenes as this, I think he would no longer follow his vicious course, but would in time become wise unto salvation.

Weds. 20 May. . . . In the afternoon Mr. Porter and myself went up to the dwelling-house of the late Mr. Piper and took a rough kind of inventory of the stock on the ground, household furniture and husbandry tackle of the deceased. Came home about 7.20. Pretty busy in the shop all day . . .

Thurs. 21 May. . . . In the afternoon down at Mr. Porter's a-writing

---

[8] Apparently some sort of generator producing a crackling light inside a glass tube from which the air had been evacuated. This device was derived from the discoveries of Francis Hauksbee (d. 1713) and the experiments of his nephew Francis the younger (1687–1763): see *D.N.B.*

of letters to the creditors of Mr. Piper to meet us at his dwelling-house on Tuesday the 9th day of June next by 2 o'clock in the afternoon, in order to settle and balance accounts . . .

Sat. 23 May. . . . My wife, poor creature, very ill again. I, who yesterday was all joy and pleasure at the prospect of my wife's being better, am now all trouble and affliction, so vain and uncertain is all worldly happiness! But, alas! why do I so afflict myself at the will of Providence, whose all-seeing eye directs all things by the power of unerring wisdom. Let me rather with an humble and truly contrite heart adore that ineffable goodness for the many blessings daily poured upon me. Oh, may I above all things live in such a manner as not to fear death whenever it comes.

Sun. 24 May. In the morn rode to Pevensey, breakfasted with Mr. Breeden and went with him to Westham Church, where I heard an excellent sermon preached by the Rev. Mr. Nicholl, rector of Westham . . . After dinner we rode to *The King's Arms* at Ninfield in order to meet Miss Sally Waller to bargain with her for a servant, but we not agreeing, Mr. Breeden and I came back and to his house and drank tea there . . .

Tues. 26 May. . . . This day sent Tho. Davy with a letter to my friend John Breeden with full power for him to bargain with Miss Sally Waller upon such terms as if it were his own affair . . .

Thurs. 28 May. In the morn my wife went home with her mother (in Mr. Coates's chaise) for the benefit of the air, but how does the thought of parting, perhaps for ever, affect me. I hardly know how to contain myself . . . In the even Tho. Durrant came back (who went with my wife) and brought me the joyful news my wife held her journey beyond what could be expected, and may she, poor creature, reap the benefit I sincerely desire, and it is my constant and most earnest wish . . .

Sun. 31 May. In the morn rode to Hartfield to see my wife, whom to my great grief I found quite ill . . .

Tues. 2 June. In the morn rode to several places to collect in some land and window tax, but could not succeed . . . Tho. Davy sat with me some time in the even. A melancholy time. Oh, how do I gush at every pore as it were with trouble.

Weds. 3 June. In the forenoon rode to several places to collect in some land and window taxes, but could not succeed . . . A remarkable wet night and all the forenoon, it having rained almost successively for 48 hours very hard, so that there are the greatest floods ever remembered in this neighbourhood, and doubtless a great deal of grass spoiled.

Fri. 5 June. . . . In the forenoon delivered to Tho. Durrant the following cash in order for him to carry to Maresfield to pay the general receiver of the land tax, *viz.*, 6 thirty-six-shilling pieces, 4 moidores, 83 guineas, 4 eighteen-shilling pieces, 2 nine-shilling pieces, 2 half *do.*, 9 shillings, 25 half guineas, [Total] £121 17*s.* 6*d.* I also gave him besides 8 half-guineas and 1 guinea to exchange if any of the money should be refused.[9] The state of the account, *viz.*,

The Collectors Dr.	
To the total amount of the land tax book for 1760	£153 18*s.* 0*d.*
To the total amount of the window tax book for 1760	£ 32 5*s.* 9*d.*
	£186 3*s.* 0*d.*
Per contra Cr.	
Jan. 9, 1761 To cash	£ 62 0*s.* 0*d.*
To salary	£ 2 6*s.* 3*d.*[10]
To cash today as before mentioned	£121 17*s.* 6*d.*
	£186 3*s.* 9*d.*

. . . This day I received a letter from my wife, wherein she acquaints me she is no better. Oh, how melancholy is that news to me. I dare say it is a weakness esteemed in me by the world to be so very fond of my wife, but let them know that a virtuous wife is an inestimable treasure, and perhaps a thing somewhat rare in this degenerate age . . .

Sat. 6 June. In the morn there was a transit of the planet Venus over the sun's disc . . .

Sun. 7 June. . . . In the even Sam. Jenner brought Sally Waller who my friend Jn. Breeden had bargained with for my servant . . .

Tues. 9 June. . . . In the afternoon went up to the dwelling-house of the late Mr. Piper where Mr. Porter and I balanced accounts with most of his creditors and debtors, and came home about 10.10 . . .

Weds. 10 June. . . . Dr. Stone bleeded me today. At home all day and pretty busy. This day was fought at Jones's a main of cocks, between the gentlemen of Hoathly and Pevensey. Query: Is there a gentleman in either of the places that were concerned?

Thurs. 11 June. . . . In the afternoon I rode to Lewes to prove Mr. Piper's will . . . Came home about 9.40. Spent . . . the court fees

[9] For the availability of coin see Appendix C. The eighteen-shilling piece was the pistole and the nine-shilling piece the half-pistole, both coins of Spanish origin.
[10] The collector was entitled to retain a salary amounting to 3*d.* in the £ collected.

£1 18s. 4d.; 1 bottle wine 2s.; spent 6d.; horses, turnpike, ostler 9d. Total £2 1s. 7d. . . .

Sat. 13 June. . . . In the even rode to Hartfield to see my wife, whom I found (poor creature) extreme ill. To describe my grief is beyond my power; therefore let a sincere and silent tear paint it in more lively colours to them that read this hereafter . . .

Sun. 14 June. In the morn rode to Uckfield and hired a post-chaise to bring my wife home in. I breakfasted there. We got back to Hartfield about 11.25, where we dined on some currant pond butter pudding . . . We then came away, my father Slater accompanying us on my horse. We got home thank God very safe about 5.40. In the even my wife was taken with an involuntary bleeding at the nose which continued near 2 hours and half, and at times the whole night through (poor dear creature). What it is owing to I cannot tell, whether from my own unhappy temper or that of my friends and relations, but in this day of trouble they seem to stand aloof and as it were staring at me like a stranger. Not one, no! not one that attempts to pour that healing balm of compassion into a heart wounded and torn to pieces with trouble. Whenever it shall please the Almighty to take from me the wife of my bosom, then shall I be like a beacon upon a rock, or an ensign on a hill, destitute of every sincere friend, and not a friendly companion left to comfort my afflicted mind and yield that pleasing comfort of consolation to a mind quite worn to the grave with trouble. Let me, oh! let me, lift my eyes and heart with sincerity to heaven for help, who alone is able to give relief. May I forever give glory to the Lord of heaven and earth in obeying the commands of his divine will in a life spent in true holiness and virtue.

Tues. 16 June. . . . My wife, poor creature, most extremely ill, nay even so ill that I think a few hours will be her last in this world.

Weds. 17 June. . . . At home all day. This day Mr. Porter administered the communion to my wife and self and servant, and as this in all human probability will be the last time we shall ever commemorate (together in this world) the death of our blessed Saviour and Redeemer Jesus Christ, so may the memory of it be a motive to spur me on through God's grace to prepare for eternity (may that awful thought be the first in my mind).

Fri. 19 June. . . . Molly Fuller sat up with my wife, who is perhaps as bad as can be and alive . . .

Sat. 20 June. Paid Mr. Howard at Hartfield 8s. in full for gingerbread and sweethearts received by him today . . . My brother came over in the evening and sat with me some time. About 11.20 my wife was taken with a strong convulsive fit which lasted some time. Dame Durrant and Eliz. Mepham sat up with my wife, and I lodged at Dame Durrant's. Oh, my unhappy misfortune. I shall lose all that is desirable to me in this world; my only true and sincere friend, a treasure of more value than all the riches this world can afford, she has ever been a virtuous and discreet woman and to me the best of wives.

Sun. 21 June.   My friend Mr. Tucker and my father Slater came to see us in the forenoon and both dined with us on a piece of veal boiled, a piece of pork, a gooseberry pudding and spinach. They both went home in the even. Dame Durrant and Bett Mepham with my wife all day and also sat up with her all night, who is so bad that it is past description and we do not expect her to live from one minute to the other. Myself lay at Master Durrant's.

Mon. 22 June.   My brother came over in the forenoon and he and Dame Durrant dined with us on part of neck of veal roasted and green salad. My brother stayed with me till about 4.10. Dame Durrant and Bett Mepham with my wife all day, who is, poor creature, so extreme bad that it would I think even draw tears from the most obdurate heart breathing to see her. Lucy Mepham sat up with my wife. I lodged in at Joseph Durrant's.

Tues. 23 June.   About 1.50 it pleased Almighty God to take from me my beloved wife, who, poor creature, has laboured under a severe though lingering illness for these 38 weeks past, which she bore with the greatest resignation to the Divine will.[11] In her I have lost a sincere friend and virtuous wife, a prudent and good economist in her family and a very valuable companion (and one endued with more than a common share of good sense). I will once more say she was virtuous even in the most strictest sense of the word virtue: she was always decent in her apparel and remarkably sweet and cleanly in her person, and had by nature a cheerful though religious turn of mind. Therefore I have lost an invaluable blessing, a wife who, had it pleased God to have given her health, would have been of more real excellence to me than the greatest fortune this world can give. Oh, may her agonizing pains and dying groans have such a constant impression on my mind that (through the assistance [of] God's grace) I may ever have the thought of death in my mind, and that by a truly religious course of life may be prepared to meet that King of Terrors; may the memory of her virtues always excite in me a love of that which is good and virtuous, and may I endeavour to copy the many excellencies she was

[11] Margaret Turner was aged 27½ when she died.

undoubtedly possessed of; therefore I may justly say with the incomparable Mr. Young: 'Let them who have ever lost an angel pity me.'[12]

We dined on the remains of yesterday's dinner. My friend George Richardson came to advise me in the even, and stayed with me all night, and both of us lodged at Master Durrant's. Dame Durrant and Bett Mepham stayed with my servants.

Weds. 24 June. In the morn my father Slater came to see me and to condole and sympathize with me in my misfortunes, a thing truly humane and friendly, and he with my worthy friend George Richardson stayed and breakfasted with me and then went away . . .

Thurs. 25 June. I dined on the remains of yesterday's dinner, with the addition of a hog's cheek boiled and gooseberry pudding. Oh, how melancholy my situation. Not a friend to pour that pleasing balm of consolation into a heart overwhelmed with grief, no, nor one enlivening object gains admittance in my distracted breast. There, oh, there is naught by melancholy and pensive sadness! What can give pleasure or present one pleasing idea to my mind since she for whom I lived and in whom I centred all earthly happiness is now no more? Oh, could I forever sing her praise and describe her virtues, it would be a task (though pleasing to my mind) far beyond the reach of my understanding. But let me call home my wandering thoughts and reflect that man, poor frail and mortal man, is born to trouble as the sparks fly upwards; then why should I who am a vile and sinful creature be exempt from that which our whole race is subject to? No, rather let me rejoice and give thanks, that Providence directeth all things with infinite goodness and wisdom, for the Lord will not forsake forever, but though he send affliction, yet will he have compassion according to the multitude of his mercies. For he does not punish willingly nor afflict the children of men.

Sat. 27 June. In the forenoon my brothers Moses, Will. and Richard came to my house, and they, together with my brother and father Slater and Dame Durrant, dined with us on a cold forequarter of lamb roasted in the oven. Yesterday about 5.50 I buried my wife at Framfield, and with her all my hopes of worldly happiness. I am now destitute of a friend to converse with or even a sincere friend on whom I can rely for advice now I have lost the dear, dear partner of my soul, with whom I could repose such confidence that I never concealed from her the most minutest circumstance of my affairs, but always found

[12] Edward Young, *Night Thoughts*, night III, line 110: 'Ye that e'er lost an angel, pity me'.

comfort in the disburthening my mind to her when it was almost even overloaded with trouble. To describe or estimate my loss is more than I can do, but still let me usurp the reins of reason and think that as everything is guided by infinite power and goodness so must it consequently be by infinite wisdom; therefore let me conclude that whatever is directed by providence is best for me and may I, oh, may I, forever be benefited by this fatal blow. Alas, I have now seen part of myself seized upon by death, relentless death; therefore may I always be careful to make my calling an election sure, through the merits and intercession of my blessed Saviour.

Tues. 30 June. . . . Bett Mepham drank tea with us, and I paid her 5s. in full for attending my wife and laying her out etc. . . .

Weds. 1 July. . . . This day I was informed of the ill-natured and cruel treatment I have privately received from malevolent tongues, who have made, propagated and spread with indefatigable industry and diligence a report that Mr. Snelling at my request (and by force) castrated my wife, which operation was the immediate cause of her death. And with such amazing swiftness is the report spread that there is hardly a child of four years old or an old woman of four score within ten miles of the place but has it at their tongue's end, and even so credulous as to give sanction to it; that is, if they do not directly believe it they will by no means let it die with them, but still continue to circulate it about, so vile and envious is man to man. Now from what occasion this palpable falsehood could take its rise I am quite at a loss to guess; as to my own part I know myself thoroughly innocent, therefore I defy and despise the malice of the vulgar multitude and if I know my own heart I sincerely forgive them; neither have I in the least any anger against them for it. No, I have not, for as ignorance undoubtedly is the mother of credulity I must ever think they deserve my pity; and, as to love and respect for my deceased wife, I want no other testimony than my own conscience, which I am sure sings peace in that affair, but however, though I have that which is beyond anything in the world besides to me, so am I not destitute of other evidence: I have even that of my wife's own handwriting, wherein she says she wants words to express her gratitude to me for my care etc. for her. I have the witness also of all about my wife during her illness; therefore I am not in the least moved with anger, though I must own I am as it were astonished to think that I can have any enemy so malicious as to propagate such a falsehood, for I can justly say there is nòt a person in the parish but has received favours at my hands, and many of them a great many; as to all my relations in general, I am sure they have sufficiently tasted of my munificence. Therefore I should have thought it a thing impossible to have been used so inhumanly and I think nothing can show more what a set of poor ignorant people I am placed among than this, for 'em to give credit to a thing which I presume cannot be done without instantaneous death, for in taking out the uterus undoubtedly the spermatic vessels must be cut, and I cannot conceive how there is to be ligatures made on the arteries, but that the person must forthwith bleed to death, or at least it is such an operation as I never read or heard of. May I, oh, may I, never think of ill, or any ways be angry at such false reports. No, let me rather bear them with joyfulness, for there is undoubtedly an overruling providence that orders everything according to infinite wisdom . . .

Thurs. 2 July. Made part of the land and window tax books . . . Tho. Davy lodged with me and to my great surprise informed me that he was shortly to be married to the widow Virgoe. Oh, I doubt he is guilty of not acting the part of an honest man, for he has kept my late servant Mary Martin company I dare say 3 or 4 years.

Fri. 3 July. I dined on some peas and bacon boiled. In the afternoon Mr. Snelling came to see me and he and I rode to pay a visit to Dr. Stone, where we stayed 2 or 3 hours. From thence we came to Stonebridge where we called and stayed about an hour. We came home about 9.40 and found my brother at my house, who came to see me in my absence, and they both stayed and supped with me and Mr. Snelling stayed all night.

Sat. 4 July. . . . At home all day and thank God pretty busy. How do I more and more daily find the loss of my wife to be great; how do I severely know the want of her in the careful and regular management of my family affairs, which are not now conducted with her conduct, prudence and good economy! No they are not; therefore her loss to me is as I have oftentimes said before an invaluable treasure—a treasure which (had it pleased God to have spared her life) would have been daily improving.

Sat. 25 July. Mr. Tomlin, hop-factor, came to see me in the afternoon . . . Down at Halland 4 times today. In the even the Duke of Newcastle and several more noblemen etc. came to Halland and stayed all night . . . Mr. Tomlin stayed and lodged at my house, and he and Joseph Fuller Jr. smoked a pipe or 2 with me in the even.

Sun. 26 July. Mr. Tomlin and myself at church in the morn . . . Mr. Tomlin and my late servant Mary Martin (who came to see me in the morn) and my brother Sam. Slater, who came in just as we were a-sitting down to dinner, dined with me on a piece of bacon and a chick boiled, a leg of lamb roasted and a plain rice pudding. After dinner we all walked down to Halland, where there was a public day, and we stayed and walked about there till near 8 o'clock. I lodged in at Jos. Durrant's, and my brother and Mr. Tomlin lodged at my house, though not one of us went to bed sober, which folly of mine makes me very uneasy. Oh, that I cannot be a person of more resolution!

Mon. 27 July. Mr. Tomlin and my brother breakfasted with me, and then we all set out together, he on his road to Withyam, and my brother and I to Lewes. I paid Mr. Madgwick 52s. in full for 4 pair breeches bought of him today. I also gave him in cash £70, for 1 bill dated today, payable to me or order, 21 days' date, on Mr. Will. Margesson. I called on Mrs. Roase. I came home so very much in liquor that I cannot tell any one thing how I came home, nor do I know if I paid my reckoning before I came away. Oh, what a poor unhappy wretch I am; I am quite distracted with trouble. Oh, how does this one piece of folly and indiscretion torment my mind. My brother went home in the even. My brother Moses came over in the even, but did not stay. Very bad all the even. Oh, my heavy and troubled mind; oh, my imprudence pays me with trouble!

Tues. 28 July. Mr. Long dined at my house on the remains of Sunday's dinner. Mrs. Coates sent me a present of a veal pasty. Oh, I am intolerable bad; my conscience quite tears me in pieces.

Thurs. 30 July. ... In the forenoon went down to Mr. Coates's and delivered to Mr. Perry the Duke of Newcastle's account. Very melancholy is my present situation. I see by repeated instances that to continue my trade as I am it will not do, for there is no dependence upon servants and to think of marrying again is what I have no thoughts of; no, not so long as the image of my dear wife is almost continual in my thoughts . . .

Fri. 31 July. ... Oh, how distracted and tumultuous are my thoughts; fearing I should by a wrong proceeding, or by the badness of trade, miscarry in my trade, and at the same being so embarrassed in my family connections that I hardly know how to act with the most prudence. I would gladly pursue that which is the most commendable and at the same time most to the advantage of myself and at the same time be a useful member of the community of which I belong.

Sat. 1 Aug. ... This day Tho. Davy's marriage with the widow Virgoe (which was solemnized at this parish church on Tuesday last) was made known, though when he dined with me, it was what I had no thoughts of, only as I know it would be soon[13] . . .

[13] For the reason for the haste see the diary entry for 27 Jan. 1762.

Weds. 5 Aug.   An acquaintance of my servant's (or rather her admirer) came to see her and dined with me . . . Almost distracted with trouble. How do I hourly find the loss I have sustained in the death of my dear wife!

Sun. 9 Aug.   After breakfast I rode to Hartfield, where I dined at my father Slater's on a coast of lamb roasted, a duck pie, a piece of pork, an apple pudding, some cabbage and cucumbers (my family at home dining on a boiled mackerel and a pudding cake and French beans). I stayed and drank tea with my father Slater and came home about 8.30 . . . How I am afflicted with grief! Oh, how great is my loss, what can equal the value of a virtuous wife?

Fri. 14 Aug.   . . . At home all day and surely a deader time for trade I never knew . . .

Sat. 15 Aug.   . . . I hardly know which way to turn or what way of life to pursue.

Sun. 23 Aug.   After breakfast I rode to Alfriston in order to take a cursory view of Mr. Sanders's stock in trade, which I am shortly to appraise for him.

Weds. 26 Aug.   . . . How severely do I feel the loss of one of the best wives . . . I may with justice say upon the whole I have lost one of the best wives and almost the phoenix of the age . . . She was religious but no bigot, discreet yet full of innocent mirth.

Mon. 31 Aug.   In the morn as soon as I had breakfasted rode to Dapp's, where I weighed up his hops, which was 1 pocket weighing 1 cwt. 0 qr. 2 lbs., and for which I paid him £2 19s. 0d. . . . In the morn sent my wool to Lewes by John Divol. In the afternoon my brother called on me in his road to Lewes in order to receive both his and my wool money . . . In the evening Joseph Fuller Jr. smoked a pipe with me. At home all day; thank God very busy. Oh, how pleasant is trade when it runs in its proper channels, and flows with a plentiful stream. It does, as it were, give life and spirit to one's actions; I think the most phlegmatic constitution must feel its pleasing and enlivening charms.

Tues. 1 Sept.   . . . In the afternoon my brother came over and drank tea with me and I received of him in cash £15 6s. 4d. in full for my wool money which he received yesterday of Mr. Tho. Friend for me . . . This day I signed a bond wherein I and Jos. Durrant are jointly bound as securities for Mr. John Long as a hop-assistant.[14] I also signed a

[14] Excise officers closely supervised the hop harvest in order to ensure the correct return to the revenue. When the harvest season was in full swing the officer needed to employ several assistants (all of whom had to give surety for their good behaviour) in order to ensure that the weights of hops picked were correctly recorded. The procedure

bond wherein we are jointly bound as securities for Tho. Durrant as a
hop-assistant. I also signed a bond wherein I and Joseph Burges are
jointly bound as securities for Tho. Davy as a hop-assistant; I also
signed a bond wherein I and John Smith of Cross-in-Hand are jointly
bound as securities for Samuel Diggens, gauger in a footwalk at
Abingdon in Berkshire.[15] The obligation of each of the bonds was
£200 and they were all witnessed by John Purdue and Lawrence
Thornton.[16] At home all day and pretty busy, but not very well . . .

Sat. 5 Sept. . . . In the afternoon a gentleman whose name was
Hastings and a humble servant of my servant's drank tea with me and
stayed all night. He brought me a present of some pigeons . . .

Sun. 6 Sept. . . . After dinner we walked to Little Horsted church,
where we heard a sermon preached by the Rev. Mr. Philips, curate of
Maresfield and Horsted, from *Philippians* 1.23, 24 . . . The gentleman
we heard appears to be not more than 20 years of age, though I
suppose he is about 26. I think I never heard a finer churchman in my
life, he being a good orator, and delivered an extreme good sermon,
both for beauty and language and piety of thought . . .

Thurs. 10 Sept. . . . In the afternoon I walked up to Mrs. Piper's and
measured a piece of land. In the even Joseph Fuller Jr. and I walked down
to Jones's, where I spent 6*d*. We came home about 8.30 . . . About noon
died a little boy of Mr. Porter's (an infant of about 9 months old)[17] and
in the evening died John French (aged 18 years) after an illness of near
6 months.

Mon. 14 Sept. . . . In the afternoon went down to Mr. Porter's and
screwed down their child's coffin. From thence I went down to Mr.
French's and served their son's funeral. Came back to Mr. Porter's
and attended their child to church, and then went and met the other
corpse and attended that to church, and came home about 6.45. In the
evening wrote my London letters.

Sun. 20 Sept. . . . As soon as I had dined, I set out for Steyning, in
order to appraise a shop of goods belonging to the widow Foreman and
now taken by my friend Geo. Richardson . . . I arrived at Steyning
about 6.20 and spent the even at *The White Horse*, in company with my

---

for weighing the harvest and the manner in which the books were to be kept are minutely
laid down in Charles Leadbetter, *The Royal Gauger; or, Gauging made perfectly easy* . . .
(4th ed. 1755), p. 367.

[15] Turner here acted as security for the good behaviour of Diggens as surveyor in
new roadworks at Abingdon.

[16] Lawrence Thornton was the local excise officer. See Appendix B.

[17] Thomas Porter had been baptized on 9 November 1760.

friend Geo. Richardson and Mr. Tho. Foreman. I lodged at *The White Horse*.

Mon. 21 Sept. I breakfasted at Mrs. Foreman's. About 10.20 we began appraising the goods; *viz.*, Mr. John Balcombe of Angmering in behalf of Mrs. Foreman and myself in behalf of my friend Geo. Richardson. We all dined at *The Chequer* on a cold duck pasty and two rabbits roasted. My brother came over in the forenoon to stand the shop for me during my absence . . . Mr. Foreman, Mr. Balcombe, Mr. Richardson and myself spent the even at *The White Horse*, where I lodged all night.

Tues. 22 Sept. We all breakfasted with Mrs. Foreman and completed valuing the stock in the forenoon, which amounted to only £82. Mr. Foreman, myself, and Mr. Richardson dined at *The Chequer* on a leg of mutton boiled and turnips and cabbage . . . I came home very safe and sober about 8.20 . . . Oh, what raptures did I use to approach home with in my dear Peggy's life when I had been out, but now how different the scene . . . No pleasing object to meet me with the smiles of approbation and all the other endearments of conjugal love and affection. Little, ah, little do the sons of riot and debauchery know how great, how far beyond description, is the pleasure that is found in the company of a virtuous wife.

Fri. 25 Sept. After breakfast walked up to Mrs. Piper's; from thence John Piper and I walked down to John Cayley's, where I bought his hops at 2 guineas per cwt. and I gave him 10*s*. 6*d*. in hand. Those hops are in joint partnership between Joseph Fuller and myself . . . In the afternoon I walked around Mrs. Virgoe's wood to examine the fences . . .

Sat. 26 Sept. After breakfast I rode over to Framfield; from thence my brother and I rode down to Savage's to look upon their hops, but did not buy them . . . Joseph Fuller acquainted me that he had bought of Tho. Reeve all his hops at 46*s*. per cwt., which we are to be joint partners in.

Tues. 29 Sept. . . . In the even Joseph Fuller Jr. smoked a pipe with me and informed me he had bought Joseph Bonwick's hops in joint partnership between us at 50*s*. per cwt.

Thurs. 1 Oct. . . . In the afternoon Mr. Thornton and I walked up to weigh Mr. Reeve's hops, but he was not home. We went from thence to George Cornwell's, where Mr. Thornton weighed his hops and I attempted to buy them, but in vain. We stayed and drank tea here, from whence we walked to Thomas Starr's where Mr. Thornton weighed his hops and I again attempted to buy them, but in vain. We called at Mr. Reeve's as we came home, and the old man would not weigh them, pretending it was too late. Came home about 6.30, where I found my brother, who came over in my absence, and stayed with me till near 8 o'clock . . . I paid Halland gardener 10*d*. for 2 gallons onions and one of potatoes received by him today. Pretty busy all day.

Fri. 2 Oct. In the morn rode up to Mr. Reeve's and weighed his hops; *viz.*, . . . 8 cwt. 0 qr. 2 lb. which at 46*s*. per cwt., the two pounds excluded, amounts to £18 8*s*. 0*d*. which I paid him. I also weighed up a pocket of Richard Hope's, marked 'R.H.' . . . for which I also paid Master Reeve £3 12*s*. 6*d*. . . .

Sat. 3 Oct. . . . In the even Joseph Fuller Jr. smoked a pipe with me and I received of him by a bank bill and cash £27 4*s*. 0*d*. in full for his half part of the money I have paid for Reeve's, Cayley's, Hope's and Bonwick's hops . . . I also exchanged a bank bill of £25 for him, giving him cash for it. Pretty busy all day. N.B. All the hops which Joseph Fuller and I have now between us are paid for and he hath also paid me his part.

Sun. 4 Oct. . . . The King's proclamation for the discouraging of vice and profaneness and for the promoting of piety and virtue was read this afternoon . . .

Tues. 6 Oct. . . . About 4 o'clock Joseph Fuller Jr. prevailed on me to accompany him to Blackboys Fair. We spent 16*d*. I came home about 10.20 but not quite as sober as I went out, but we drank very little . . .

Thurs. 8 Oct. . . . My father Slater . . . came to see me . . . My father Slater and I balanced accounts . . . In the even went down to Mr. French's, where I received of him in cash £2 in full for a coffin for his son on account of Mrs. Browne and also £4 3*s*. 4*d*. in full for the funeral of his son on my own account. I came home about 9.30 . . . The reason of my father Slater's visit today proceeded from the rumour that now prevails at Hartfield of Mr. Snelling's castrating my wife by force at my request. What would the credulous mob (for so I can justly call them) have with me? If I do injury to any one I would not I am sure, did I know it, and as to this affair I am sure I not only know my own innocence, but at the same time, it is a thing unprecedented in the practice of surgery, a thing I believe never done, so that it must betray the height of ignorance, which undoubtedly is the mother of credulity, and shame, which generally are twin sisters and oftentimes are linked together. But however it is, the rumour it seems finds people simple enough to give credit to it, and even so let them—I no ways envy their happiness and sure I am if I know my own heart I sincerely forgive them (and at the same time all mankind) and was it in my power to confer a favour upon any of them I would do it, and that without upbraiding them, and I think no man ever loved a woman better than I did my wife, for which I want not the mob for witnesses. No! I do not. I have a better, even that of my own heart . . .

Fri. 23 Oct. . . . This day John Jones and Mary Heath were married . . .

Mon. 26 Oct. Paid John Piper 18*d.* for a bushel oats received by him today. In the forenoon Mr. Porter and I walked down to the sign of *The Chequer* at Whitesmith, there being a court-baron and leet[18] held there for this manor. We dined there in company with near 20 or more on a rump of beef boiled, a brisket of beef and a leg of mutton boiled, 2 geese roasted, a giblet and plum pudding (my family at home dining on the remains of yesterday's dinner). Mr. Porter and I was admitted to the copyhold estate late Mr. Will. Piper's, according to the tenor and effect of his last will. The expenses I paid today were: to a heriot compounded for at £5; to the beadle's fee for seizing the heriot 6*s.* 8*d.*; to the beadle's fee etc. for admission, 2*s.* 0*d.*; expenses 2*s.* 0*d.*: [Total] £5 10*s.* 8*d.* I came home very sober about 10.45; I wish I could mention the same of my companion. Pretty busy all the forenoon and in the morn wrote my London letters.

Tues. 27 Oct. In the forenoon Mr. Porter and I walked down to Halland where I paid Mr. Michell the steward £2 3*s.* 4*d.* for his fees for our admission yesterday, and now there remains £21 for us to pay for a fine, which sum we have agreed to pay in three months . . .

Thurs. 5 Nov. At home all the forenoon. I dined at Mr. Porter's on a sirloin of beef roasted, a buttock and a piece of brisket boiled, raisin suet pudding and turnips, in company with a great many more of the parishioners, it being the tithe feast day . . . I paid Mr. Porter 8*s.* in full for 1 year's tithe due on Michaelmas last. Came home very sober about 5.10. Very little to do all day— a most dismal, dull time for trade . . .

Sat. 7 Nov. . . . In the even paid Joseph Fuller Jr. in cash £36 17*s.* 0*d.* in full for his half part of the money which I had received of Mr. Tomlin for the hops he sold, which was bought in partnership between myself and Joseph Fuller . . .

[18] The court baron was the manorial court which dealt with matters relating to the tenancy of land within the manor, the holding of that land, and the regulating of agricultural matters. It was in that court that Mr. Porter and Turner would have been admitted to Piper's copyhold as trustees. The leet or view of frankpledge was held in conjunction with the court baron normally twice a year, and determined matters formerly dealt with by the sheriff including misdemeanours, minor assaults, watch and ward, and the regulation of the assize of bread and ale etc.

Weds. 25 Nov. . . . At home and very busy all day. Oh, what a pleasure is business! How far preferable is an active busy life (when employed in some honest calling) to a supine and idle way of life, and happy are they whose fortune it is to be placed where commerce meets with encouragement and a person has the opportunity to push on trade with vigour . . .

Thurs. 3 Dec. . . . In the even went down to Whyly in order to balance accounts with Mr. French, but did not. I stayed smoking a pipe with Mr. French till near 10.20. Mr. Porter informed me today that Burrage, who some two years ago absconded and left his wife and 6 small children to this parish, was heard of again, and advised the parish officers to send and get him home.

Fri. 4 Dec. After breakfast Joseph Fuller and I walked down to Halland in order to consult Mr. Coates in regard to Burrage's affair . . . Jos. Fuller Jr. smoked a pipe with me in the even, as did Mr. French who, poor creature, is I think quite stupid through drinking.

Fri. 11 Dec. . . . In the even down at Mr. Porter's a little time. This day was brought home by two men (whom our parish had sent on purpose) Will. Burrage, who had absconded above 5 years ago and left a wife and 6 small children as a burden to the parish. Now as the affair makes a great noise and confusion in the place and the inhabitants seem much divided in their opinion about the treatment which he deserves to meet with, I shall for the future satisfaction of anyone who may happen to see my memoirs deliver my sentiments in the affair. First, Mr. Porter, Mr. Coates and Mr. French are desirous he should suffer the punishment due to so atrocious a crime as deserting his family, by which means they have been an expense to the parish of upwards of £50, and the poor woman become a lunatic through grief, in the most rigorous manner. The rest of the people are all desirous he should escape without any further punishment. Neither of which methods are I think agreeable to reason or justice, for as to the first I think it is too severe, as many things appearing in his favour, and as to his first going away I doubt he had some faint reason for so doing (though none adequate to answer the leaving his family) such as his wife and he living unhappy through the instigation and unhappy temper of her mother, who lived with them, and the many attempts made by the first-mentioned gentlemen to reduce the price of day labour by bringing into the parish certificate-men for that purpose, which undoubtedly must be very disheartening to an industrious man and what I think hardly just and human.[19] Then, since he has been in

---

[19] Burrage had absconded in October 1756 (see above, p. 69) and at that time Turner

custody he has behaved extreme well, giving the men no trouble, and coming home with a seeming cheerfulness, and at the same time promising to maintain and do the best for his family in his power. These things duly considered should, I presume, be some mitigation of the rigorous punishment due to his crime (which must be allowed great). The greatest number or body of the people who are for releasing him immediately plead that it will be of no service to the parish to confine him in the house of correction, and that the interest to the parish barely considered should be a motive sufficient to release him. This I think savours too much of a contracted and self-interested mind, or rather a pusillanimous disposition, neither of which methods I approve of. I would then advise justice should take place in such a manner that a strict eye may be had to mercy, and not in the height of executing justice to forget that benign virtue. No, I would not but that it might be so tempered together as not to interfere or clash with each other, for undoubtedly he ought to suffer in some manner, either by confinement or corporal punishment, or both, in order to deter others from the same offence, but then I would have mercy so far take place that he might be convinced his punishment did not proceed from choice but as it were from real necessity in order to keep a due subordination in the parish and that the mitigation of the punishment proceeded altogether from the lenity of the officers and gentlemen of the parish. This if I can form any judgment of the affair would seem the most eligible way, as justice would not be perverted and yet mercy shine through the whole in the most beautiful and delightful colours. However as I write or think not to direct others, only to deliver my own sentiments in the affair, if my arguments were brought to the test and strictly scrutinized they might perhaps be found quite insignificant and of no weight. Therefore I shall with all due deference submit to the opinion of them who are better judges.

Sat. 12 Dec. . . . *N.B.* By way of a postscript to what I wrote yesterday relating to Burrage's affair, I doubt not but to every human and well-disposed person (especially in this land of liberty and freedom) there must at the first thought something appear shocking to humane nature at the prospect of a person's suffering punishment even when their crimes justly merit it, and we are, as it were, instantaneously struck with a fellow-feeling for the delinquent, and at

made the point which he repeated now: that the wealthier men in the parish were making it almost impossible for a pauper resident in East Hoathly to get work by bringing into the parish cheap labour in the form of paupers from elsewhere—'certificate men'—who could, as soon as the work was finished, be transported back to their place of settlement, testified by their settlement certificates.

the same time sympathize and as it were bear a part of his punishment, but so soon as the first starts of the impression made on our minds by the representation of outward objects (if I may be allowed the expression) are over and reason resumes its place, we shall then consider and reflect that, were not human laws in some measure executed, there could be no security of property, and when a person is become truly obnoxious to the laws of his country, totally to pervert the executing of justice would be as great a crime as wholly to exclude mercy, so that what before appeared as compassion and pity we shall now look upon rather as pusillanimity and as proceeding from fear, rather than from a true principle of mercy. Therefore in my opinion justice with humanity should be first executed and then let mercy and benevolence open their extended wings and close the scene.

N.B. Were the executive part of our laws quite laid aside (as I fear they are in many cases too much) then mercy would cease to be a virtue, and become a real vice.

Sun. 13 Dec. ... After churchtime Tho. Durrant and myself walked down to Whyly, where we stayed and spent the even, and to my eternal shame be it spoken, came home not thoroughly sober. Oh, my imprudence! What frailty and perverseness is there in human nature! If I could have the resolution to drink only water, as my brains are so weak, happy would it be for me, for I think the very smell of liquor intoxicates my brain and when reason is lulled to sleep, then, then, undoubtedly appetite reigns triumphant, and we no longer can be said (with any propriety of reason) to be rational creatures, but rather brutes ...

Weds. 16 Dec. ... A continual rainy day and at home all day and as little to do all day as it is possible to imagine. Oh, heavy and dismal trade. I am quite in a strait not knowing what to do. Sometimes I think I will set myself clear in the world and leave Hoathly, as I am so embarrassed in my affairs, what with my friends, and together with the debts due on my trade. At other times I think it would be looked upon as a discontented mind that occasions me to do it, so that I am at loss which measure to pursue, being unwilling to act imprudent and yet at the same time desirous to act in such a manner as might be most consistent with reason and at the same time not to distrust the wise distributions of providence. My inclinations seem most inclined to leaving this place, thinking it is the most prudent method as my circumstances are at present ...

Fri. 18 Dec. ... After dinner I rode to Lewes upon my brother's horse ... Spent the even at Mr. Geo. Verral's in company with my brother and lodged at *The White Hart*. I cannot say I went to bed quite sober, so frail, so wicked a creature am I, and as I have transgressed today, so may I perhaps again tomorrow . . .

Sat. 19 Dec.   Came home about 2.10. My brother went away immediately. My family at home dined on a piece of pork, an apple pudding and turnips . . . Oh, could I say I went sober to bed. Oh, how does my tormented mind suffer anguish—but let me quite, oh, let me for ever, conquer this easy temper; let me fix my resolution and with divine assistance let me pluck as it were out a right eye; that is, let me endeavour never more to be guilty of this folly. Oh, how do I detest myself; let me repent in dust and ashes, and once more try to obtain a victory over my unruly passions.

Sun. 20 Dec.   . . . In the even went down to Halland where I supped and stayed all night in order to assist Mr. Coates in distributing of a gift left by some of the ancestors of the Pelham family forever to be given yearly on St. Thomas's Day to every man and woman 4*d.* each and every child 2*d.* each, together with a draught of beer and a piece of bread . . .

Mon. 21 Dec.   About 3.10 we began our work and distributed upwards of £7 and I conjecture to between 5 and 6 hundred people, as also a sack of wheat made into bread and at the least a hogshead of good fresh beer. I stayed and breakfasted with Mrs. Coates and after receiving my groat came home about 11.10. I dined on part of a bullock's heart fried . . . Relieved [29] poor people today and 1*d.* and a draught of beer each . . .

Thurs. 24 Dec.   . . . At home all day and not very busy. How melancholy a time I have. Not that I would be thought anyways to murmur at the wise distributions of providence; no, far from it . . . I doubt not if I was to look around me upon the rest of mankind, I might see many more wretched than myself, and that I enjoy many blessings denied to a great number of my fellow creatures, and if not, I can with propriety say I enjoy more than I deserve; but still, as I am fearful my trade does not answer, and am at a loss to get into any other branch (nor do I think my capacity equal to carry on a great trade), how can I help being uneasy, and more particularly as I have not a friend to whom I can entrust the management of my affairs to, so that when I am from home my affairs suffer at home, and when I am at home my affairs are neglected abroad, for wanting seeing after; so that I have as it were a curb put upon any industry I might exert myself with.

Fri. 25 Dec.   This being Christmas day, myself and one servant were at church in the morn . . . We both stayed the communion; I gave 6*d.* . . .

The widow Marchant, James Marchant, Bett. Mepham and Philip dined with me on a buttock of beef roasted, a raisin suet pudding and turnips . . .

Sat. 26 Dec. In the morn Mr. Long walked with me to many places to collect in some land tax, but I found myself unsuccessful in the attempt, not getting one farthing . . . In the afternoon went down to a public vestry at Jones's . . . the surveyors for the year ensuing are Edw. Foord and Thos. Fuller; electioners, Jn. Vine and Edw. Hope . . .

Weds. 30 Dec. . . . Somewhat indisposed today with a sore throat occasioned as I presume by the lodging of a raisin stone in my throat the 21st instant.

Thurs. 31 Dec. In the forenoon my brother and I walked out in order to get in some land taxes, but I could not succeed. We then walked to Dr. Stone's in order to ask his opinion on my throat; we came home about 2.20. My brother dined with me on a piece of pork, a piece of beef, and a raisin suet pudding and turnips boiled . . . In the evening went down and balanced accounts with Mr. Porter . . . Stayed at Mr. Porter's till about 11.40. My brother stayed all night with me. One of my servants out all night a-dancing . . .

1762

Fri. 8 Jan. . . . After breakfast I rode to Maresfield (calling at my brother's as I went) to meet Mr. Tho. Gerry, deputy receiver-general of the land tax, to whom I paid in cash £71 in part of the two first quarters of the land tax . . .

Sun. 10 Jan. In the forenoon Mr. Stone's apprentice came and let my blood and one of my servants' . . .

Mon. 11 Jan. Richard a-cutting my vine in the forenoon, for which I gave him 6d. . . .

Tues. 12 Jan. . . . All day (and even from three o'clock yesterday in the afternoon) we had a very high wind and several very hard showers of hail, with several claps of thunder and many flashes of lightning, and really I think it a very great hurricane . . .

Thurs. 14 Jan. I walked down to Halland about 1 o'clock, where I dined on 2 chickens boiled, 1 roasted, a sirloin of beef roasted, a ham boiled, 2 ducks roasted and some mince pies (in company with Mr. James Michell, Mr. and Mrs. Porter, Mr. and Mrs. French, and Mr. and Mrs. Gibbs) . . . I stayed and drank tea and coffee there and also stayed and supped there, in company with the aforesaid, Mr. James Michell excepted, on part of a cold sirloin of roast beef, a neck of mutton roasted, a cold duck pie, a gooseberry tart and potatoes. We played at brag in the even; I won 15½d. and gave Mr. Coates's servant 12d. We came home about 10.50, and I came home sober. Oh how comfortable does the word 'sober' sound in my ears. Let it be my earnest and constant endeavour always to do the same. Then will my thoughts in this particular yield a pleasure (not to be expressed in words) in this retrospect view.

Sun. 17 Jan. Myself and both servants at church in the morning . . . The thoughts of that most amiable creature, the best of wives, has this day overspread my mind with a melancholy gloom. Oh, how severely do I feel the loss! To give her a character which her merits justly deserved and which they have a just claim to demand requires a pen more nervous than my own. She was undoubtedly superior in wisdom, prudence and economy to most of her sex and I think the neatest and most cleanest woman in her person I ever beheld . . . I think words can convey but a faint idea of the pleasure and happiness that a husband finds in the company of a virtuous, prudent and discreet woman, one whose love is founded not on the basis of sensual pleasures but on the more solid foundation of friendship and domestic happiness, whose chief delight is to render the partner of her bosom happy. For whatever the libertines may say or think of marriage, I believe they never felt that secret pleasure which is to be found in it, notwithstanding their boast of freedom and I know not what . . .

Sat. 23 Jan. . . . At home all day and really hardly anything to do in the shop, so dull is trade. A fine day and frosty, and I think the pleasantest day we have had for some months. Employed myself part of the day in sawing wood.

Sun. 24 Jan. Myself and both servants at church in the morn . . . We had a brief[1] read for rebuilding the parish church of Penal in the county of Merioneth in South Wales, the expense of which amounts to the sum of £1270 and upwards; I gave 1d. . . .

Mon. 25 Jan. . . . At home all day; a cold sharp frosty day; very little to do all day. In the even wrote my London letters. Employed myself today part of the day in sawing of wood. Oh, melancholy time—what to do I hardly know.

Tues. 26 Jan. . . . At home all day and really very little to do all day. In the day sawed some wood. In the even myself, Jos. Fuller Jr., Master

---

[1] For briefs see above, p. 158, note 37.

and Tho. Durrant smoked a pipe or two with Mr. Long. We came home about 11.20. It is impossible to say how dull trade is, for I have hardly anything to do, and I am sure I must daily lose money, so that I am as it were come to a resolution to leave this place . . .

Weds. 27 Jan. . . . This day died the wife of Will. Alcorn after being married about 10 months. The wife of Tho. Davy was this day delivered of a girl after being married only 6 months, two people whom I should the least have suspected of being guilty of so indiscreet an act. But what can be said of this passion? How careful should we be of ourselves in this particular when we daily see people of the strictest virtue (apparently) guilty of it.

Mon. 1 Feb. In the forenoon Mr. Porter and myself rode to Maresfield, there being a meeting of the commissioners of the window tax for the last day of appeal.[2] Mr. Porter appealed and had a relief of the last half year of 3*s.* 6*d.* We dined at *The Chequer* . . . Came home about 5.50 . . .

Sun. 7 Feb. No service at our church in the morning on account of the death of Mr. Porter's brother[3] . . . After churchtime Mr. French, Jos. Fuller and myself gave away among the poor of this parish the £4 I received yesterday of Mr. Porter and John Piper, it being the interest money of a hundred pounds left by Mr. Sam. Atkins to this parish forever, the interest of which is yearly to be distributed among the poor of this parish in such manner and to such persons as the minister and churchwarden for the time being shall think proper . . .

Tues. 9 Feb. . . . My brother came over to inform me that his carrier would go with a team to Maidstone Fair on Saturday next, and we together with Sam. Jenner packed up my rags in order to go to the fair by him. They weighed 5 cwt. 0 qr. 22 lbs. net weight . . .

Weds. 10 Feb. In the day my brother sent me over word his carrier had disappointed him in going to the fair, for the roads were so bad he could not get along . . . A most remarkable cold day; a sharp piercing wind and a very hard frost. At home all day, though not very busy. I hardly remember a colder day, I think, at this season of the year.

Thurs. 11 Feb. . . . At home all day; a good deal of rain in the even; a sharp frost . . . Joseph Fuller smoked a pipe with me in the even. We

[2] The window tax, introduced in 1696, had been recast in 1747 and increased in 1758 and 1761 to help finance the Seven Years' War. W. R. Ward, 'The administration of the window and assessed taxes, 1696–1798', *English Historical Review*, 67 (1952), pp. 522–46.

[3] This was Richard Porter who had been Thomas Porter's predecessor as rector of East Hoathly and with whom he has often been confused. See Appendix B.

have had almost this 3 months past a deal of rain, and yet a frost almost every day or two.

Fri. 12 Feb. . . . In the even copied a letter for Mr. Thornton to send into Kent to a gentleman who seems to espouse the part of a fellow now confined in jail for clandestinely taking several things from him some time ago.

Weds. 17 Feb. . . . A truly melancholy time; hardly anything to do, and it is impossible to get any money owing to me, so great a state of poverty abounds amongst us. I think there is too great reason to fear that it proceeds from a too free indulgence of that bane of private property LUXURY.

Sat. 20 Feb. This morn about 1.30 died (very suddenly, though after a long and lingering illness) Tho. Fuller, aged 58 years . . . I dined on the remains of yesterday's dinner. At home all day. In the evening about 6.30 I walked down to Tho. Davy's (by whom I had been earnestly solicited to come), his infant daughter being baptized in the afternoon. I stayed and spent the even there in company with Tho. Durrant, Ann Dallaway, James Marchant, Elizabeth Mepham and Mr. Jn. Long. I supped there on some bread, cheese and plumcake. Came home about 12.30, sober. I gave the nurse 6*d*. Tho. Durrant stayed and laid at my house; the people being all abed in his house. Oh, melancholy and dismal time: trade dull and money more so. How doth such misfortunes break a person's temper and render him too often an unsociable creature.

Tues. 23 Feb. . . . In the morn I rode to Lewes to meet Mr. Stephen Fletcher, servant to Mr. Sam. Ridings, in order to buy some Manchester goods. I breakfasted with him at *The White Horse* in company with my brother. I called and did some business with Mr. Madgwick and came home about 12.15 . . .

Thurs. 25 Feb. . . . After dinner I went to the funeral of Tho. Fuller, where I read his will to his friends and relations. He was buried about 5.30, and aged 58 years. This day I, my late and present servant witnessed the will of Jarvis Bexhill of the parish of Chiddingly which I had made for him. Very little to do all day. Nay, I may almost say hardly anything at all to do, so dull and melancholy is trade, and I believe it to be the same throughout the county.

Fri. 26 Feb. . . . In the forenoon I, at the request of Tho. Fuller's executors and Mr. Porter, in company with Joseph Fuller, went to the house of the deceased, where I examined his writings and looked over his papers, and also read such of them to his relations as was necessary . . .

Sun. 28 Feb. . . . Myself and both servants at church in the afternoon . . . at home all the day; read part of Drelincourt on death and in the even one of Tillotson's sermons. How is the thoughts of the best of wives come as it were afresh again to my memory; the idea of her spotless virtue is present with me . . .

Mon. 1 Mar. Sent to Mr. Robt. Olive by the post 5*s.* to pay my fine to excuse my serving steward to the club[4] . . .

Weds. 10 Mar. . . . At home all day and I think the very worst day's trade I have ever known since I have been at Hoathly . . .

Fri. 12 Mar. This being the day appointed by proclamation for a general fast and humiliation before Almighty God to be observed in a most devout and solemn manner by sending up our prayers and supplications to the Divine Majesty for obtaining pardon of our sins and for averting those heavy judgments which our manifold provocations have most justly deserved, and imploring His blessing and assistance on the arms of His Majesty by sea and land, and for restoring and perpetuating peace, safety and prosperity to himself and to his kingdoms, myself and both servants was at church in the morning . . . In the morn we had a very crowded audience at church, and undoubtedly a very good sermon.

Thurs. 18 Mar. . . . In the even went to Jones's, where we made a poor rate for the relief of the poor at the rate of 6*d.* to the pound, which was signed by everyone present . . . I came home about 9.20 and neither spent anything nor drank a drop of liquor, so that in all probability I was thoroughly sober . . . Delivered to T. Babcock 4*s.* 6*d.* for him to pay for 28 lbs. gingerbread tomorrow in Lewes for me. Pretty busy all day, though a most dismal melancholy time to get in any money which is owing to me on book or notes. Nay, so difficult is it that I find it impossible.

Fri. 19 Mar. . . . In the even Mr. Long sat with me a little time, I being instructing him in some rules of arithmetic . . .

Thurs. 25 Mar. . . . At home all day and really very little to do considering the season of the year. Joseph Fuller Jr, Tho. Durrant and Mr. Long smoked a pipe with me in the even. Oh, how does the memory of that ever valuable creature my deceased wife come over my thoughts as it were a cloud, may I (not) with truth say daily! For who is that man among mankind that has once been in the possession of all this world can give to make him happy and then lost it but must ever and again think of his former happiness—which is my case . . .

Sun. 28 Mar. In the morn about 5.40 I set out for Hartfield where I

---

[4] The Mayfield Friendly Society, see above, p. 99, note 25.

arrived about 8.20. I breakfasted with my father Slater and also dined there on a beef pudding, a shoulder of mutton boiled, a piece of pork boiled, carrots and greens (my family at home dining on a piece of beef boiled and some turnip greens). I came home about 6.50, though I cannot say thoroughly sober. Yet I think it almost impossible to be otherwise with the quantity of liquor I drank. But however so it was, and notwithstanding the many resolutions I have taken, and which I have hereto as often broke as made, I hope once more to assume so much fortitude and resolution as to conquer the weakness of my brains by an entire abstinence from any liquor strong . . . But however much in liquor I was, my reason was not so far lost but I could see a sufficient difference at my arrival at my own house between the present time and that of my wife's life, but highly to the advantage of the latter. What I mean is: one I always found at home and everything serene and in order; now one or both servants out and everything noise and confusion. Oh, it will not do, it cannot do! No, no, it never will . . .

Fri. 2 Apr. After breakfast Joseph Fuller and I walked up to the dwelling house of the late Tho. Fuller where we took an inventory, or rather a list of his effects; the stock-in-trade, utensils, furniture in the house, husbandry tackle, stock etc., debts excepted, amounted to £152 19*s*. 7*d*. And I dare say we valued it at £50 under the real worth, was it to be sold . . .

Sun. 4 Apr. . . . Myself and both servants at church in the afternoon . . . We had a thanksgiving prayer read today for the success attending His Majesty's arms in the reduction of that important island of Martinique, which has lately surrendered unto His Majesty's general etc. employed in the expedition against that island[5] . . .

Easter Monday, 12 Apr. . . . In the afternoon went to the parish meeting at John Jones's, where I made up the yearly accounts between Jer. French and John Vine . . . The overseers for the ensuing year are my own self, for the doing of which I am to have two guineas, [and] Richard Hope; the electioners, John Watford and Jos. Durrant; churchwarden, Jer. French, electioner, Jn. Vine Jr. . . . Thank God pretty busy all day; nay, I may say very busy! Oh, how pleasant did business use to be in my dear Peggy's life when the fatigue of the day was over! Then I had the opportunity of unbending my mind in the company of a most engaging and agreeable companion . . .

Sun. 18 Apr. . . . In the afternoon put out Anne Thomas Vinal to

---

[5] Martinique capitulated on 15 February 1762. Julian S. Corbett, *England in the Seven Years' War* (1907), II, p. 225.

her mother Mary, the wife of Richard Parkes, to keep at 9*d*. a week, her husband being present and agreed thereto.[6]

Tues. 20 Apr.   After breakfast my brother and Tho. Durrant and myself walked down to Halland to see the gardens . . . My brother, Tho. Durrant and Mr. Long dined with me today on a leg of mutton roasted and some onion sauce. My brother went away soon after dinner. Oh, how I blush to mention what I would for ever wish to conceal; that is, from going about from house to house today with my brother I got really very drunk. Oh, that hateful and to me truly odious name—that I can ever be so foolish! Sure I am the most irresoluteless creature breathing, a thing I am sure I sincerely detest and yet have been so foolish as to be guilty of it . . .

Mon. 3 May.   . . . In the even went down to Jones's, there being a vestry to consult about repairing the inside of the church, when it was unanimously agreed by all present to repair the same according to the estimate delivered in by Mr. John Vine and Mrs. Browne, with such alterations as the rector and churchwarden shall think proper.

Thurs. 13 May.   . . . In the afternoon Mr. Porter's children drank tea with me. This day was played at the common belonging to this parish a match of cricket between 11 boys of this parish and 11 boys of Laughton Parish, which was won by our youth with great ease.[7] In the evening Tho. Durrant and I walked out a little way for a walk, it being a very fine even . . .

Fri. 14 May.   . . . Mr. Terry, rider to Messrs. Kendall and Rushton, called on me, to whom I gave a small order. In the forenoon my father Slater came to see me in his road to Lewes, and I rode with him to Lewes where I paid Mrs. Roase in cash £6 4*s*. 6*d*. in full . . . My father Slater being in a hurry, I dined nowhere . . . I borrowed of my father Slater in cash 1 guinea . . . Came home about 10.10 (my father staying at Lewes), but wish I was so happy as to have in my power to say I came home sober. For what with going without victuals and calling with my father at so many places, the liquor soon crept into my poor shallow brains. Oh, how do I lament my want of resolution in this particular! It is not, I am sure, the love of liquor that entices me to be guilty of this folly; no, it is not, but an easy and foolish modesty that occasions it. But

[6] Anne Thomas Vinal was the illegitimate daughter of Mary Parkes, born some time before her marriage to Richard Parkes which had caused the parish officers such trouble five years earlier (see the diary entries for 23–25 October 1757). Here the parish paid Mary Parkes to look after her own child. A year later the child was put out as a parish apprentice to Mrs. Browne.

[7] See above, p. 9, note 9.

7. Thomas Turner in account with the East Hoathly overseers for goods supplied from his shop, 1764–5: *E.S.R.O. PAR 378.*

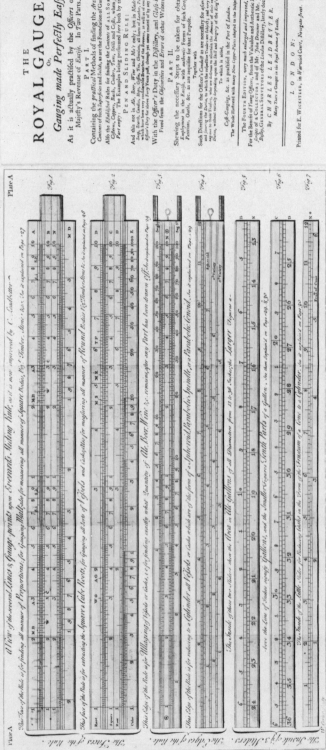

8. Title page and illustration of the slide rule from Charles Leadbetter, *The royal gauger . . .* (1755). Turner read this book and taught others the use of the slide rule. *Bodleian Library, Oxford.*

however once more I will try and see if I cannot get the better of it, for I will drink nothing strong.

Mon. 15 May. . . . Thank God pretty busy all day. Oh, how pleasant is the season of the year. All nature wears the livery, as it were, of gaiety.

Sun. 16 May. . . . No service at our church in the afternoon. One of my servants went to Chiddingly church, and myself and the other walked to Little Horsted church in company with Joseph Fuller Jr., Sam. Jenner and Joseph Durrant. We had a sermon preached by the Rev. Mr. Philips, curate of that parish and Maresfield, from *Song of Solomon* 2. 10, 11, 12, . . . from which words we had, I think, as good if not the best sermon I ever heard, both for elegance of language and soundness of divinity, the gentleman discoursing on the words in a very spiritual manner; so I really think it quite a masterly performance. After churchtime we went into Mr. Lewer's, where we smoked a pipe or two and came home about 8.10. A very fine pleasant day for the season of the year . . .

Thurs. 27 May. After breakfast I rode with Thornton to *The Black Boy* where Joseph Gibbs and his wife signed a certificate, or at least a paper, whereby they yielded their consent for him to marry Mary their daughter, a spinster under age.[8] From thence we went to Lewes where he took out a licence to marry the young girl, myself being his bondsman . . . Spent nothing today, Mr. Thornton paying my expenses . . . A very dull time for trade, but what is that when compared to the melancholy gloom that is in my mind in my calm and tranquil hours. There, there is a scene which would, I believe, move a heart if it was hard as the nether millstone.

Mon. 31 May. . . . In the even went to see Mr. Thornton and his bride where I spent the evening in company with Joseph, Molly and Bett. Fuller, Mrs. Browne, T. Davy, Eliz. Mepham, Jas. Marchant, Mr. Shoesmith, Mr. Long, and Sawyer Moon. I came home about 12.30, sober . . .

Thurs. 3 June. This being the excise sitting day at Lewes, I rode there to swear to a coffee and tea book[9] . . .

[8] For the significance of this marriage see Appendix B, *sub* Thornton. Thornton and Mary Gibbs were married the following day at Framfield: E.S.R.O., PAR 343/1/1/4.

[9] All shopkeepers who dealt in coffee and tea were required to keep an accurate record of all their purchases and sales. This record was kept in a book which was produced for the excise officer when he came on his rounds. The shopkeeper made an oath, or affirmation, to the truth of the entries in the book and was given a new book. The regulations are laid down in Charles Leadbetter, *The royal gauger; or, gauging made perfectly easy* . . . (4th ed., 1755), a book which Turner possessed.

Fri. 4 June.   After breakfast I rode to Maresfield to meet the general
receiver of the land tax for this county. I paid Tho. Gerry, his deputy,
in cash £87 14s. 0d. and by a bill on Mr. Will. Margesson, dated today,
14 days' date, payable to him or order, value £23 4s. 3d. for value
received of His Majesty's money, no. 522, which together makes the
sum of £110 18s. 3d. and is in full for one year's arrears of land and
window tax from this parish due at Lady Day last, *viz.*,

> Dr.
>
> | | |
> |---|---|
> | To one year's land tax | £153 18s. 0d. |
> | To one year's window tax | £ 30 6s. 0d. |
> | | £184 4s. 0d. |
>
> Per contra Cr.
>
> | | |
> |---|---|
> | Jan. 8 1762 in cash | £ 71 0s. 0d. |
> | Today in cash | £ 87 14s. 0d. |
> | *Do.* by bill | £ 23 4s. 3d. |
> | Salary | £ 2 5s. 9d. |
> | | £184 4s. 0d. |

Sat. 5 June.   After breakfast rode to Lewes to ask Mr. Michell the
favour to serve the nails for the Keeper's house now a-rebuilding . . .
Came home about 12.20 . . .

Mon. 7 June.   After breakfast I rode to Maresfield to carry in the
window and land tax books,[10] there being a sitting of the commissioners
of the land tax. I came home about 2.20. Spent only 2d. There being a
main of cocks fought at Jones's today between the gentlemen (if any
such there be) of this parish and the gentlemen of Lewes, Mr. Ben.
Hudson of Hailsham who came to see it dined with me on a fillet of
veal roasted, a gooseberry pudding and some green salad . . .

Tues. 8 June.   In the morn Mr. Long and I went to John Cayley's,
where we measured a piece of wood ground of about 5 acres. Mr.
Long came back and breakfasted with me. In the forenoon I papered
out a bag of nails. I dined on the remains of yesterday's dinner. About
4.30 Jos. Fuller Jr., Mr. Long, Mr. Thornton, Tho. Durrant and
myself went to Laughton to pay a visit to Mr. Ed. Shoesmith, with
whom we drank tea and stayed and supped on a veal pasty. Came home
about 1.40, not thoroughly sober, nor very much in liquor, walking
home very well without so much as a fall or anything like one.

Thurs. 10 June.   . . . At home all day. Busy a-putting my shop in
order.

---

[10] These were the books of assessment. See above, p. 216, note 3.

Mon. 14 June. . . . In the even wrote my London letters. But very little to do all day, and I think trade and money never was so dull before. Oh, what can I do? My hands are as it were tied, having no person whom I can trust with my business when I am from home. And by my being so much confined at home, undoubtedly my affairs abroad must be neglected.

Thurs. 17 June. I dined on the remains of yesterday's dinner with the addition of a piece of bacon boiled and a light pudding. In the afternoon Joseph Fuller and myself went to Chiddingly to see a match of cricket played between that parish and Laughton, when Chiddingly won the same, having 5 wickets to go down.[11] Came home about 9.05, having spent nothing during my entertainment. But very little to do all day. In the forenoon took a ride to collect in a debt, but did not receive it.

Fri. 18 June. . . . At home all day and but very little to do. Oh, melancholy time! At work in my garden part of the day.

Sat. 19 June. . . . In the morning read part of a book entitled *A Defence of Plurality of Church Benefices*, but I cannot be persuaded by his reasons that it is anyways beneficial to promote our most holy religion.

Sun. 20 June. . . . In the even we had a fine refreshing shower, there being a great drought for some time past, so that in many places the grass is almost burnt up.

Fri. 25 June. . . . At home all day. In the afternoon drawed up articles for a club in this parish to enter into to indemnify any of the society from serving in the militia by raising a fund to pay the money charged upon anyone by Act of Parliament who shall be chosen by lot and refuse to serve or to provide a substitute to serve in their stead[12] . . .

Sun. 27 June. . . . After dinner Mr. Long, myself and Sam. Jenner rode to Lewes and all went to the church of All Saints where we heard a sermon preached by the Rev. Mr. Lund, rector of the said parish, from *Galatians* 6.2. . . . I spent the afternoon in company with my brother, Sam. Jenner and myself (Joseph Fuller staying all night). Came home about 11.20 sober. Spent today upon myself, horse and turnpike 11*d*. A fine calm and serene day and I think never to go so far to church again, for company dissipates the thoughts so much that the good impressions which the service of the church should have on the mind is, I doubt, too much obliterated.

[11] See above, p. 9, note 9.
[12] The militia was raised on a county basis, and in each parish men were chosen by lot to serve for three years or else to provide £10 for a substitute. Here the parishioners of East Hoathly were organizing a benefit society to provide the £10. *Tate*, pp. 180–1.

Fri. 2 July. . . . In the even Mr. Buller and I ran a race of 20 rods for a bottle of cider, and I had the good fortune to beat him; and him, myself and Joseph Fuller Jr. went to Jones's in the even where I spent 4*d.* and came home sober. Oh, sure there never was so dull a time for trade or money. I think I shall be quite distracted for money and cannot anyways get any in.

Sun. 4 July.   No service at our church in the morn, Mr. Porter preaching at Ripe . . . After dinner my brother called on me, and we rode to Lewes together. I went to consult an attorney about a bill of sale I have of Master Darby's effects, but had not the opportunity of seeing the gentleman I wanted to see . . . Not at church in the afternoon, but for this simple reason only, its being wet weather.

Mon. 5 July.   In the morn rode to Lewes where I met the attorney I went to see yesterday. Came home about 1.50 . . . In the even played a game of cricket[13] . . .

Tues. 6 July. . . . Who can think the anxiety of my mind at the thought of distressing[14] poor Darby! But what can I do? This seems the only chance I have for recovering a debt justly due to me of about £16. The first law of nature I think will tell me I am doing no injustice. And should I neglect this opportunity, it appears as if I should never have such another. Therefore I think if I do not do it now, I am doing a piece of injustice both to myself and creditors. So I am constrained by near necessity to put my bill of sale in execution. Notwithstanding which, there is something in the thought of distressing a fellow creature that my mind as it were recoils at.

Thurs. 8 July. . . . This day Charles Cooper entered with my bill of sale upon the premises of Tho. Darby and made a distress, but I am afraid it will come out that I shall lose the whole debt, which is about £18 18*s.* 0*d.*

Fri. 9 July. . . . After breakfast I rode to Lewes about Darby's affair, but it still wears a gloomy aspect and I make no doubt I shall lose the whole sum . . .

Fri. 16 July.   After breakfast rode to Lewes concerning Darby's affair. Came home about 11.10 . . . In the even my worthy friend and acquaintance Mr. Snelling called on me and stayed with me some time. Very little to do all day and truly a very melancholy time. What, oh, what shall I do, or what will become of me? Work I cannot; steal I hope I never shall once think of; therefore let me try and submit with patience to any circumstances it shall be my lot to endure.

---

[13] See above, p. 9, note 9.   [14] Distraining on his goods.

Mon. 19 July. . . . Very little to do all day. Sure a more duller time for trade I never knew in all my life. What shall I do? I cannot think. There is no such thing as getting in any money, so that I think I shall be ruined.

Tues. 20 July. In the forenoon I went to church with Mr. Carman's daughter, for which I stood godfather in company with Mrs. Vine the younger and Mrs. Piper. We then walked down to Mr. Carman's, where I dined in company with Mr. and Mrs. Porter, Mr. Aaron Winton and his wife, two Mr. Clarks, Mr. Piper and Mrs. Vine, on a leg of lamb boiled, a roasted pig, 4 chicken roasted, a boiled ham, a currant suet pudding and a plain flour pudding, carrots and cabbage (my family at home dining on the remains of yesterday's dinner with some French beans). We stayed and drank tea at Mr. Carman's, and I came home in company with Mr. and Mrs. Porter about 8.15, as sober as when I went from home. Gave the maid 12*d.* and the nurse 2*s.* . . . N.B.: The child was named Susanna.

Tues. 27 July. . . . Delivered the sum of £47 for Benjamin Shelley to pay in London for me this week . . . *viz.*, one thirty-six shilling piece, four moidores, twenty-seven guineas, twenty half-*do.*, one nine-shilling piece, one six-shilling-and-ninepence, three shillings, [6] halfpence[15] . . . This day received a note of hand from Will. Burrage for him to pay to the churchwardens and overseers of this parish and their or either of their successors three shillings weekly and every week as a part of the expense this parish is now burdened with in maintaining his wife and family . . .

Fri. 30 July. . . . This day received of Ben. Shelley by his brother a receipt from Mr. Will. Margesson for £45 19*s.* 0*d.* and also a guinea returned, it being not weight, which together makes the sum of £47 and is in full for the £47 delivered to him the 27th instant . . . In the even Mr. Richard Sterry called on me and supped with me and also stayed all night. In the even we balanced our accounts . . .

Sun. 1 Aug. . . . There being only prayers at our church in the afternoon (and that more than we knew of till we came from church), Sam. Jenner and I took a ride to Seaford where we took a walk by the seaside and took a view of the two forts newly erected there, one of which has 5 24-pounders mounted and the other 5 12-pounders. We drank tea at the sign of *The Tree*. Came home by Alfriston where we overtook on the road my servant and Tho. Durrant. We came home about 11.10—oh, could I say thoroughly sober. I think I am the most unfortunate fellow

[15] See above, p. 150, note 26, and p. 226, note 9; and also Appendix C.

in the world, for only a few glasses of wine intoxicates my brains. I was not so far intoxicated today as to be guilty of any indiscretion, but still though we only took a ride with no other design than innocent inoffensive amusement, and with an intention of reaping the advantage of serious and improving conversation from each other, yet being guilty of this one folly, the whole of our journey must become contaminated, which otherwise could not have been more than a breach of an active obedience. Each of us spent today as follows: to tea 6*d*., wine 6*d*., beer 2*d*., horses 4*d*., a-coming home 7½*d*.

Fri. 6 Aug. . . . Paid Ben. Shelley in cash 2*s*. 4½*d*. and by money he received of Mr. Ben. Treacher in full for rabbitskins I sent him 9*s*. 7½*d*. which together makes the sum of 12*s*. and is in full for the same sum he paid Dr. Godfrey in full for three dozen of his cordial this week . . .

Sun. 8 Aug. Myself and one servant at church in the morn . . . Myself and both servants at church in the afternoon . . . How much more pleasure it is to be at home all day of a Sunday and to attend the service of the church than to be rioting about as I have too much been of late. But may I never more offend in that point.

Thurs. 12 Aug. . . . In the even His Grace the Duke of Newcastle came to Halland where he lodged all night in his road for the races on Lewes Downs tomorrow. This morn about 7.30 Her Majesty Queen Charlotte, consort of our most gracious sovereign, was safely delivered of a prince, and are both like to do very well.[16] At home all day and pretty busy a part of the day.

Fri. 13 Aug. . . . This day the King's plate of £100 was run for on Lewes Downs by Lord Grosvenor's horse Boreas and Mr. Howard's crop mare Suskey, which was won by Boreas, he distancing the mare in the first heat.

Sat. 14 Aug. . . . A very rainy day and but very little to do all day . . . Sent to Mrs. Coates by my servant Hannah Marchant the Duke of Newcastle's bill for goods delivered this time of his being at Halland, amounting to 18*s*. 2*d*. This day the purse of £50 was run for on Lewes Downs, when Mr. Wildman's horse Lincoln, Mr. Tod's gelding Junior, Mr. Blackman's mare Slouching Sally and Mr. Wilson's mare Harmless started for the same, which was won by Lincoln, he getting the first two heats. There was a third heat for the stakes between Junior and Slouching Sally, which was smartly contested between them, Junior getting it not but above half a length . . .

---

[16] This was the birth of Prince George Augustus Frederick, later George IV.

Sun. 15 Aug. . . . This day paid Sally Waller my house-keeper cash and goods £4 in full for her servitude due to this day, and in the forenoon she left my service in order to go to Catsfield to keep her uncle Mr. May's house, who came for her. Her leaving my service was, as she protested, without any dislike or the least reason for dislike and contrary to her inclinations and the most earnest persuasions of her friends, but this uncle, being a widower and having two children, over-persuaded her to live with him in order that she might see the children well done for . . .

Mon. 16 Aug. . . . My brother called on me in his road to the Dicker where he was going to see a cricket match played between the parish of Wadhurst and the parishes of Ringmer, Ripe, Chalvington, Arlington, Herstmonceux, Wartling and Hailsham. At home all day. In the even my brother called on me in his return home and informed me the parish of Wadhurst was beat 10 runs[17] . . .

Thurs. 19 Aug. . . . A very dull time for trade, and 'Money, money!' is the general cry of everyone.

Sun. 22 Aug. . . . This day a prayer of thanksgiving was read for Her Majesty's happy delivery, as was also the young prince prayed for in the service. A melancholy time—no pleasant or agreeable companion to spend even a few minutes with. All, all is melancholy and dismal sadness.

Mon. 23 Aug. . . . In the afternoon went down to Halland where I received of Mr. Coates in cash 18*s.* 2*d.* in full for a bill delivered the 14th for goods delivered to Halland for the use of the Duke of Newcastle this last time of his being there. I also received of Mr. Coates 18*s.* in full for poor tax assessed on him by a rate made the 9th instant. Received of Joseph Fuller in cash 24*s.* in full for the same sum assessed on him by a poor rate made the 9th instant. Dame Roase, buying many things in the shop, drank tea with me. In the even wrote my London letters. Thank God I have been very busy all day, and I may say the busiest day I have known this many a day. Oh, what pleasure it is to be busy! It quite cheers the spirits and chases away the gloom that hangs on a melancholy brow . . .

Fri. 27 Aug. I paid 2*d.* for 6 plaice bought at the door today. I dined on some boiled plaice with the remains of yesterday's dinner. Paid Ben Shelley in cash £1 14*s.* 11½*d.* which with £32 2*s.* 0*d.* I sent by him the 24th instant makes together the sum of £33 16*s.* 11½*d.* and is in full for the same sum he paid in London for me this week, *viz.,*

---

[17] See above, p. 9, note 9.

To cash paid Mr. Will. Margesson for which Mr. Shelley hath given
me no receipt £32 2s. 0d., to a scale beam 1s. 9d., to a shaving
box and brush 2s. 6d., to the porter and a quart bottle 5½d., to
money paid Mrs. Mary Chandler in full on account of James Fuller
£1 8s. 6d., to 6 papers ink powder 2s.
At home all day and but very little to do . . .

Mon. 30 Aug. . . . At home all day but not very busy. I think I never
knew so general a complaint for money in the little circle of my
knowledge as there is at this present time. It is hardly to be met with in
the common course of trade, and those few that have money be too
often apt to look down with an eye of disdain and even contempt on
those who know the want of it.

Sun. 5 Sept. After breakfast I set out for a place called Ninfield
Stocks in order to meet my late servant Sarah Waller, agreeable to an
appointment previously made, where we both arrived about 10 o'clock.
I stayed and talked with her about two hours and came for home where I
arrived about 2.20. My servant dined at home on the remains of
yesterday's dinner, and as for myself, I ate not any dinner . . . Now
should those minutes of my journey ever come to any person's eye, or
should my journey come to be known publicly to the world (that I took
upon me such a journey, and for no other reason than purely to see an
old servant), why I doubt not but they will very readily conclude she
was his sweetheart, or if not so favourably disposed in their censure,
perhaps the sagacious eye of scandal may see in it something worse.
But however, they will be wrong for once in both conjectures. For I can
with most sincerest truth assure them it was neither, but as my servant
had not received so good usage from her uncle, whom to oblige and
serve, as she thought, she left my service, therefore in a letter I
received from her since she left my service, she begged I would so far
condescend as to come to Catsfield. Which I, not approving of,
appointed this place. For as she was destitute of any friends to consult
or advise with, she sent to me and now beg[ged] she might come to my
service again in the spring if I was not provided with a servant before
that way). I positively declare I have not, since that fatal day which
this? Why, if I may truly speak, just none at all. But now I am
endeavouring to vindicate my character from any aspersions that the
malevolent tongue of envy or ill-will may strive to blacken it with (as I
have had demonstration they are not wanting in their best endeavours
that way). I positively declare I have not, since that fatal day which
deprived me of all, all, this earth can ever give to make me happy (I
mean that melancholy day which took from me my wife), ever once

made my addresses to any one of the fair sex. No! I have not: notwithstanding the busy world have made it their business to proclaim the contrary—and that with some vehemence of clamour, for such only it really was. Not that I have taken up any resolution to celibacy, for I can with truth declare marriage to be the only state that I found any happiness in and at the same time assert I hardly think I have had one minute's peace of mind since I have been a widower, if I have been sober and in the calm possession of my reasoning faculty. But as to what liberty the world takes with my fame, I think it not worth my notice, so long as my conscience does not join in declaring their censures are just. Therefore let the vain and giddy world talk on as freely as they please of me, and I hope to have grace not to merit anything odious enough to hurt my mind; and if not, I shall think myself no ways the worse for their censure, neither shall I think all their clamour worthy my notice, but permit them to talk on till they are tired, which perhaps it may be said will never be. Why then even let them talk on *ad infinitum*.

Tues. 7 Sept. . . . Mr. Coates's maid (who clamours of the world—but without any, even the most distant, of truth—call my sweetheart), the two Miss Porters and their servant drank tea with me . . .

Sun. 12 Sept. . . . Myself and servant at church in the afternoon . . . In the afternoon Mrs. Coates's maid drank tea with me. This! this is the girl that the world proclaims is to be my wife. But oh, 'tis an egregious mistake, a thing that yet has never entered my thoughts. Nor did I ever give the girl the least reason to think of any such thing, for I am sure I have not kissed her, except once, since Whitmonday. But during the time I lived in Lewes she and I, being almost next-door neighbours, were very intimate, but then there was nought of love. And now she coming to live at Mr. Coates's at Lady Day, and as there is some alteration in the state I move in now and then; that is, I was then only a servant, I think that was I not to take the same notice and use her with as much civility as I did then, I should be guilty of a great deal of pride and ignorance and show a great deal of folly. This, this is the reason and the only one that induces me to take notice of her (that is, I mean there is no love between us), thought I greatly regard the girl as an old acquaintance and very sincerely wish her an uninterrupted state of happiness. And was it in my power to serve her in anything that might tend to her advantage, I would very gladly and cheerfully do it (but still no love affair, no *amour* a-going on). My good neighbours, I have not taken up a vow of celibacy, but I am in no hurry. Love yet has made no impression on my mind since I lost the only woman that I imagine I shall ever love, or at least that I ever shall love with that energy and sincerity I did her. After Pat went away, Mr. Tipper,[18] Tho. Durrant and myself took a walk for some air.

Tues. 14 Sept. ... At home all day and pretty busy. In the afternoon employed myself a-writing. In the even Mr. Tipper read to me part of a—I know not what to call it but *Tristram Shandy.*

Thurs. 16 Sept. Paid Joseph Fuller Jr. 4s. 6d. in full for 2 stone 2 lb. beef bought of him today. Sam. Virgoe, a-writing for me, dined with me on a cold skirt pie and carrots. James Marchant, Eliz. Mepham, Dame Vallow and daughter, buying some things in the shop, drank tea with me. In the even went down to Mr. Porter's, where I stayed and supped and smoked 2 or 3 pipes. Came home about 10.30. Received of Richard Prall by the payment of his daughter Ann 7s. 6d. on account. Very busy all day. I wish I was employed in this manner every day.

Fri. 17 Sept. Sam. Virgoe, a-writing for me, dined with me on part of a cold beef pie. Paid Mr. Ben. Shelley in cash £14 10s. 10d. in full for the same sum he paid in London for me this week ... My brother came over in the afternoon ... I received of him in cash £2 11s. 9d. in full for a parcel of silk handkerchiefs bought of me today ... My old acquaintance Mr. Coates's servant accidentally happened to be at my house this afternoon and drank tea with me again. Oh, a most delicious and savoury morsel for the gossiping part of my neighbours to chew and bandy about from house to house. Surely it must be vastly entertaining and pretty, as well as improving to introduce Mr. and Mrs. Turner (that is to be) in chat at their tea tables. 'Why really, who could have thought Mr. Turner would have courted such-a-one! Surely Miss such-a-one or B----such-a-one would be a much better match for him, but there, he is so very difficult that nothing will go down with him but a woman that can talk fine and one who he thinks he can be master of. Well, for my part I don't envy her happiness. I am sure I would not have him was I in her place. No, I wouldn't. Only think how he used his last wife, poor creature. But there, I warrant you this poor girl, as she has had a misfortune as they say, is willing to have anyone that will have her. Well, I don't doubt that he'll be paid off for his old sins, and I really wish he may.' And is not all this vastly

[18] Thomas Tipper was a fairly frequent visitor. He was a well-known Newhaven brewer and was in East Hoathly at this time in his capacity as excise officer during the hop harvest. He was a cultivated man, and of about the same age as Turner. He died aged 54 in 1785 and his tombstone records him as a brewer of 'Stingo' and as a man who 'knew immortal Hudibras by heart'. The tombstone is illustrated in *S.C.M.* 12 (1938), p. 279.

pretty? But I can assure you, my good friends, neighbours, acquaintances, intimates, gossips, lovers, haters, foes, farters, friskers, cuckolds and all other sorts of Christians of what name or denomination soever, that there has not [been] one word of courting yet passed between us, or ever will be. And if this can't satisfy your consciences, even think and say as you like, for it will still be one and the same thing to your humble servant.

Mon. 20 Sept. . . . In the even went down to Jones's, where we had a vestry holden to consult on proper ways to raise a sum of money to pay for new-pewing and beautifying the church etc. It was agreed to raise or hire any sum requisite to pay off the whole expense, and the interest and part of the principal to be paid every year . . .

Fri. 24 Sept. . . . At home all day and not very well in the even, proceeding, as I imagine, from leading so recluse and sedentary a life together with the anxiety and uneasiness of mind I labour under . . .

Thurs. 30 Sept. . . . Mr. Jos. Hartley at Lewes (coming to take measure of me for a wig), together with Dame Cushman and Mary Durrant, who bought many things in the shop, drank tea with me. In the even Tho. Durrant and I walked down to Mr. French's to inquire how Molly did, she being very ill . . .

Sat. 16 Oct. . . . My old acquaintance, Mr. Coates's servant, drank tea with me, but still not one word of courting, no, not even a kiss as a preparative thereto. Oh, sad disappointment this must be to the busy censorious world to find their conjectures are never like to prove true . . .

Sun. 17 Oct. No service at our church the whole day, the inside being all taken up for to be new-pewed . . . The Rev. Mr. Warnford, Rector of Horsted, which I heard today, is I think as good an orator as I ever heard and really made a good harangue.

Thurs. 21 Oct. . . . This day John Jones in this parish surrendered (at a court baron held at Whitesmith for this manor) all that his messuage, tenement etc. in this parish into the hands of the lord of this manor by the acceptance of his steward Will. Michell Esq. Mr. Thomas Neatby, distiller in Southwark, was admitted thereto by his attorney Mr. Tipper, according to the form and custom of this manor. Now Jones owing Mr. Neatby on bond £32 10s. 0d. and he thinking it not safe agreed to take this house etc., for the bond; and I being an acquaintance of Mr. Neatby's, he implored me to act for him, but as I could not attend the court myself, I got Mr. Tho. Tipper to be the supposed attorney (as they call it). But however, Jones would not surrender unless the bond was produced in court, which I could not do, Mr. Neatby having that himself. But to make all easy and to get Mr. Neatby admitted, I gave Jones my note of hand, payable to him or order for the sum of £32 10s. 0d. upon condition that I did not deliver him up the bond within one calendar month from this day.

Fri. 22 Oct. . . . Paid Mr. Tho. Tipper in cash £1 4s. 1d. in full for the same sum he paid as Mr. Neatby's half part of the expense yesterday at the court . . . At home all day but not very busy.

Sat. 23 Oct. . . . At home all day, but not very busy. In the even Mr. Long and Mr. Tipper gave me a few tunes, one on the violin, the other on the German flute. A very wet even.

Sun. 24 Oct. No service at our church today, it being a-repairing. A prodigious wet forenoon so that it was almost impossible to get to any other church . . .

Fri. 29 Oct. Jarvis Bexhill, buying some goods in the shop and it being a very wet forenoon, dined with me on a piece of beef boiled and cabbage with a piece of pork. Dame Durrant made me a present of a goose, and she, Tho., Mr. Tipper and Sam. Jenner drank tea with me. I gave the good woman also a little of that which she esteems of more value than gold, ay, of that delicious cordial, Mrs. Nant's,[19] a thing truly of greater value than a goose. Oh, that sweet delicious relish! How it enlivens the spirits, gives one all the pleasing sensations that are so agreeable to our nature. And above all, when there is too much taken, it renders the most agreeable part of the creation mere brute creatures, as is too often this poor woman's case . . .

Sat. 30 Oct. This day agreed with Mr. Harmer the stonemason at Heathfield to write our altar-piece and completely ornament and paint it, the letters to be the best leaf gold, and he to have seven guineas for doing it[20] . . . A very dull heavy time with me, trade bad, money very scarce and hardly any friend in the world that can or will be a friend to me. But many of my relatives quite the reverse, which daily brings to my mind the memory of that sincere and virtuous friend whom I have not, my wife . . .

Sun. 31 Oct. No service at our church in the morn. I dined on a goose roasted and apple-sauce. No service at our church in the afternoon. I walked down to see Mr. Carman. I drank tea with him and his family; came home about 8.20 very sober. This is not the right use that Sunday should be applied to. No! it's not. Therefore what an

[19] Brandy: a corruption of the French place-name Nantes.
[20] The bill submitted by William Harmer of Heathfield for this work is in the E. S. R. O., AMS 5841/8. It was amongst the papers discovered in Turner's house in the late nineteenth century. See below, p. 269, note 9.

unhappy man must I be to continue to practise things out of compliance and complaisance to the custom of the world, which my reason condemns as wrong and irreligious.

Weds. 3 Nov. . . . Paid Mr. Blackwell, paper-maker at Hawkhurst, in cash and a sort of rags called hand-stuff £4 4s. 6d. in full for 20 reams of paper received of him today . . . At home all day and but very little to do . . .

Sun. 7 Nov. . . . Paid Mr. Joseph Hartley in cash and goods £1 1s. 0d. full for a wig received of him today . . .

Thurs. 11 Nov. . . . A very cold day and a great deal of snow fell, but there being no frost, it did not lie long on the ground . . .

Sat. 13 Nov. . . . A very sharp frost, and in the even or some part of the night there fell a great deal of snow.

Sun. 5 Dec. . . . In the even my brother came to my house in order to stand in the shop for me during my absence from home to appraise a shop for Mr. William Bennett (which he has now taken) late Mr. Hunter's in Lewes. My brother lodged at my house.

Mon. 6 Dec. After I had breakfasted, I set out for Lewes in company with Thomas Durrant. I dined at Mr. Bennett's (whom I was at work for) on a leg of mutton boiled, samphire sauce and turnips . . . I supped at *The White Horse* in company with Mr. and Mrs. Bennett, my brother Dicky, Tho. Durrant and Peter Bowra on some veal cutlets fried. As soon as the moon arose, Tho. Durrant and I came for home, where we arrived about 1.30 . . . After I came home wrote my London letters. My antagonist in the appraisement was Mr. John Browne of Ditchling, really a very hard man to deal with, one that has not a very extensive judgment, but so careful withal that it was with great trouble and difficulty that we went on; he really asking more than the intrinsic value of many things was which delayed our getting on in such a manner that we did not value above half the stock; so that I must be obliged to attend again tomorrow. My brother stayed at my house all night.

Tues. 7 Dec. After breakfast I set out for Lewes. I dined at Mr. Bennett's on the remains of yesterday's dinner . . . We worked very busily all day and in the even adjourned to *The White Horse* to cast up our books and fix the price of the fixtures, which we concluded about 11.10, and the sum total not more than £56. I supped and spent the evening at *The White Horse* and lodged there all night, going to bed very sober.

Sun. 12 Dec. Almost just as I was going out to church in the morn, very unexpectedly came along my brother Richd. and Mr. Will.

Bennett, who prevented my going to church. They both dined with me off a fowl and piece of bacon boiled, a piece of beef roasted and some potatoes. They both stayed and drank tea with me and then set out for Lewes . . . Lent my brother in cash 10s. 6d. Oh, the pleasure I found in my brother's company was very great, he seemingly become a very sober prudent young fellow. But how would it have been heightened had it not prevented my attending the church! . . .

Mon. 20 Dec. . . . My brother came over in the even in order to stand in my shop tomorrow during my absence, I having promised Mr. Coates to assist him in the morn in distributing a gift left by some of the Pelham family for ever and annually given upon St. Thomas's Day; that is, 4d. each to every man and woman, and 2d. to every child, come from where they will, and a piece of bread and draught of ale to each . . . I walked down to Halland where I supped and stayed all night. My brother lodged at my house as did Tho. Durrant.

Tues. 21 Dec. About 3.20 I arose and began to pay the people and finished about 10.20, having paid away £11, and I presume there had been 9 hundred people. I stayed and breakfasted with Mr. Coates, received my groat and came home about 11.40. My brother went home immediately . . . Relieved [thirty] persons with one penny each and a draught of ale . . .

Weds. 22 Dec. . . . This day died Robt. Baker in this parish after a few days' illness, aged 88 years. He had been parish clerk [21] years and upwards.

Sat. 25 Dec. . . . This being Christmas Day, I was at church in the morning . . . I stayed the communion; gave 6d. The widow and James Marchant, Eliz. Mepham and 2 girls dined with me on part of a buttock of beef boiled, a raisin suet pudding and turnips and potatoes. James Marchant, Eliz. Mepham and one of the girls drank tea with me. At home all day. Read part of Young's *Estimate of Human Life*.

Mon. 27 Dec. . . . In the even went down to John Jones's, there being a vestry to choose surveyors for the coming year. I stayed and drew up the last year's surveyors' accounts and then came home. The surveyors chosen for the coming year were Thos. Carman and James Fuller, and the electioners John Vine and Edw. Hope . . .

[21] Blank in the manuscript.

1763

Tues. 4 Jan. . . . I having a pretty large quantity of cash in the house, got Sam. Jenner to lodge with me. At home all day . . .

Weds. 5 Jan. . . . In the even Joseph Fuller Jr., myself and Tho. Durrant walked down to Whyly where we supped at Mr. French's, in company with Bett. and Molly Fuller, Fanny Hicks, Richd. Fuller, Mr. Edwd. Shoesmith and James Fuller on 2 ducks roasted, 2 chickens roasted, 1 *ditto* boiled, a cold shoulder of mutton roasted, a plate of cold ham, and plum and gooseberry tart. We played at brag in the even; I won 7½*d*. Gave the maidservant 6*d*. Came home about 11.20 very sober . . .

Thurs. 6 Jan. . . . Paid Wm. Jenner, the constable for this half Hundred in cash £6 8*s*. 3*d*. in full for the same sum assessed on this parish as their quota for the county stock[1] . . .

Fri. 7 Jan. . . . This being the day appointed by authority to meet the general receiver of land tax at Maresfield for the collectors to pay the two first quarterly payments of the said tax, I sent by Sam. Jenner in cash £76, which with 19*s*. outset for the salary for collecting makes £76 19*s*. 0*d*. and is in full for the first 2 quarters' payment of the land tax. The money delivered was 71 guineas, 1 moidore, 2 shillings[2] . . . Very little to do and very cold.

Sun. 9 Jan. Myself and servant at church in the morn . . . My old acquaintance (and I wish I could with propriety say 'worthy acquaintance') Mr. John Tucker, came to see me after churchtime and dined with me on a bullock's tongue boiled and turnips . . . A very cold sharp frosty day, and so has been every day since the 23rd or 24th of December. The frost is now so severe that I believe there is ice 5 inches thick; the dust in the highways steams as in summer, and really it is as good travelling for carriages in most places.

Thurs. 13 Jan. . . . Master Hope's servant being big with child, I walked to his house to talk with her about it, Tho. Durrant accompanying me. We came back about 7.30. I supped and spent the remainder of the even at Master Durrant's. Very little to do all day. A

---

[1] Turner was here acting in his capacity as overseer of the poor, and raising, as an extension of the parish poor rate, a sum laid on the parish by the justices as its contribution to a county rate. The procedure had been finalised by Act of Parliament in 1739: 12 George II cap.29. See J. V. Beckett *Local taxation: national legislation and the problems of enforcement* (Standing Conference for Local History, 1980), p.10.

[2] See above, p. 150, note 26, and p. 226, note 9; and also Appendix C.

very cold sharp frosty day, as there has been every day since the 24th ult., the ice in many places being 7 inches thick.

Fri. 14 Jan. In the forenoon Mr. French and I walked up again to Edwd. Hope's, but could do nothing, the master insisting the servant should not leave his service, and the girl will not consent to swear the father . . . Fanny Hicks, James Marchant, Fanny Weller and Bett. Mepham drank tea with me, and they stayed and spent the even with me and played at brag. They all met by accident, coming to buy goods in the shop, Fanny Weller excepted. I won at cards 2s. 2½d.

Tues. 18 Jan. . . . Just as I was undressed and going [to bed], Mr. Ed. Shoesmith of Laughton and Mr. Edwd. Stiles of Brightelmstone[3] called to see me. And really we sat so long that I am so unhappy to say I went to bed very much in liquor. I have once more taken a resolution, and I hope that a firm one and which through divine grace I hope steadfastly to adhere to; that is, never to drink anything stronger than wine and water. For the least quantity of anything so soon intoxicates my weak brains that I do as it were loathe myself upon the account of it. At home all day and very little to do . . .

Fri. 21 Jan. . . . At home all day, pretty busy . . . A very severe frost, which has continued now very severe, and that without almost any even the least thaw for a month. The ice in many places is not less than 8 or 9 inches thick.

Mon. 24 Jan. . . . In the even extreme busy a-casting up the measurements of the church as given me by a person employed to measure the same in behalf of the parish . . .

Tues. 25 Jan. . . . In the forenoon finished the squaring of the dimensions of the church work and paid the measurer, Mr. Jarvis at Chailey, 12s. 6d. for his time . . . At home all day and really very little to do. The frost continues very severe, and this day there was such a rime as I hardly ever knew . . .

Sat. 29 Jan. . . . The frost began to thaw today after having continued very severe for 5 weeks . . .

Sun. 30 Jan. In the morn I walked home with Sam. Jenner, with whom I breakfasted, and then we set off for Alfriston, I wanting to talk with a person whom I have some thoughts of putting my nephew Philip out an apprentice to. We dined at Wm. Wrothfield's, an acquaintance of Sam. Jenner's, on as fine a roast pig as I ever see. We came home about 8.20, but not thoroughly sober, though I think never so fatigued in all my life. Spent today 21d. What with my fatigue and drinking but

---

[3] Brighton.

little, it got up in my head. But as I have so weak a brain and have so often set a resolution not to be guilty any more of this vice, a vice I really detest, I hope yet to have a resolution strong enough to refrain from drinking anything strong for the future. Oh, may I have strength of mind and resolution firm enough to make good my real and designed intention! . . .

Thurs. 3 Feb. In the morning my late servant Mary Martin came to see me and . . . stayed all night . . .

Fri. 4 Feb. . . . My late servant stayed with me all day and aired all my wife's clothes. Sam. Jenner and I played a few games of cribbage in the even; I neither won or lost. He stayed all night. At home all day and pretty busy, but yet my mind is quite tumultuous, hardly knowing which way to pursue or what way of life to engage in. To keep house with servants in the business I am situated in is not either agreeable to my natural inclination or advantageous to my interest. No, it is not, for to the one it is quite different from, and the other I doubt it is greatly prejudicial to; and at the same time I doubt my trade will hardly answer the expenses of a family which might in all probability be the consequence of marriage. Though undoubtedly could I meet with a virtuous and discreet girl (one whose person I could like), who had taken more care to adorn the internal than external parts, I say with such an one I should prefer marriage before a single life, a life in my opinion agreeable to the dictates of nature, reason and religion. For agreeable to the Apostle marriage is honourable in all men and the bed undefiled, though marriage I doubt in this licentious and libertine age is too oft entered with lucrative views or to gratify some inordinate passion, both of which I presume are contrary to the real instituion of marriage. For in my mind virtue and a sincere love or friendship for each other seems the only basis to build a lasting happiness upon in the marriage state. Then so long as life continues, so long will happiness and love continue to increase. Happy, for ever happy, beyond the power of imagination to describe, are they whose lot it is; happy, may we presume, to all eternity.

Mon. 7 Feb. In the morn James Marchant and Elizabeth Mepham was married, and I was what is commonly called father, and also together with Thomas Davy signed the register book. My servant dined with her brother and sister, the new married couple, and I and my late servant dined on the remains of yesterday's dinner. In the even went down to Jones's, where we had a public vestry. Stayed there till about 10.10. Then I went to the wedding house and sat and smoked several pipes in company with Mr. Robert Hook, Tho. Durrant, Tho.

Davy, Jos. Fuller Jr., Fanny Hicks, my servant, and the new married couple; came home about 2.20. My late servant stood the shop for me during my absence. Sam. Jenner, coming in the even, stayed with my servant for company and also took part of my bed. Pretty busy in the day, but oh, the torment of my mind—I hardly know what to do!

Weds. 9 Feb.   After breakfast my late servant went home . . .

Sun. 13 Feb.   . . . Received of the Rev. Mr. Porter £4 in full for the interest of £100 due from him to this parish the 20th Nov. last for the same sum left to this parish by the will of Samuel Atkins Esq. deceased, which interest is yearly to be distributed among the poor of this parish . . . We accordingly disposed of the said interest money [to 27 people] . . . As the sum given away did exceed the interest money, Mr. Porter made it up out of the money given at the sacrament . . .

Weds. 23 Feb.   My brother came over to stand the shop for me . . . In the afternoon I went to Lewes in company with John Watford; took out a warrant to bring Catharine Jenner before Mr. Rideout to swear the father of the bastard child she was delivered of the 25th ult. I drank tea at Mr. Bennett's and he, myself, John Watford and my brother Richd. spent the evening at *The White Horse* . . .

Thurs. 24 Feb.   Instead of going to bed my brother and I went and called up Mr. French and from there we went to the miller's, who was headborough, and gave him the warrant, and then we went and took up the woman, and I went with her to Lewes where she swore the child on Edwd. Hope, a man of upwards of 70 years of age. I just stayed and breakfasted in Lewes and took out a warrant to apprehend the father, which I did about 11 o'clock and kept [him] in hold till he thought proper to give a note of hand for the payment of £40 on demand unless he enters into a bond to indemnify the parish from all expense that may attend the said bastard. My brother went away about 2.20. Spent at Lewes upon the woman and the man that carried her, horses, ostler and turnpike 4s. 6d. Paid for the warrant and examination 3s.

Fri. 25 Feb.   . . . At home all day and very little to do, ah! very little indeed. I think it makes me quite by myself. I am so confined with family connections and some large debtors in my parish that I hardly know which way to act or extricate myself out of so great a dilemma. For as I live now every, nay, all my effects and affairs are in a ruinous condition. My household affairs suffer waste and my business both at home and abroad receives injury from neglect.

Fri. 11 Mar.   . . . In the even Tho. Durrant and I walked up to James Fuller's to pay him a wedding visit. We stayed till near

11 o'clock and came home very near sober. (I forgot to mention we were
not asked to eat a mouthful of bread) . . .

Sat. 12 Mar. . . . At home all day, but very little to do. The wind
vastly high and prodigious cold, it being so excessive piercing that
people could hardly bear it, and the frost very sharp. Paid a needle-
maker from Chichester 9s. 3d. in full for the following needles bought
of him today: 12½ hundred Travilors⁴ needles 3s. 9d., 1 thousand *do.*
4s. 0d., 100 chapel *do.* 1s. 0d., ¼ hundred looping 6d.

Sun. 13 Mar. . . . I went to Little Horsted in order to meet my
brother and another person whom we had thoughts of putting out our
nephew Philip an apprentice to. But he having lately taken a lad, our
walk was to no purpose . . .

Mon. 21 Mar. . . . Sam. Jenner coming in the even, I persuaded him
to stay all night in order to help kill a hog for me tomorrow, which I am
to have of Robt. Hook. Paid Mr. Gilbert in Waldron 9s. for: 1 doz.
wooden hand dishes 6s. 6d., 1 dozen porridge *do.* 1s. 6d., ½ dozen
flitting *do.*⁵ 1s. 0d. received of him today.

Tues. 22 Mar. . . . Sam. Jenner breakfasted with me and stayed and
assisted me in the killing of my hog . . .

Weds. 23 Mar. I dined on some hog's liver and rashers of pork
fried. In the afternoon Joseph Fuller Jr. cut out my hog which weighed
23 stone 5 lb, and which I am to give Master Hook 22d. per stone
for . . .

Thurs. 24 Mar. Jarvis Bexhill dined with me on some hog's liver
and kidneys fried. At home all day and pretty busy. In the even went
down to Jones's in order to make a poor rate (there being a vestry held
for the same intent at which there was Mr. Porter, Jos. and James
Fuller, Edwd. Foord, Richd. Hope, Jos. Durrant and Mr. Carman).
We stayed till near 1 o'clock quarrelling and bickering about nothing
and in the end hardly did any business. The design of our meeting was
to have made a poor rate in which every one that was taxed was
intended to be assessed to the racked rent,⁶ that everyone might pay his
just quota (in proportion to his rent) of the money expended in
maintaining and keeping the poor. But how do I blush to say what
artifice and deceit, cunning and knavery there was used by some (who
would think it cruel and unjust to be called dishonest) to conceal their
rents, and who yet would pretend the justness of an equal taxation was
their desire. But however great their outward zeal for justice appeared,

---

⁴ The meaning of this word is not obvious. It was perhaps a brand name.
⁵ A flitting or fleeting dish was a flat dish used for skimming cream from the milk.
⁶ A raised rent: a rent more nearly equal to the full value of the land.

that cankerworm of self-interest lay so corroding in their hearts that it sullied the outward beauties of their would-be honesty. I say 'would-be' honesty because I look upon that man, be him who will, that endeavours to evade the payment of his just share of taxes to be a-robbing every other member of the community that contributes his quota, and also withholding from the poor what is their just right, and above all sinning against a positive command of our Saviour of doing to others as we would be done unto . . .

Tues. 29 Mar. . . . I dined on a hog's tongue and tail boiled and some greens. At home all day and pretty busy, but not in my shop any other than in marking up of goods. A good deal out of order with my cold . . .

Thurs. 31 Mar. . . . I dined on some cold pork bones roasted in the oven yesterday . . . Lent William Harmer, the person who paints our church, in cash £3 3s. 0d. which he gave me his note of hand for . . .

Tues. 5 Apr. I dined on a hog's clad piece[7] boiled and some greens. Paid Robert Hook in cash £3 in part of his parish bill. Molly and Bett. Carman, Pat Pollard and Lucy Emery drank tea with me, and the two Carmans and Sam. Jenner and myself played a few games of whist in the evening; I won 3½d. At home all day and thank God very busy. Sam. Jenner stayed and took part of my bed. Oh, what pleasure does a busy active life give! It quite enlivens the spirits and adds fresh vigour to nature.

Weds. 6 Apr. . . . So busy that I ate no dinner. Sarah and Ann Inkpin drank tea with me. This day Edward Hope signed and sealed a bond to indemnify this parish from all expenses touching and concerning the bastard of Catharine Jenner which she charged him with begetting in her examination taken in writing before Richd. Rideout Esq. The obligation of the said bond is £40 and dated the 1st day of March 1763, witnessed by Robt. Hook.[8] I then gave him the note of hand he gave me as a security until the said bond was properly signed and executed. He also gave me his note of hand payable to me or order on demand for 24s. in full for all the expenses I was at in taking him up, the woman's examination etc . . .

Thurs. 7 Apr. . . . In the afternoon went down to the yearly vestry at Jones's where I made up my accounts as overseer . . . The overseers chosen for the year ensuing are myself and Mr. Tho. Carman, the

[7] Probably clod-piece: the coarse part of the neck nearest the shoulder.
[8] The actual bond given by Edward Hope is now E.S.R.O., AMS 5841/11. It is amongst the group of papers described opposite, p. 269, note 9.

electioners Jas. Fuller and Jn. Watford; the churchwarden Jer. French, electioner Rd. Hope[9] . . .

Mon. 11 Apr. This day balanced accounts with Mr. John Vine Jr. and made even on both sides to this day, which is the first time we have been even for upwards of 11 years . . .

Sun. 17 Apr. In the morn I set out for Ticehurst with James Marchant in order to carry him to his parish, which he had sworn upon the said parish, and which Mr. French had yesterday taken out an order to remove him to. I arrived at Ticehurst about 9.30 and delivered him to Mr. Fuller, a glover there. I went to church and heard a sermon preached by the Rev. Mr. Medlicott, vicar of the same parish, from *Matthew* 5.17: 'Think not that I am come to destroy the law, or the prophets: I am not come to destroy, but to fulfil.' I dined at Mr. Noakes's on a calf's head and bacon boiled. I must pass over anything more than this: I got so very drunk that I came no farther than *The Half Moon* where I lodged all night, and in coming there I received a fall and hurt myself very much. Had it not been for an overruling Providence I must I think inevitably have been lost. May this, oh, may it for ever prevent my drinking anything stronger than water as my brains are so weak that even the smell of liquor almost makes me drunk.

Mon. 18 Apr. Came home about 10.30. I am such a terror and burden to myself I can say no more than this: was I possessed of the Indies, I think I would dispose of them in exchange for that peace of mind I enjoyed on Saturday evening more than now.

Thurs. 21 Apr. At home all day. I dined on some boiled sausages. Oh, how bad are I! Nothing, no, nothing but tumult and outrage is in my breast. May it please God to bless me with a true and real piety that I may never more be guilty of that sin of intemperance. Mrs. Fuller and her 2 daughters drank tea with me.

Sat. 23 Apr. Received of Mr. Porter in cash and a bank note £32 11s. 0d. as the following gratuities towards repairing our church: his

[9] A group of vouchers and receipted bills paid on this day by the overseers and churchwardens of East Hoathly were found (with some similar documents from the years 1775–8 and 1781–2) in the roof of Thomas Turner's house at the end of the nineteenth century. They were the subject of an article by L. B. Smith and F. A. Hadley, 'Turner trove. A link with a famous Sussex diarist', *S.C.M.* 9 (1935), pp.546–50, and were subsequently given to the Sussex Archaeological Society by C. P. Ranger. They are now deposited in E.S.R.O. where their reference is AMS 5841. They include bills for goods supplied by Turner himself, his brother Moses, Jeremiah French, Elizabeth Browne, and Richard Hope. Other bills and vouchers from 1761 onwards are to be found amongst the East Hoathly parish records, in E.S.R.O., PAR 378. The earliest overseers' account book, for the years 1762–79, (PAR 378/31/1/1) is entirely in Turner's hand: see Plate 4.

own gratuity £10 10s. 0d., the Duke of Newcastle's £21, Mr. Sam. Durrant's £1 1s. 0d. . . .

Mon. 25 Apr. . . . Lent Mr. Tho. Carman the £32 11s. 0d. I received of Mr. Tho. Porter the 23rd instant on the parish account and received his note of hand for the same payable to Mr. Porter, Jer. French and myself, or any of us, our or any of our order on demand and with interest for the same after the rate of three pounds percent per annum in trust for the parish . . .

Weds. 27 Apr. . . . Paid Tho. Davy in goods and cash 4s. 9d. in full for the same sum he paid yesterday at the feast of our Society at Mayfield which, as a member thereof, is in full of my arrears[10] . . .

Sun. 1 May. . . . Mrs. Burges, widow in the Street, being taken with an apoplectic fit today, Mr. Stone was sent for to her and called on me and let my blood . . . Tho. Durrant, having taken physic today, sat with me during service. Sam. Jenner drank tea with me, and to whom in the evening I read two of Tillotson's sermons. It being late when I was bleeded, and being as it were alone, Sam. Jenner stayed all night with me. In the afternoon died the widow Burges, aged 80 years and 11 months.

Mon. 2 May. This day William Bristow, Fanny his wife, and William and Samuel their children was brought home by an order of two of his Majesty's justices of the peace to this parish by the overseer of the poor of the parish of Heathfield . . . At home all day and pretty busy. In the evening wrote my London letters. The man's wife who was brought home today drank tea with me . . .

Thurs. 5 May. This was the day appointed by authority for a general thanksgiving for the late peace. No service at our church in the morn. Mr. Porter being on a journey . . . we had a sermon preached on occasion of the thanksgiving day by Mr. Roger Chalice, Vicar of Mayfield . . . We have had no kind of rejoicing in this place though it is the day for proclamation of peace and general thanksgiving throughout the nation. I think almost every individual seems to be dissatisfied with the peace, thinking it an ignominious and inglorious one.[11]

Sat. 7 May. . . . In the afternoon my late servant Sal. Waller and her lover Mr. Hastings called to see me and drank tea with me. I received of my late servant 14s. in full. My servant and paramour went away about 6.20. At home all day and very busy. Sam. Jenner, being to go a

[10] See above, p. 99, note 25.
[11] The Treaty of Paris which ended the Seven Years' War was signed on 10 February 1763.

journey with me, took part of my bed. Who can account for the inconsistent behaviour of some women? During the time this girl was my servant, the person who was with her made his addresses to her, but alas! all in vain. She even appeared to loathe, nay, hate the sight of him, but now I am inclined to believe she will soon make him her husband. What pity it is that the most beautiful of the creation is so fickle and unconstant.

Sun. 8 May.   About 3 o'clock Sam. Jenner and I set out for Newhaven to pay my friend Tipper a visit where we arrived about 6.10. We breakfasted with him and also dined with him . . . Came home about 9.10, really very sober . . .

Mon. 9 May.   Sam. Jenner dined with me on the remains of yesterday's dinner. At home all day, vastly stiff and tired with my journey yesterday. In the evening wrote my London letters and read Shakespeare's *As you Like It* and *Taming a Shrew*, both of which I think good comedies.

Mon. 16 May.   . . . Sam. Jenner, coming in late in the even, took part of my bed. I am in my mind very uneasy. For as I have no person to whom I can trust the management of my affairs with, I have many things that sustain loss for want of a more strict examination. Neither is my life as regular as I could wish and desire it was . . .

Fri. 20 May.   Delivered to Tho. Durrant in cash £109 11s. 3d. in order for him to carry to Maresfield for me to pay the general receiver of this county who collects there today, and which is in full for land and window tax due from this parish at Lady Day last . . . I also gave him 3 guineas over and besides the aforesaid sum in order for him to exchange any of the guineas . . . that might be complained of for want of weight . . .

Whit Sunday 22 May.   . . . Myself and servant at church in the afternoon; the text *Luke* 24.49: 'But tarry ye in the city of Jerusalem, until ye be endued with power from on high.' . . . The sermon we had this afternoon I have heard Mr. Porter preach 7 times with very little or any alteration.

Weds. 25 May.   . . . In the even went down to Jones's, there being a vestry holden to consult about appealing to an order with which William Bristow, Fanny his wife, and William and Samuel their children, were brought home from the parish of Heathfield; when it was unanimously agreed by all who were there (except myself) to appeal to the order at the next general quarter sessions at Lewes. The persons who constituted the vestry were Mr. Thomas Carman, Mr. Jer. French, Mr. Burges, Richd. Hope and myself. We came home about 8.40 all very sober. But in my humble opinion, notwithstanding we have had the opinion of a counsellor concerning it, that so certain as we stand trial, or at least appeal to the order, as certain we shall be cast, or lose our suit. No, I hardly think we have a bare probability of winning, so great is the advantage against us. Sam. Jenner, staying with my servant during my absence, took part of my bed.

Thurs. 26 May. . . . This morn James Hutson and Sarah Stevens were married at our church, and I doubt, a very unpromising match. Both parties, I doubt, have little else in view than to gratify passion, a sordid motive truly to build happiness upon in a married state.

> Wedlock's an evil [the] human race
> Are fond with open arms t'embrace.[12]

Not that I would be thought to have any dislike to matrimony. No, far from it. I think it a state designed for our happiness, or at least to make us as happy as our natures are capable of in this world. And as to my own self I can say by experience it is the only state in which I ever knowed any thing that might with any degree of propriety be called happiness. I only speak as matches are now commonly made, which I doubt are too often from lucrative and avaricious views, or else to gratify a base and inordinate appetite . . .

Sun. 29 May. In the morn Mr. Carman and I set out for Newhaven in order to retain counsellor Humphrey on a cause to be tried at the next quarter sessions to be held at Lewes on an appeal of this parish against Heathfield. Tho. Durrant accompanied us for pleasure. We got to Newhaven about 9.20 and waited upon Mr. Humphrey, and so soon as we had done our business, Mr. Carman set out for Lewes. Tho. Durrant and myself and Mr. Tipper dined at *The White Hart* on a veal pudding, a knuckle of veal boiled, a piece of pork and greens. After we had dined, Mr. Tipper and we rode to Seaford where we drank tea and parted. Tho. Durrant and I came home very safe and sober about 8.20. Spent today . . . on the parish account: my dinner 6*d.*, horse 7*d.*, ferry 1*d.*, tea at Seaford 6*d.*, liquor 1*s.* 3*d.* In my absence today my old acquaintance Mr. Tucker came to see me and brought me a present of some fine mackerel.

Sat. 4 June. . . . At home all day and very busy . . . After the fatigue of the day was over, I read part of Shakespeare's *Works*. What can equal the pleasure of half an hour's study after the fatigue of a busy

---

[12] A couplet evidently derived from the aphorism in the *Monostichoi* of Menander generally translated as 'Marriage is an evil that most men welcome'.

day is over! How does it unbend the mind and make it suitable for business again with a fresh alacrity!

Weds. 8 June. . . . In the even read part of Shakespeare's *Works*, which I think extreme good in their kind . . .

Mon. 13 June. In the forenoon my brother came over to stand the shop for me in my absence today, this being the day appointed for the assessors of the land and window tax to carry in their assessments to Maresfield, and there being a meeting of the commissions of the land and window tax. I set out for Maresfield about 11.20.[13] I got my assessment properly signed and allowed. I dined at Maresfield in company with George Medley Esq., Mr. Peckham, Mr. Tapsell and gentlemen to the number of 18, exclusive of myself, and really most of them were men of considerable property . . . I spent 3s. 8d. but it really was in the company of my betters. Came home about 8.30, very sober . . .

Fri. 17 June. . . . My brother being kind enough to stand the shop for me in the afternoon, I took a ride to a cricket match about 4 miles distant, there being a match played between the parishes of Uckfield and Isfield and the Borough of Lewes and ten miles around. It was decided in favour of the former, having only 14 runs to get in the last innings.[14] I came home about 9.20, very sober. Spent upon myself, horse and turnpike 10d. My brother stayed and supped with me and then went home. Very tired and somewhat ill in the even with a violent pang in the head. My brother very busy today in my absence.

Mon. 20 June. . . . This is the day on which I enter into the 35th year of my age, and may I (through the assistance of divine grace), as I grow in years, every day increase in virtue and piety. Then and not otherwise shall I live the life designed for me to lead on earth in order to my happiness in heaven to all eternity.

Thurs. 23 June. . . . This day two years ago was the day on which it pleased Almighty God to take from me my dear wife, and in the loss of her I sustained a very great one. During which interval of time the world has many times discovered I have been on the point of marriage. But I am clear in this that I have never yet made any offers of love to any one woman (no, not anything like courting) notwithstanding the voice of the world has been so much of the contrary opinion. Not that I have made any resolution to live single. No, I have not, for I really think could I meet with a woman whom I really loved, it would be much more conducive to my interest to marry than live single—and I

[13] See above, p. 216, note 3.   [14] See above, p. 9, note 9.

am certain a great deal more to my internal peace and satisfaction of mind, and I should hope a means to forward me in a state of piety and religion. The reason, if I know my own mind, of not seeing a proper object proceeds from that true (and I will boldly say) sincere love I had for my dear Peggy, who has almost continuously been in my mind since her death. Such an effect has the remembrance of her had upon my mind that I have thought it next to impossible I could ever like a woman well enough to make her my wife. For methinks when I compare the virtue and graces of my late wife with any of the female sex I have had the opportunity to be conversant with, they so far exceeded any I have found that they quite obliterated the thoughts of marriage that might otherwise run in my mind. But if I ever do marry again, I am sure of this, that I shall never have a more virtuous and prudent wife than I have already been possessed of. May it be the will of Providence for me to have as good a one—I ask no better. For a more virtuous and deserving woman I never did, or do I think ever shall, be so happy as to be acquainted with.

Sat. 25 June. . . . In the afternoon we had a cricket match played upon the common between the married men of this parish and the bachelors (or at least them that pass under that denomination) which was won by the former by 6 wickets[15] . . .

Sun. 26 June. . . . My late servant and her sister and my present servant went in the afternoon to hear a clergyman (lately curate of Laughton but now become field preacher) where there was I understand a pretty large congregation . . .

Tues. 28 June. . . . In the even Joseph Fuller and myself played a game of cricket with Mr. Geo. Banister and James Fuller for half a crown's worth of punch, which we won very easy. But in being hot, and drinking a pretty deal of punch, it got into my head; so I came home not sober[16] . . .

Weds. 29 June. I dined on two tench stewed, given me by Tho. Fuller. At home all day, but very little to do. Very stiff and disagreeable to myself upon my game of cricket last night.[17] In the even read part of Beveridge's *Thoughts*.

Tues. 5 July. . . . In the afternoon there was a game of cricket played between this parish and an eleven picked out of Framfield. It was played in the park and my servant went to see it, but it was not played out[18] . . .

Weds. 6 July. . . . In the even went up to the common where Richd.

---

[15] See above, p. 9, note 9.   [16] See above, p. 9, note 9.
[17] See above, p. 9, note 9.   [18] See above, p. 9, note 9.

Fuller and I played a game of cricket with James Fuller and Sam. French for half a crown's worth of punch, and to our disappointment we were beat.[19] We went down to Jones's in the even and spent the money in company with Joseph Fuller, Samuel Jenner and Tho. Durrant. Came home about 11.20 very sober.

Sun. 10 July. Myself and servant at church in the morn . . . Mr. Banister's child being to be baptized, and the sponsors not being all come, it could not be baptized before dinner. As T. Durrant and myself were to be sponsors, we dined at Mr. Banister's on a piece of cold roast beef, some cold roast chickens, tart, custards and a piece of cold bacon in company with Mr. T. Martin and his wife, Mr. T. Colbran and his wife, Molly Fuller, James Fuller and his wife and Miss Hall (my servant at home dining on the remains of yesterday's dinner). Myself and servant at church in the afternoon . . . After churchtime myself, Tho. Durrant, Miss Hall and Molly Fuller appeared at the font as sponsors for Mr. George Banister's daughter, which was baptized by the name of Molly (simple enough!). We retired to Mr. Geo. Banister's, where we drank tea and spent the afternoon. Came home about 9.40, very sober. Gave the nurse 12*d.*; *ditto* the servant 6*d.*

Mon. 11 July. . . . In the afternoon Mr. Jos. Smith, attorney at law, and Mr. Turner, clerk to Mr. Burtenshaw, drank tea with me. The former came to take my further orders concerning two trials we have the ensuing sessions, the other to summons me to bring the land and poor rates to the sessions . . .

Weds. 13 July. . . . In the even read several political papers called *The North Briton*, which are wrote by John Wilkes Esq., member for Aylesbury in Bucks, for the writing of which he has been committed to the Tower, and procured his release by a writ of *habeas corpus*. I really think they breathe forth such a spirit of liberty that it is an extreme good paper[20] . . .

Thurs. 14 July. My servant's father came to appear as an evidence on our parish account tomorrow at the sessions. He dined with me on some beans and bacon. Paid James Crowhurst in cash and goods £2 7*s.* 0*d.* in full for a parcel of earthenware received of him today. This day received a subpoena from Mr. Jos. Smith, the attorney our parish has

---

[19] This reference is not included in the article in *The Cricket Quarterly* referred to above, p. 9, note 9.

[20] John Wilkes (1725–97) the radical demagogue and pamphleteer began *The North Briton*, a weekly, in June 1762. Issue no. 45 (23 April 1763) criticised the King's speech and for this Wilkes was imprisoned in the Tower. He was released in May following mass demonstrations under the slogan 'Wilkes and Liberty'. Turner evidently approved of Wilkes's journal.

employed, for Eliz. Marchant, which I accordingly served her with . . .
At home all day. My servant's father lodged at my house.

Fri. 15 July.   In the morn Mr. Carman and I set out to attend the
sessions in Lewes. My brother came over to stand the shop for me
during my absence. We accordingly consulted our attorney relating to
our trials in everything necessary in order to be prepared in every
point. We then attended the court, and the second trial which came on
was one of ours where the parish of Ticehurst were appellants and the
parish of East Hoathly removants. The case was this: James Marchant,
the son of Tho. Marchant, an inhabitant legally settled in the parish of
Ticehurst, did by and with the consent of the said Thomas his father
put himself out as an apprentice to his uncle Richd. Marchant, an
inhabitant legally settled in this parish, for the space of 7 years, during
which time he was faithfully to serve him as an apprentice (and the
pauper always deemed himself as an apprentice). The uncle declared
upon making the agreement he would have no indentures because he
[James] should not gain a settlement in this parish. The said pauper
served 5½ years of his time then, some difference arising between
them, he left his service and never after obtained any subsequent
settlement (by any act of his whatever). This was the purport of his
examination, upon which our order of removal was granted.[21] What
motives Ticehurst had for their appeal I know not, but there
appeared none in the course of the trial, and it was, after a short
hearing, given in our favour and the order confirmed. The next cause
that came on was also ours, where we was not so fortunate as in the
former. Indeed we had now another part to act, being in this cause
appellants. The case was this: About the year 1740 the parish of
Waldron granted a certificate, duly executed and allowed by two
justices, with John Bristow and Ann his wife, John, William (the
pauper), Edwd. and several more sons and daughters named in the
said certificate, to this parish. William the pauper about the 11th year
of his age, after having lived with his parents in the parish of East
Hoathly under the certificate several years, went back into the parish of
Waldron and lived several years as a servant. From thence he went into
the parish of Isfield and obtained a legal settlement in Isfield by

[21] The justices' decree that James Marchant's place of settlement was Ticehurst did
not, of course, mean that he was forced to reside there. It meant that the parish officers
of East Hoathly were within their rights in insisting that Marchant must arm himself with
a settlement certificate from Ticehurst so that he could be removed there should he
become a burden on the rates. He duly obtained such a certificate, dated 8 October
1763, which is still amongst the East Hoathly parish records (PAR 378) in the East
Sussex Record Office. See Plate 6.

servitude. He then came into the parish of East Hoathly again and lived as a yearly servant several years, hiring himself by the year, and served a year subsequent to the hiring. By some means he got into the parish of Heathfield and married. His family coming on, and no very promising person to maintain them, they began to be uneasy and talked of removing them. He then applied to this parish for a certificate, and it being a case somewhat difficult, our parish had the case drawn up, stated and laid before counsellor Humphrey for his opinion, which was that William the pauper did not according to law belong to this parish. Upon this authority we refused him a certificate, and after some short space of time, perhaps 20 weeks, he was brought home with an order from the parish of Heathfield. We then applied to Mr. Humphrey again, who still insisted he was right and that the man could not belong to this parish and advised us to give notice of an appeal. We did and the trial, as I observed, came on; which, after being learnedly debated by the counsel on both sides for I believe an hour, it was the opinion of the court the order should be confirmed, and it was stated special. This ended our business.[22] Mr. Carman and I dined at *The White Horse* in company with 9 more on a quarter of lamb roasted, a leg of lamb and a piece of beef boiled, carrots, turnips and French beans. My brother and servant at home dined on bacon and beans. I drank tea at Mr. Madgwick's and came home safe and sober about 10.30. My brother went home in the even. Spent today upon the witnesses and all expenses that I have paid out of my pocket about these trials £1 10s. 3d. My servant's father lodged at my house.

Thurs. 21 July. This morning, about 3.20, died my neighbour, and in a measure my landlady, Mrs. Weller, after about 4 days' illness, aged 69 years[23] . . .

Sat. 23 July. . . . His Grace the Duke of Newcastle came to Halland and stayed all night, as did the Earl of Ashburnham and several gentlemen. Very busy all day.

Sun. 24 July. . . . After churchtime Sam. Jenner, Joseph Fuller, Tho. Durrant and myself walked down to Halland (there being a

[22] The case of William and Fanny Bristow and their two children illustrates the complications of the law of Settlement. The justices, despite Mr. Humphrey's arguments, presumably held that Bristow's last legal place of settlement was East Hoathly by virtue of his having had employment as a yearly servant in the parish prior to his move to Heathfield—he having fulfilled none of the conditions necessary to gain legal settlement in his new parish since moving there. The conditions under which settlement by service was obtained or not obtained were extremely complicated and occupy nine pages in *Burn*, pp. 520–28.

[23] This was the mother of Francis Weller who owned the shop occupied by Turner. He subsequently bought it from Weller in 1766: see Appendix B.

public day) at which was the two judges on the circuit, *viz.*, Lord
Chief Justice Pratt and Justice Bathurst; Lord Gage; the Earl of
Ashburnham; Mr. Pelham of Stanmer; Rose Fuller Esq; Sir John
Bridger Kt; Thomas Chowne Esq; William Plumer Esq. and a great
many gentlemen of all ranks.[24] We neither ate nor drank, but took a
survey of the company and saw most of the gentlemen take their
coaches and horses and then came home about 7.20 . . .

Mon. 25 July. . . . In the afternoon attended the funeral of Mrs.
Weller and served the same. We carried her to church for interment
about 5.40. There was a sermon preached suitable to the occasion
from *I Corinthians* 15.19: 'If in this life only we have hope in Christ, we
are of all men most miserable'. I gave away . . . 6 crepe hatbands, 10
pairs women's gloves, 13 pairs men's *do.*, 1 pair maid's *do.*[25] . . .

Tues. 26 July. . . . In the even walked down to Halland having an
invitation from Mr. Coates. I supped there on a piece of cold roast beef
and a cold venison pasty in company . . . We stayed and spent the even
there till about 12 o'clock and then came home, but not one of us
thoroughly sober. Mr. Francis Weller took part of my bed.

Thurs. 28 July. . . . At home all day and pretty busy. This day the
King's Plate was run for on Lewes Downs when the following horses
started, *viz.*, the Duke of Grafton's bay horse Havanna, who won the
two first heats; Mr. Beaver's bay horse Young Blank, distanced the
first heat; Mr. Osbaldestone's bay mare Dismal, who came in 2nd both
heats . . .

Sat. 10 Sept.[26] . . . Paid Thomas Tulley 14*s.* in full for 1 dozen
pattens and 1 dozen clogs received of him today . . .

Sat. 17 Sept. . . . At home all day and very little to do. In the
afternoon about 5.30 died Mr. French of this parish after a long and
lingering illness, which, it's to be doubted, was first brought on by the
too frequent use of spirituous liquors and particularly gin. If it was

[24] Lord Chief Justice Charles Pratt (1714–94) later 1st Earl Camden, and Henry
Bathurst (1714–94) 2nd Earl Bathurst were both subsequently to become Lord
Chancellor. For Lord Gage, the Earl of Ashburnham and Mr. Pelham see above, p. 156,
note 33. Rose Fuller, of the Heathfield gun-founding family, was MP successively for New
Romney, Maidstone and Rye: Perceval Lucas, *Heathfield Memorials* (1910) p.86. Sir
John Bridger of Hamsey died in 1816. William Plumer of Blakesware had in the previous
February become MP for Lewes. Thomas Chowne of Alfriston was the antiquary who,
two years later, was to see his residence, Place House, and all his manuscripts and
antiquarian collections destroyed by fire. The story is told in T. W. Horsfield, *The history
and antiquities of Lewes and its vicinity*, II (Lewes, 1829), pp. 6–7.
[25] See above, p. 7, note 5.
[26] The volume of the original diary covering the period between 29 July and
9 September is missing.

possible to make any estimate of the quantity he drank for several years past, I should think he could not drink less on a moderate computation than 20 gallons a year. Let me, from such instances, fly the habit of drinking and think upon the final consequence.

Sun. 18 Sept. . . . I went down to Whyly to speak to Mrs. French about the funeral of her husband . . .

Tues. 20 Sept. . . . Drew up the minutes of an agreement between Mr. Fra. Weller and Mr. Geo. Banister wherein Mr. Weller agrees to let Mr. Banister that end of his house,[27] late his mother's, and all the furniture as it stands, an inventory of which was taken and annexed to the agreement. Mr. Banister is to give 2s. 3d. each week . . .

Weds. 21 Sept. My brother came over in the forenoon and dined with me on the remains of yesterday's dinner with the addition of a piece of pork boiled and some carrots. My brother stayed and waited in the shop while I went to serve Mr. French's funeral. I served the funeral and came home about 7.20.[28] My brother went away immediately. A very remarkable wet forenoon, it raining almost incessantly all the forenoon. Mr. French was aged 55 years. An account of the gloves given away today, *viz.*,

Men's black ribbon-bound gloves:

1 Mr. Jn. French	3 John Baker
2 Mr. Jn. Thatcher	4 Tho. Tester

5 Tho. Davy

Women's black ribbon-bound gloves:

1 Mrs. French	6 Mrs. Durrant
2 Molly French	7 Dame Cornwell
3 Mrs. Thatcher	8 Dame Prall
4 Mrs. Virgoe	9 Dame Trill
5 Mrs. Davy	10 Eliz. Henly

11 Mrs. French's maid.

Boy's black ribbon-bound gloves:

1 Sam. French

---

[27] The other end of the house from that occupied by Turner.
[28] See above, p. 7, note 5.

Men's looped gloves:

1 Mr. Porter

Total of gloves:

5 pair men's 2*d.* chamois
11 pair women's *do.*
1 pair men's looped
1 pair boys' 2*d.* chamois or black ribbon-bound.

Mon. 26 Sept. . . . I think I never in my life knew any place so much gone off for trade as is this place since I have lived in it. Most of the principal inhabitants, as we esteemed them, being dead and the remaining reduced, trade is got to be very trifling. The occasion of poverty's being so frequent proceeds from luxury and imprudence, I fear, too often. For custom has brought tea and spirituous liquors so much in fashion that I dare be bold to say they often, too often, prove our ruin. For by the frequent and continual use of them we increase our expenses, bring on idleness and render ourselves less capable to struggle with the world and above all hurt our health and I doubt often, by the too frequent use of both, entail a weakness upon our progeny.[29]

Weds. 28 Sept. . . . Not at all busy today. My spirits quite low, though I know not for what, unless it be for want of company, hardly ever seeing anyone but those who come to and from the shop.

Thurs. 6 Oct. . . . In the afternoon, it being an extreme fine day, I and Joseph Fuller took a walk to Blackboys Fair. I spent 15*d.* and came home about 7.20 not thoroughly sober, though I drank very little indeed at the fair.

Sat. 8 Oct. . . . At home all day and very busy. A very windy and wet day. Mrs. Carman, not Mr. Carman, made me a present of a fine duck. Let not this, my readers, be a matter of speculation. There is no criminality in it. For I never kissed the woman but once, and that was the day I stood sponsor or godfather for her daughter. So I presume her kindness (though I believe unknown to Mr. Carman) proceeds from a spirit of gratitude.

Sun. 9 Oct. . . . After churchtime I went up to Jos. Fuller's (being invited), where I drank tea in company with Tho. Fuller and his wife (newly married and come to pay their friends a wedding visit), John Fuller and his wife, Mr. Shoesmith and his wife, Miss Hammond, Tho. Hammond, Bett. Fuller and their own family. I also stayed and supped there and came home about 10.10, sober . . . I must

[29] See above, p. 159, note 38.

acknowledge I was entertained this even in a very polite and elegant manner at Mr. Fuller's.[30]

Sat. 5 Nov. . . . In the even went into Joseph Durrant's, where I stayed till about 12 o'clock and spent the even in company with their own family and Mr. Thornton, formerly an officer of excise in this round, and their boarder, but now an officer of excise at Spondon in the county of Derby. I cannot say but my stay was too long, for the liquor was too powerful for my brain, so that I was somewhat in liquor, though not very much, but however too much, and in reality more than is consistent with the Christian religion, and what I greatly lament as a great weakness in myself.

Mon. 7 Nov. . . . In the forenoon I, in company with John Piper, rode to Maresfield, there being a sitting of justices for the surveyors of the highways' business, and Mr. T. Carman, one of our surveyors, being a man young in office and I presume willing to show his power, had purposed to return several people in this parish as defaulters in their work on the roads. And, as many of the number were poor indigent people, I, at the desire of Mr. Porter and several more of the principal inhabitants, went in order to plead in their behalf. Some of them I persuaded him not to return, and upon the whole there was a warrant granted for only three[31] . . .

Thurs. 10 Nov. . . . Let me once more describe my uneasy situation, but at the same time acknowledge the many blessings of Providence which I enjoy far superior to many of my fellow creatures perhaps far more deserving of them than myself. But alas! what afflicts me is the loss of my dear Peggy, though in this I in no ways repine at the allwise dispensations of Providence. For I am well assured whatever is by the appointment of heaven is best for us, and I think it my duty to cheerfully submit . . . No one but a servant to trust the care of my concerns to or the management of my household affairs, which are now all confusion. My affairs abroad are neglected by my confinement at home, and were I to be more from home, my affairs there perhaps might still suffer a greater injury from my absence . . . And for want of the company of the more softer sex and through my overmuch confinement I know I am become extreme awkward, and a certain roughness and boisterousness of disposition has seized on my mind. So the want of those advantages which flow from society and a free intercourse with the world and a too

[30] The volume of the original diary covering the dates between 14 October and 29 October is missing.

[31] For the duties of the surveyor and his supervision by the justices see *Tate*, pp. 242–50.

great delight in reading have brought my mind to that great degree of moroseness that it is neither agreeable to myself, nor can my company be so to others.

Thurs. 24 Nov. . . . Mr. Banister, having lately taken from the smugglers a freight of brandy, entertained Mr. Carman, Mr. Jos. Fuller and myself with a bowl of punch in the even at my house.

Fri. 25 Nov. Mr. —————,[32] the curate of Laughton, came to the shop in the forenoon, and he, having bought some things of me (and I could wish he had paid for them), dined with me and also stayed in the afternoon till he got in liquor. And with being so complaisant as to keep him company, I was quite drunk. How do I detest myself for being so foolish! As my brains are so weak they will bear nothing that is strong, sure I am highly culpable for ever attempting to drink anything stronger than small beer or water, and more especially as it gives my conscience so much uneasiness afterwards . . .

Tues. 6 Dec. . . . Very little to do all day. In the even I, in company with Joseph and Richard Fuller and Thomas Durrant, went to the sign of *The Chequer* in Framfield parish in order to spend the even, the people who keep the house having laid out a good deal of money with me. We supped there on some Welsh rabbits and stayed till about 10.20. Spent 19*d.* each and came home about 11.20. I cannot say I was thoroughly sober, but not so much in liquor but I could walk very well. But however I was the worse for liquor.

Weds. 7 Dec. In the morn Mr. Banister and I walked to Chiddingly, he in order to survey a victualler and I for a walk, it being a fine morning . . . I think since I have lived at Hoathly I never knew trade so dull or money so scarce, the whole neighbourhood being almost reduced to poverty.

Thurs. 8 Dec. This morn were married at our church (with licence) Joseph Durrant and Eliz. Reeve, both of this parish, and they, together with Richd. Hope, happening to come in just as I was at breakfast, breakfasted with me. This match appears to me to be one of our country sort; that is, the woman is pregnant if her own word is to be taken for it, and I dare say in this affair a woman's word for once may be taken for fact . . .

Tues. 20 Dec. . . . My brother came over in the even in order to stand the shop for me while I assist Mr. Coates tomorrow in

[32] There was a succession of curates at Laughton during the mid-eighteenth century and it is not possible to identify this one with certainty. W. A. Pearson, in his history of Laughton entitled *The village of the buckle* (1931), pp. 55 mistakenly implies that it was Thomas Porter. It was most likely, however, to have been Thomas Fletcher.

distributing a gift left by some of the ancestors of the Pelham family to be distributed yearly on St. Thomas's Day . . .

Weds. 21 Dec. In the morn about 2.40 we began our work, Mr. Coates paying the money and I taking down the names and parishes of the people. About 11.20 we completed our work, having given away about £8 6s. 0d., 4 bushels of wheat made into bread, and I imagine not less than 100 gallons of beer. I breakfasted with Mr. Coates and received my groat and then came home about 11.40. My brother went away immediately. In my absence today my brother and servant relieved [32] poor inhabitants of this parish with a penny each and a draught of ale . . .

Sun. 25 Dec. Myself at church in the morning . . . Just after we came from church, my old and worthy acquaintance Sam. Jenner came to see me and dined with me, as did also the widow and James Marchant and his wife and child and Joseph Burges Jr., on part of a rump of beef roasted, a raisin suet pudding and gooseberry pie . . .

Mon. 26 Dec. . . . In the even went down to Jones's, there being a public vestry to choose surveyors for the ensuing year as also to pass and balance the accounts of the last surveyors. Upon balancing the same there was due to the parish the sum of 11s. 7½d. We then rechose the two old surveyors and the two electioners, *viz.*, Tho. Carman and James Fuller surveyors and John Vine and Edwd. Hope electioners. And I think had we tried all Sussex, or even England, for to have found two such, we could not have done it. These people are very improper to serve the office of surveyors, being litigious to the extremest degree and withal very ill-natured, executing law to the utmost. But in the midst of law they quite forget justice, equity or charity, and in permitting those two to continue their office I think the parish has and will sustain a very great injury, as well to the parish in general as also to every individual thereof . . . We stayed at the vestry and spent 6d. each and came home about 3.20, not very sober. Oh, liquor, what extravagances does it make us commit! . . .

1764

Mon. 2 Jan. In the forenoon was married at this parish church William Williams and Lucy Mepham, both living in this parish, and I was, as it's commonly called, father; that is, the person that gives the woman to the man. It being a parish job, I paid Mr. Porter 5s. for his fee and the clerk 18*d*. for his fee. I also gave the girl 1 guinea as a wedding portion . . .

Tues. 3 Jan. . . . This day by a warrant from George Courthope Esq. we took up the man that was married yesterday in order to swear his parish; and Mr. Carman accordingly went with him to Uckfield where he swore himself on the parish of Beaumarais in the Isle of Anglesey. Paid Mr. Carman 6*s*. 3*d*. for his expenses attending the same[1] . . .

Fri. 6 Jan. Sent by John Watford to the general receiver of the land tax for this county, who collects today at Maresfield . . . the sum of £76 and is in full for half a year's land tax for this parish (the salary outset) due at St. Michael last . . .

Weds. 11 Jan. . . . This morn about 7.30 in the morn was found dead in our parish Will. Lidlow, a person belonging to Chiddingly, supposed to drop on account of his being in liquor and perished by the inclemency of the weather. At home all day and very little to do. How should such instances as these teach mankind to shun that hateful vice of drunkenness, a crime almost productive of all other vices.

Fri. 13 Jan. In the forenoon walked down to the Nursery where the body of Will. Lidlow was, who was found dead in our parish the 11th instant, the coroner's inquest being to sit touching the same. Accordingly Mr. Attree, one of the coroners of this county, came and swore the jury, *viz.*, myself the foreman of the jury and the other 12 . . . The verdict brought in, as it appeared plain by the examination of the people who found him, that he was in liquor and laid himself down on the ground and perished through the inclemency of the weather . . .

Sat. 14 Jan. . . . Received of Thomas Overing one guinea for the expense our parish has been at on account of the death of Will. Lidlow, and he accordingly took his body to Chiddingly in order for interment.

Mon. 16 Jan. . . . Paid John Brazer and Will. Starks 8*s*. for their trouble laying out etc. of Will. Lidlow, found dead in our parish the 11th instant[2] . . . I think to the best of my memory I never remember so wet a time as the present, there having been hardly 48 hours of fine weather for this 2 or 3 months past, it raining almost continually.

[1] The East Hoathly parish register, E.S.R.O., PAR 378/1/1/4, gives his place of origin as 'Rixson' (i.e. Wrexham), Denbighshire. The expenses in connection with this parish wedding are shown in the accounts of the overseers of the poor (PAR 378/31/1/1) which are in Turner's hand. It is interesting to note that in the accounts Mr. Porter is shown as having received 6*s*. and Mr. Carman 9*s*. 3*d*. See Plate 4.

[2] Plate 4 shows the parish expending 7*s*. 6*d*. on 13 January over this affair, but shows no expenditure on 16 January.

Tues. 17 Jan. Paid Mrs. Browne by her servant Henry Godley in cash 8*s.* 6*d.* for the coffin had for Will. Lidlow. In the forenoon Mr. William Goldsmith the constable, Mr. Tho. Carman and myself assisted John Watford in distraining the goods of Sarah Thunder for three pounds for three years rent due the 5th instant. Mr. Goldsmith as constable swore Mr. Carman and I to appraise the said goods, which we did at 46*s.* John Watford delivered her a list of the same and a notice to either pay the rent or replevy[3] the said goods or he would at the expiration of 5 days sell the same . . . My brother Moses sent me a present of a turkey.

Sun. 22 Jan. Myself at church in the morn . . . In the time of singing the second time I went out and searched Jones's but found no persons proper for reproof . . .

Tues. 7 Feb. . . . At home all day, and very little to do. In the even Richd. Prall came to Mr. Banister's end of the house, he being in bed, and called him all the knaves, villains etc. he possibly could think of, which so far enraged Mr. Banister as to occasion him to come down into the street and indiscreetly enough gave him the discipline of the horse-whip. I cannot but own Prall very justly deserved what he had a pretty sufficient dose of, but at the same time I dare say as it's a breach of the peace it will cost my neighbour Banister some money.

Thurs. 9 Feb. All the forenoon I was endeavouring to make up the affair between Mr. Banister and Mr. Prall, and both of them being very headstrong, I could not prevail. Mr. Carman being headborough served the summons on Mr. Banister . . . After dinner they set out for to appear before the justice in order to abide by his determination. As I heard it, Mr. Banister desired me to accompany him as an evidence, which I did. Upon their appearing before Luke Spence Esq., the justice who granted the summons, he seemed to think it a very great breach of the peace and told them the best way for the mutual interest of both parties was to agree it up, which they agreed upon. Accordingly we went to *The Cats* where we made it up after three or four hours' squabbling and wrangling about nothing. Mr. Banister gave him a guinea and was to pay the horse hire of Prall's horse to Lewes. We came home very sober about 1.30. I spent nothing, Mr. Banister paying my expenses . . .

Fri. 10 Feb. . . . At home all day, and very little to do. Mr. Pepper, the supervisor of excise for this district, coming to survey with our officer, and it being in the even and our public house affording no

---

[3] Obtain the return of the goods by giving security to have the matter tried in court.

entertainment either for man or horse, he took part of my bed and his horse a part of Mr. Banister's horse's.

Weds. 15 Feb. . . . I dined on a piece of beef roasted in the oven and a batter pudding under it . . .

Thurs. 16 Feb. . . . I dined on some cold beef but not the remains of yesterday's dinner for my servant through carelessness let the dogs get into the buttery and eat up all my beef . . .

Fri. 17 Feb. . . . In the even Mr. Banister gave myself, Jos. Fuller, T. Durrant and Mr. Long a treat (at my house), he having lately made a seizure of some brandy from the smugglers. Mr. Long took part of my bed.

Sun. 19 Feb. Myself at church in the forenoon . . . My friend Sam. Jenner dined with me on part of a sparerib (made me a present of by Dame Durrant) roasted and some applesauce . . . In the even I read to my friend a sermon preached at the last Visitation held at Lewes, written by Mr. Nicholl, Vicar of Westham in this county, and part of three discourses written by James Walder, a Baptist preacher, the last of which I esteem the best performance, it being in my judgment written with a true spirit of piety and in a pretty modest style,[4] And what may, I presume, be proper for to be read by any sect whatsoever, there being nothing more in it than what is the duty of all Christians both to practise and believe.

Mon. 20 Feb. . . . About 5.20 myself, Jos. Fuller Jr., Tho. Durrant and Robt. Hook set out for *The Black Boy* in order to meet some more company, there to spend the evening with our old acquaintance Mr. Thornton, who was formerly excise officer in this place, but has lately relinquished the excise and taken *The Black Boy* . . . Came home about 12.20 very sober . . .

Tues. 21 Feb. . . . Tho. Durrant and my friend Sam. Jenner breakfasted with me, and then we set out on our journey, there being a sale of condemned goods to be sold at the Custom House at Newhaven, some of which I went with an intent to purchase . . . I bought at the sale 56 lbs of currants at only 50*s.* per hundredweight (which I paid for) and came away from Newhaven immediately . . . Thank God came home to my own house about 11.30, very sober and not greatly fatigued.

Sat. 25 Feb. . . . My worthy friend Sam. Jenner came in the even, and the weather being very bad, he took part of my bed. Sam. Jenner and I played a few games of cribbage in the even. I lost two games, being very incapable to play, having just heard of a book debt of £40 I

[4] For both works see Appendix D.

am like to lose—and that in a measure through the knavery of the man and my own too great indulgence, showing him more favour than I am now by woeful experience like to find he deserves. But who can tell as the world goes the man that is now honest? For what by our extravagant living and an indolent way of life we are got into, makes custom so prevalent that rather than retrench our expenses, we too often see people run out of their estates and defraud their creditors.

Sun. 26 Feb. . . . In the forenoon my old and I wish I could say worthy acquaintance Mr. Tucker of Lewes came to see me and dined with me on a bullock's mint pudding, part of a bullock's cheek hashed and some boiled tripe . . . During churchtime my foolish and drunken visitant went home . . .

Sun. 4 Mar. . . . Just as I was got out of my doors intending to go to church, my brother came over in order to stand the shop tomorrow in the forenoon for me. I immediately set out on foot for Lewes in order to meet Mr. John Ridings and Mr. Stephen Fletcher, son and servant to Mr. Sam. Ridings of Manchester, with whose horses they are now at Lewes. My brother took a walk with me and brought me going on my road to Lewes, above half way, where I arrived about 5.20. I drank tea with my very worthy acquaintance Mr. Madgwick, and he, myself and Mr. Richardson, Mr. Tho. Woodgate and Mr. John Ridings and Mr. Fletcher spent the evening together at *The White Horse* till about 10 o'clock. We supped at *The White Horse* on some Welsh rabbits. I lodged at Mr. Madgwick's. A very fine pleasant day, but extreme cold. My brother came back and lodged at my house.

Mon. 5 Mar. In the morn I arose and went up to *The White Horse* where I bought a pretty large parcel of goods of Mr. Ridings and Mr. Fletcher for myself and brother. I paid Mr. John Ridings in cash £3 4s. 0d. in full on my account with his father Mr. Sam. Ridings to this day, and also paid him in cash £5 2s. 0d. in full on my brother Moses' account with Mr. Sam. Ridings.

Weds. 7 Mar. . . . Lent Eliz. Akehurst in cash £1 1s. 0d. for which she stands debtor, and also I received a note of hand for the same from Joseph Akehurst; so that between both parties I think there will be no fear of losing it, though really the generality of people are so poor and withal so full of tricks that it is hardly any knowing whom we can trust.

Fri. 9 Mar. . . . Sam. Jenner at work for me all day, dined with me on a light pudding, a piece of beef boiled and some Savoy greens. At home the whole day . . . A very cold day. Sam. Jenner went away in the even. Perhaps it may appear odd, Sam. Jenner's being so much at my house, but he is a good-natured willing person and oft does my gardening etc. for nothing, and he is undoubtedly a worthy companion.

Sat. 10 Mar. . . . Mr. Long, late writing-master of this place, but now under instructions for the excise, coming into my house in the even, took part of my bed . . .

Sun. 11 Mar. . . . In the morn Mr. Stone called on me and bleeded me . . . Waldron singers came and sang at our church in the afternoon. After churchtime Mr. Porter and myself gave away the interest of a hundred pounds, being a legacy given to this parish for ever by Mr. Sam. Atkins, formerly gardener and house steward at Halland to his Grace the Duke of Newcastle, the interest of which was annually to be distributed among the poor householders of this parish; and I presume them to be such (according to the testator's intention) as had no monthly or even any relief at all out of the parish book. But be that as it will, we now deviate from that rule . . .

Mon. 19 Mar. I dined on the remains of yesterday's dinner. At home all day; posted my day book. Molly Hicks drank tea with me. In the even wrote my London letters. Very little to do all the day.

Thurs. 22 Mar. . . . Sam. Jenner at work for me all day and dined with me on a beef pudding and some rashers of pork boiled. At home all day and not at all busy. Sam. Jenner, staying the even, took part of my bed. A very cold day, but a very fine one. Samuel Jenner and I played a few games of cribbage; I lost 2 games.

Fri. 23 Mar. . . . In the even Sam. Jenner, coming late, took part of my bed. But very little to do all day. In the even, read some *Universal Magazines*. It may perhaps be thought by the censorious and malevolent part of my neighbours to appear odd that Sam. Jenner should be so much at my house, but I assure them I have always found him a very sincere friend, a worthy man, a learned and agreeable companion, a sincere and good Christian, and at any time willing to do any business for me, and that without any gratuity. Therefore I cannot think him an improper person for a companion.

Sun. 25 Mar. . . . After dinner I walked up to Mr. John Vine's in order to converse with him about his affairs, he being insolvent and keeps out of sight . . .

Sun. 1 Apr. . . . This day about 10.10 began an annular eclipse of the sun, but as it was not clear weather during the time, it yielded but an indifferent prospect to the spectator with the naked eye, though I heard some of the curious say, who viewed it with proper glasses, that they saw a star or two appear; so we may conclude the sun must be a good deal darkened. But I am pretty certain had I not been apprized of it before, I should never have discovered anything about it . . .

Tues. 10 Apr. . . . At home all day and, thank God, very busy. In the even went down to Jones's, there being a public vestry held to make the poor rate. But what through the ignorance of some and the avaricious views of others, we made none, notwithstanding this being the second vestry called for that purpose. I am sorry to say it, my partner Mr. Carman was the chief, nay only, instrument of its not being made, and that for no other reason than to serve his own private ends and to be assessed lower according to the law than anyone alse—a sad principle for an overseer or any other person who is concerned in a public office. For I think whenever the public good and that thing, self-interest, stand in competition with each others, the last should always submit and give way to the former. For in all communities I presume were the public good more studied and practised, the consequences which would ensue must be that of the private interest of every individual . . .

Weds. 11 Apr. . . . In the afternoon Mr. Robt. Cleaver, a woollen draper in London, called on me and drank tea with me. I gave him an order . . .

Thurs. 12 Apr. . . . In the afternoon Thomas Durrant and I walked to Uckfield to pay my friend Mr. Elless a visit, with whom we drank tea and spent the even, and came home very safe and well and pretty sober about 1.10. And I think I was never entertained in so polite and genteel a manner by any one person I ever paid a visit to, everything being conducted with the greatest politeness imaginable, and yet with the greatest freedom and friendship . . .

Tues. 17 Apr. . . . In the even went down to Jones's, there being a public vestry held to make a poor rate (which was the third meeting), and we did accordingly make a rate at 4s. in the pound. It was allowed of and agreed to and also signed by everyone present except Mr. Tho. Carman, who opposes it, as it will greatly affect him . . .

Weds. 18 Apr. After breakfast Mr. Joseph Fuller and I went up to take an inventory of the effects of Mr. John Vine (now a prisoner in his own house for debt) in order to have the same made over to 2 or 3 of his principal creditors by virtue of a bill of sale, and to be by them distributed equally among the whole number of his creditors in proportion to their respective debts . . . We came home about 6.20, and after we had done our business, we sat down and smoked our pipes and drank a bottle or two of cider, which I must own got up in my head so that I cannot presume to say I came home sober . . . Upon taking the inventory of Mr. Vine's effects, I find not anything of value, neither do I think the whole of the household goods are worth above £60, and hardly any stock of value.

Good Friday, 20 Apr. . . . In the even walked down to Whyly, and from thence Mr. John French and myself walked up to Mr. John Vine's in order to settle the account between Mrs. French and Mr. John Vine. But it appearing a very intricate affair, we were obliged to postpone it till further consideration . . .

Sun. 22 Apr. In the morn Mr. Vine called on me, and we, together with Mr. Jos. Burges, walked to Lewes in order to settle Mr. Vine's affairs, when we called upon Mr. Baley and Mr. Michell to consult them, and we agreed to have all his effects turned over (by virtue of a letter of attorney or some other deed that Mr. Michell should think more proper) to Mr. Abraham Baley, Joseph Fuller and myself to sell and dispose of for the benefit of ourselves and his other creditors, and to distribute it among his other creditors in proportion to their respective debts[5] . . .

Mon. 23 Apr. . . . In the afternoon went down to Jones's, there being a public vestry held for making up the present overseers' accounts and choosing new officers, when Mr. Carman and I as overseers made up our accounts with the parish, and there was due to the parish £6 7s. 1d. The officers nominated for the ensuing year are: Thomas Turner, churchwarden, Edward Foord, electioner; myself again, overseer (as well as churchwarden), Thomas Carman, overseer; James Fuller and John Watford, electioners . . . We stayed till near 3 o'clock in the morn before we broke up and spent 10s. allowed out of the poor book and a halfpenny each . . .

Thurs. 26 Apr. In the morn I set out for Yalden to see my old acquaintance and cousin (and I ought to have reason as well to say worthy friend). I called at *The Hand and Sceptre* on the road to Tunbridge where I dined on some cold beef and baited my horse. I got to Yalden about 3.10. After I came there Mr. Hill and I took a walk and called and spent an hour or two with a gentleman of his acquaintance. We came back about 7.20 . . .

Fri. 27 Apr. After breakfast Mr. Hill and I set out for Maidstone where he gave me a bond for £60, dated today, bearing interest at £5 percent per annum, the same witnessed by Tho. Pope Jr. The above bond is in lieu of the £30 for which he gave me his note of hand the 5th day of November last, as also the £30 lent him the 15th of January last, both of which notes of hand I gave up to Mr. Hill. He paid me the interest, *viz.*, £1 2s. 6d. due on the said notes to this day. I bought of Mr. Tho. Pope Jr. a parcel of thread etc. amounting to £5 15s. 0d. for

---

[5] Abraham Baley's draft letter-book for the period 1763–73 survives (E.S.R.O., S.A.S. HA 313). It contains many letters concerning the attempts to settle John Vine's affairs.

which I paid him, and he is to send it to *The White Hart* Inn, Southwark, carriage paid. We baited ourselves and horses at Maidstone and came back to Yalden about 6.20. In the even we took a walk and came back and supped with Mr. Hill on some sausages in company with two gentlemen his acquaintances, the one a writing-master, the other a wheeler and philosopher. I stayed at Mr. Hill's all night. Spent today 13½d.

Sat. 28 Apr. In the morn, it being wet, I concluded to stay another day. In the forenoon I took a walk with Mr. Hill to visit a patient. Came back and dined with Mr. and Mrs. Hill on some pork bones boiled and greens. In the afternoon we took a walk to see Mereworth Place and church, the first the seat of the late Earls of Westmoreland, but now Lord Despencer's, and I think the seat as beautiful a little seat as I ever saw, there being a great deal of extreme good painting, some very fine marble and everything of ornament very noble. The church is modern-built and excessively handsome, but small. We came back to Yalden where I supped with Mr. Hill and stayed all night. Gave the servant who showed the Place 12*d.*, Mr. Hill paying the person who showed us the church. We spent about 7*d.* each at the public house.

Sun. 29 Apr. After breakfast I intending to come away, I gave Mr. Hill's servant 12*d.*, the man who cleaned my boots 6*d.*, and to my very great surprise, Mr. Hill permitted, nay almost forced, me to pay the expense of my horse at Yalden, which was 5*s.* Now this is what I could not have once thought would be so. I [have] always given Mr. Hill, when at Hoathly, both entertainment for himself and horse, though I might use the same excuse Mr. Hill did, of not having a stable of my own. And then I went entirely about Mr. Hill's business to Yalden, which was to talk to him about the money he owed me. Because had I wrote about it Mrs. Hill would in all likelihood have discovered it, and as it was unknown to her, so he desired it might remain a secret. Oh, ingratitude, ingratitude! thou common but hateful vice, a vice which in my opinion clouds all our other virtues, and I think no man guilty of it in a higher degree than Mr. Hill. Mr. Hill accompanied me on my road as far as Tunbridge where we dined on a calf's head boiled, a piece of bacon and greens, and then we parted, he for Yalden and I for home . . .

Mon. 30 Apr. . . . In the morn I rode to Maresfield (there being a sitting of justices) in company with Mr. Porter, where Mr. Carman endeavoured to appeal to our poor rate, but could not obtain any redress. I also verified the last year's account of the parish upon oath. I paid for a new warrant and summons 5*s.* . . . .

Tues. 1 May. In the morn went up to Mr. John Vine's, and after breakfast I rode to Lewes upon Mr. Vine's business. I dined at *The White Hart* in company with about twenty more upon a fillet of veal roasted, a ham boiled, a fore-quarter of lamb roasted, 2 hot pigeon pasties, 2 raisin and currant puddings, greens, potatoes and green salad. The reason of my dining on so elegant a dinner was on account of my having business with Mr. Baley, steward to the devisees and heirs of the Rt. Hon. Hen. Pelham Esq. deceased, who held an audit there today . . .

Weds. 2 May. . . . This day was fought a main of cocks at our public house between the gentlemen of East Grinstead and the gentlemen of East Hoathly for half a guinea a battle and 2 guineas the odd battle, which was won by the gentlemen of East Grinstead, they winning 5 battles out of six fought in the main. I saw three battles fought, but as I laid no bets, I could neither win nor lose, though I believe there was a great deal of money sported on both sides . . . This day Mr. John Vine signed a deed whereby he consigned over to Mr. Abra. Baley, myself and Mr. Joseph Fuller all his effects whatsoever both real and personal, in order to dispose and make the most of it we can for the benefit of ourselves and other creditors . . .

Fri. 4 May. In the morn I set out for Lewes in order to attend the sessions (as I did in a manner expect Mr. Carman would appeal to the poor rate) . . . Mr. Carman, agreeable to my expectation, lodged an appeal to the poor rate, and as it was a broken day, we stayed and heard the sessions trials. Came home about 10.10, very sober . . .

Sun. 13 May. Myself, Mr. Dodson and servant at church in the morn . . . During the time we were at church my friend Mr. Geo. Richardson and my brother came to see mê and they dined with me on a calf's heart in a pudding, a piece of beef boiled, greens and a green salad . . . During divine service Mr. Joseph Hartley came to bring me a new wig . . . Paid Mr. Joseph Hartley in cash 8*s.* 6½*d.* by goods 16*s.* 5½*d.* [total:] £1 5*s.* 0*d.* in full for a new wig received today £1 1*s.* 0*d.* and new mounting an old one 4*s.* 0*d.* . . .

Thurs. 17 May. . . . Very busy today in making the new land and window tax books, as also settling other accounts, that I was almost the whole day continually busy a-writing . . .

Fri. 18 May. . . . After breakfast I rode to Maresfield to meet the general receiver of the land tax, where I paid his deputy Mr. Tho. Gerry the following cash: 100 guineas, 10 half *do.*, 1 quarter *do.*, 2 shillings, [total:] £110 12*s.* 3*d.* which, with the salary allowed for collecting is in full for one year's window tax, and house duty,[6] and half a year's land tax, due from this parish at Lady Day last . . .

Sun. 20 May. . . . We had a vestry called, and we stayed in the churchyard to consult whether we should lend Francis Turner the sum of six guineas on the parish account, in order for him to discharge a debt with which he is threatened with an arrest if the same is not paid tomorrow; when it was the unanimous consent of all present to lend him the said sum . . . After churchtime Mr. Dodson and I walked down to ask Sam. Jenner how he did, with whom we stayed and drank tea and came home about 8.20. A very fine pleasant day, but when I consider the nature of my circumstances, that is no one person to whom I may entrust the management of my affairs, it almost drives to distraction.

Tues. 22 May. . . . This morn about 5 o'clock died Joseph Hutson aged 19 years, of a diabetes.

Thurs. 24 May. . . . Received of Mrs. Eliz. Browne by the payment of her servant Henry Godley 24s. in full for the same sum assessed on her by a poor rate made the 17th ult. (The above was not in cash, but a coffin, for which I received the money of the heirs of John Vernon).[7] Today there being a court-baron held at Whitesmith for this manor, I walked there in company with Mr. Joseph Burges, who was admitted to a house, lately erected in this manor by Mr. John Vine, but to which he was never admitted. However (upon the steward's promise that he should have it as his property) Mr. Burges bought it and as I observed before, was this day admitted to it for the first admittance. The day of payment agreed upon between Mr. Vine and Mr. Burges was Midsummer. Mr. Burges, not having got his money for payment, immediately after he was admitted surrendered it to me and my heirs (in the way of a mortgage), but upon condition he, the said Joseph Burges, pays the sum of £50 to me upon Midsummer Day next. Then the said mortgage is to be discharged, Mr. Burges paying all the court fees, except 2s. I paid, the agreement being for him to pay 30s. which was the whole charge (except my 2s.).[8] I dined at *The Chequer* at

[6] By 'house duty' Turner probably meant the fixed duty payable on all houses with fewer than ten windows. Houses with more than ten windows paid tax on them *pro rata* over and above the fixed sum. W. R. Ward, 'The administration of the window and assessed taxes, 1696–1798', *English Historical Review*, 67 (1952), pp. 522–46.

[7] Vernon had been the gardener at Halland and had died on 9 March. Mrs. Browne had evidently supplied the coffin, leaving Turner to extract from Vernon's heirs payment which would cover Mrs. Browne's poor rate.

[8] This was the public house now called *The King's Head*, though at this stage it was still referred to officially as 'all that new-erected messuage or tenement and garden . . . commonly called the Schoolhouse'. Until 1768, according to Turner's yearly notices of vestry meetings recorded in the East Hoathly overseers' account book (E.S.R.O., PAR

Whitesmith, the house where the court was held, on a piece of the
brisket of beef boiled, a leg of mutton boiled, a fore-quarter of lamb
roasted, 3 chickens roasted, 2 raisin and currant suet puddings, green
salad etc. in company with about 20 more (my family at home dining
on a piece of beef boiled, a suet pudding and greens). I stayed till about
5.20 when, being sent for to go to Mrs. Browne's, where she gave
me some further orders relating to the burying of her brother
tomorrow . . .

Fri. 25 May. . . . After dinner myself and Mr. Dodson went down to
the funeral house, I to attend the funeral, and he as an underbearer. I
gave away the following goods: . . . 1 mode[9] band, 4 Italian crepe *do.*, 1
pair men's white kid gloves, 2 pairs men's white lamb *do.*, 1 pair
women's *do.*, 4 pair women's black 2*d.* chamois *do.*, 4 pair men's *do.*, 1
pair maids' *do.*, 13 favours. We brought the corpse to church and
buried it about 8.20. Afterwards I and Tho. Durrant walked down
again to Mrs. Browne's to bring home the things left. Came home
about 9.40 . . .

Sat. 26 May. In the forenoon my brother Moses came over to
acquaint me of the death of Philip Turner, natural son of my half-
sister Eliz. Turner (the boy we had the care of, as also his maintenance
according to the will of my father). He lived with my brother in order
to learn the trade of a tailor, and died this morning about 5 o'clock of a
scarlet fever, aged 15 years . . .

Tues. 29 May. . . . In the afternoon there was played at Hawkhurst
Common in this parish a game of cricket between this parish and that
of Ringmer, but is was not played out, Ringmer having three wickets to
go out and 30 notches to get, so that in all probability had it been
played out it would have been decided in favour of Hoathly.[10] Mr.
Dodson played . . . At home all day and really very little to do; trade
excessively dull. In the afternoon busy a-putting my shop to rights.

Fri. 1 June. . . . In the even Mr. Banister and myself smoked a pipe
or two with Tho. Durrant, purely to keep Mr. Banister from
quarrelling, his wife big with child, lame of one hand and very much in
liquor, being out in the middle of the street at tennis among a parcel of
girls, boys etc. Oh, an odious sight, and that more so to a husband . . .

Sun. 3 June. Myself at church in the morn . . . We had a brief read

378/31/1/1), it was called *The Maypole*. Thenceforward he recorded it as *The King's
Head*. Burges subsequently sold it to James Fuller in 1768. Fuller sold it to Turner in
1772. The details are all to be found in the Laughton manor court books E.S.R.O.,
S.A.S. A 663–4.
    [9] A band made of alamode: a glossy, black silk.   [10] See above, p. 9, note 9.

for rebuilding the parish church of Sittingbourne in Kent, burnt down by the carelessness of workmen, the expense of rebuilding which, exclusive of the old materials and the parson's chancel, amounted to the sum of £2086 and upwards. The brief was to be gathered from house to house in the several counties of Kent, Surrey and Sussex.[11] My friend T. Tipper dined with me . . .

Mon. 4 June. . . . After dinner I rode to the sitting of the commissioners at Maresfield to get the land and window tax books signed and also the distress warrant for each . . . I also took out a distress warrant for poor tax[12] . . .

Fri. 8 June. In the morn went up to Mr. Vine's, where I attended a sale of part of his effects; I wrote and took the cash . . . My old acquaintance Mr. John Long (now an expectant[13] in the excise) called on me in the evening and took part of my bed. The goods sold of Mr. Vine's today and money received were as follows:

1 cart mare to Will. Goldsmith, not paid	£ 6 6s. 0d.
Geo. Siggs at Waldron: 1 turn-wrist plough,[14] paid	£ 1 0s. 0d.
Sam. Stace at Berwick: the grass in two fields till Michaelmas, paid	£ 2 2s. 0d.
Mr. Ansell Day: 1 cow, and 1 calf, 1 twelve-monthing bull and heifer, paid	£10 15s. 0d.
John Piper: 1 collar, paid	6d.
Robt. Cornwell at Framfield: 1 skive,[15] 1 piece of wanty,[16] paid	6d.
Thomas Osborne at the Dicker: 2 hogs, not paid	£ 2 6s. 0d.
John Lee at Ringmer: 1 cart, and hind horse harness, paid	£ 2 12s. 0d.
Mr. Tho. Parker at Hellingly: 1 hind harness, paid	13s. 6d.
Sam. Stace at Berwick: 1 horse, 2 harnesses, 1 round plough, paid	£ 7 0s. 0d.
Do. 1 hog, paid	£ 1 0s. 0d.
	£33 15s. 6d.

[11] For briefs see above, p. 158, note 37.
[12] A warrant signed by the justices allowing Turner to distrain on the goods of those not paying their taxes or rates.   [13] A candidate for office.
[14] Turnwrest plough: one on which the mould-board may be shifted from one side to the other at the end of each furrow, so that the furrow-slice is always thrown the same way.   [15] Probably a pared piece of leather.
[16] The rope used to fasten the pack or load on the back of a horse, or the girth under a draught-horse's belly which prevents a cart from tipping back.

Mr. Robt. Turner at Chiddingly was salesman, but he is not paid for his work . . .

Whit Sunday 10 June. . . . Myself at church in the morn . . . I stayed the communion; gave 6*d*. I dined on a shoulder of lamb roasted, and green salad. Myself and servant at church in the afternoon . . . Sam. Jenner and Tho. Durrant drank tea with me and after tea we, together with Joseph Fuller Jr, walked over the Chiddingly to see a house which was a-repairing there. We walked into Mr. Robert Turner's, where we stayed and smoked two serious pipes, and then came home about 9.40. As pleasant an evening as I ever walked in my life.

Thurs. 14 June. . . . Today I signed and sealed a bond wherein Jos. Fuller and myself were joint obligers in a bond of £200, the condition, the good behaviour of Mr. Long, now employed as an officer of excise . . .

Fri. 15 June. . . . About 11.30 I set out for Newhaven, where was to be a sale of foreign brandy at the custom house there. I dined at *The White Hart* at Newhaven (my friend Tipper not being at home) in company with 5 gentlemen (or at least other men) on a shoulder of mutton roasted, a plain hard pudding, a currant butter pudding cake, cabbage and lobster. The sale came on about 3.50, where there were put up for sale to the highest bidder 78 half-ankers[17] of brandy, 3 in a lot, which is 26 lots, each lot containing 13½ gallons of foreign brandy. Each lot was sold . . . After the sale was over I paid as follows, *viz.*,

Lot No. 1, 13½ gallons at 6*s*.	3*d*.	£ 4	4*s*.	4½*d*.
Do. 22	5*s*. 10*d*.	£ 3	18*s*.	9*d*.
Do. 26	5*s*. 9*d*.	£ 3	17*s*.	7½*d*.
		£12	0*s*.	9*d*.

I then rosined down my casks and spent about half an hour with my friend Tipper, who came to bring me going on my road home almost as far as Lewes. I came home about 9.20 very safe and sober . . .

Sun. 24 June. In the morn about 5.30 Tho. Durrant and I set out for Newhaven to see my very worthy friend Mr. Tipper. At Ripe we met Sam. Jenner, agreeable to appointment, and proceeded on our journey to Newhaven where we arrived about 7.50 and breakfasted with my friend Tipper, after which we walked down to the sea where we entertained ourselves very agreeably an hour or 2. We also had the pleasure to see a lunette battery[18] erected there to guard the entrance

---

[17] A measure or size of keg containing that amount. An anker was commonly in England 8⅓ gallons. By Turner's calculation here the half anker was 4½ gallons.

[18] A particularly shaped fortification consisting of two faces and two flanks.

of the harbour. It consists of give guns, 18-pounders, mounted and everything ready for action. There is a very neat house and magazine belonging to the fort and a gunner resident there. We dined with my friend Tipper on a leg of lamb boiled, a hot baked rice pudding, a gooseberry pie, and a very fine lobster, green salad and fine white cabbage . . . We stayed with my friend Tipper till about 4.30 and then came away, he accompanying us on our journey home as far as Lewes. Came home safe and well about 9.30. Spent today upon myself, horse, ostler, turnpike and ferry 9*d*.

Tues. 26 June. After breakfast I rode to Lewes, in company with Mr. Porter (this being the Visitation held by the Rev. Dr. D'Oyly, Archdeacon of the Deanery of Lewes). I paid the deputy collector of the brief money in cash £1 6*s*. 11*d*. in full for the money collected in this parish on the briefs (being 5) since the last Visitation.[19] I then went and attend church service at St. Michael's, where the Visitation sermon was preached by the Rev. Mr. Robert Austine, curate of St. Michael's, St. Mary and Ann Westout, as also master of the free grammar school at Lewes, from *Ephesians* 4.11, 12, 13 . . . I was then sworn into the office of churchwarden, which cost me 4*s*. 6*d*. and paid the apparitor[20] 12*d*. for a form of a thanksgiving prayer sent to our clergyman last year on the safe delivery of our most gracious Queen Charlotte of the Prince of Osnaburgh[21] . . .

Sat. 30 June. . . . After breakfast John French and I set out for Eastbourne. The reason of my journey was this: Mrs. French's wagon, with her son and servant was yesterday a-bringing a cord of wood to my house, and just at that instant of time as they was before my door, came by Mr. Sam. Beckett's postchaise and 4 horses in their road from Uckfield to Eastbourne (their home); and in driving a great pace and together with a sufficient degree of carelessness and audacity, they, in their passing the fore-horse in the team, in order to get into the road again before the other horses, drove against him and (I presume by accident) drove the shaft of the chaise into the rectum of the horse about 9 inches, and then it pierced through the gut into the body, of which wound the horse died in about 7 hours. Now as I see the accident, Mr. French desired I would go with him to talk with Mr. Beckett about it. We called at Mr. Fagg's in our journey, Mr. French wanting his advice, he being a justice of the peace, who soon informed

---

[19] For briefs see above, p. 158, note 37.

[20] The officer or messenger of the ecclesiastical court.

[21] Frederick Augustus was born on 16 August 1763, the second of Queen Charlotte's fifteen children. He became Duke of York and a field marshal.

him what he had been before told, that it was not justice business. We then continued our journey to Eastbourne, where we see Mr. Beckett, who behaved extremely civil and agreeable; Mr. French and he agreed to leave it to Mr. Fagg and Mr. Porter to appoint what he should pay for the damages etc. sustained. Mr. Beckett keeping a public house, we dined there on some pork and beans and a beef pudding . . . We came home very safe and sober about 7.30. Spent nothing today, Mr. French paying the whole expense of the journey . . . Mr. Beckett made me a present of a few mackerel. I bought of him 6 pieces of clear lawn, to be delivered to East Hoathly for £12.

Tues. 3 July. . . . In the morning I walked up to Mr. Vine's, where I witnessed a conveyance from Mr. Vine to the Rev. Mr. Thomas Porter of the house, stable, etc. now in the occupation of William Vinal . . .

Thurs. 5 July. . . . In the after[noon] drawed up the minutes of an agreement between John Goldsmith of Laughton and John Piper of this parish, wherein John Goldsmith has sold to John Piper his freehold and copyhold land in this parish, with all the hedgerows, woods, underwoods, shaws, timber trees, tillows,[22] barn, house and stable, with all appurtenances as it now stands, for £380, to be paid upon the conveyance of the freehold and admittance of the copyhold, which is to be at the court-baron for the manor of Laughton, held next after St. Michael next; John Piper to pay for all the writings and expenses necessary for conveying to him the freehold land and his part of the expenses of the admittance to the copyhold, and John Goldsmith the other part; and Goldsmith to leave £50 in Piper's hands to pay an annuity of £2 *per annum* till such time as the land shall be fully discharged from the said annuity; and Piper is to enter upon the said premises at St. Michael next, old style; Goldsmith to sign all writings whatsoever that shall be necessary to make a good and warrantable title to the said estate. Both of which parties signed the same and I as witness attested it and the agreement is in my hands. At home all day and but very little to do.

Mon. 9 July. . . . In the afternoon went to Master Durrant's hayfield, where I stayed and worked a-haying an hour or two . . . A violent hot day . . .

Weds. 11 July. . . . Mrs. Fuller, widow, buying some things in the shop in the morn, breakfasted with me, as did Miss Fanny in the afternoon, and drank tea with me. Received of Mrs. Highwood by Mr. George Banister 30s. in full for 1 cask of brandy. At home all day; busy

---

[22] For shaws and tillows see above, p. 221, note 5.

in my garden all the forenoon, and in the afternoon and even read part of Burnet's *History of the Reformation* which I esteem a very impartial history, as the author has everywhere treated his subject with moderation and coolness, which is in my opinion always a sign of learning and virtue.

Tues. 24 July. . . . In the even walked over to Mr. Joseph Burges's, there being an affair to be made up between my neighbour Mr. Banister and Jacob Parkes of Hellingly, for assault of Mr. Banister upon him, which was accordingly agreed. Came home about 9.25: spent nothing nor drank anything. Sam. Jenner was one of the arbitrators, who stayed and took part of my bed. At home all day and really very little to do in the shop. I think I never knew trade so dull in my life, or money so scarce.

Sun. 29 July. . . . In my absence today Mr. Simonds Blackman and his bride made their first appearance in church[23] . . .

Mon. 30 July. . . . About 2.30 we had a tempest of thunder and lightning, and a great deal of rain. The thunder had been heard all the morn from about 5.20. The storm here was not very severe, neither the thunder or lightning so sharp as I have often seen or heard, but it was excessive dark, and I dare say was very severe both to the southwest and southeast of this place. But it passed over here pretty light. I do not remember ever to have seen the heavens in so seeming a tempestuous situation as they was today, the whole element seeming in a commotion. It did not last a great while here. Sam. Jenner drank some coffee with me, it being a wager I lost . . .

Weds. 1 Aug. . . . A very wet evening. Sam. Jenner, coming in and the rain continuing, took part of my bed. But very little to do in the shop all day. In the evening instructed Samuel Jenner on the 'Branan Rules'.[24]

Fri. 3 Aug. . . . At home all day and pretty busy. In the even read part of the *London Magazine* for July, in which I find a great many excellent pieces, more than I ever remember to have seen in any one magazine. Perhaps I may be partial in my opinion, and only think them excellent as they agree with my own sentiments, for we are apt to be partial in our judgment of men and books as they agree and are similar to our own thoughts, few having so sound a judgment as to think and act impartial when their interests or sentiments are the topic.

Fri. 10 Aug. . . . This day Mr. Porter's house, now a-building, was

[23] Blackman was the new tenant of Whyly Farm previously occupied by Jeremiah French. The marriage was an interesting on see below, pp. 308–9.

[24] The meaning of this sentence is obscure.

reared, and in the evening there was an entertainment at Mr. John Vine's occasion, to which I went and came home about 11.20, but not quite sober.[25]

Sat. 11 Aug. . . . Very unpleasant and irksome to myself today; the punch taken in too great a quantity last night occasions my head to ache violently today. A very fine pleasant day.

Mon. 13 Aug. . . . Mr. Merrick, a grocer in Southwark, called on me today and dined with me on the remains of yesterday's dinner with some cucumbers, and stayed and smoked a pipe or two with me after dinner. At home all day and pretty busy. A great deal of rain fell in the afternoon. In the even wrote my London letters, and afterwards went down to Joseph Fuller's, where I stayed and spent the evening till near 1 o'clock in company with their family, Mr. Banister and Tho. Durrant. But I cannot say I came home sober. How do I lament my present irregular and to me very unpleasant way of life, for what I used to lead in my dear Peggy's time—then was home of all places the most agreeable, but now the most unpleasant and irksome. Then did I not know the want of a virtuous and pleasant companion, whose good sense was always pleasing and made life in that respect agreeable: but now, alas, I know not the comfort of an agreeable friend and virtuous fair. No, I have not spent hardly one agreeable hour in the company of a woman since I lost my wife, for really there seem very few whose education and way of thinking is agreeable and suitable with my own.

Mon. 20 Aug. . . . Mr. Banister's wife being delivered this morning, he took part of my bed, his own house being full . . .

Thurs. 23 Aug. . . . Mr. Banister dined with me on some hashed venison and after dinner we set out together for Lewes Races where His Majesty's purse of £100 was run for on Lewes Downs, when Sir John Moore's grey horse Cyclops and Mr. Bowles's horse Cyrus started for it, which was won by Cyclops, he winning the first heat (after which Cyrus [with]drew) so that we had what may be called bad sport, though I don't know I ever remember the King's Plate being run in less time, they performing it in 8 minutes and 15 seconds. Cyrus won the King's Plate at Salisbury, and Cyclops at Canterbury and Winton. At the latter he beat Cyrus before. Came home about 9.30, but happy should I be could I say sober. Oh, my unhappy—nay, I may say, unfortunate —disposition, that am so irresolute and cannot refrain from what my

---

[25] A rearing-feast was usually given to the workmen when the roof was 'reared' or put on a house. W. D. Parish, *A dictionary of Sussex dialect and collection of provincialisms in in the county of Sussex* (Lewes, 1875).

soul detests . . . Saw several London riders upon the Downs, with whom I drank a glass or two punch.

Fri. 24 Aug. . . . This day the five-year-old plate of £50 was run for on Lewes Downs, when Mr. Scott's horse Alphonso, Messrs. Scrase and Verral's mare Swishabout, Mr. Smith's horse Jack o' the Green and Mr. Bickham's mare Venus started for the same, which was won by the former, he winning the 2 first heats (easily). The stakes were strongly contested in a third heat between Swishabout and Jack o' the Green, and after the strictest scrutiny by the gentlemen on the stand it was given in favour of Swishabout by the length of her nose only. N.B.: The last three belonged to gentlemen etc. in and about Lewes.

Sat. 25 Aug. . . . My brother-in-law Sam. Slater called on me in his road to Lewes Races, dined with me on some mutton chops broiled, a cold lamb's heart and cucumbers, and at his desire after dinner I went with him to Lewes Races where the give-and-take plate of fifty pounds was run for on Lewes Downs. The 5 following started for the same *viz.*, Mr. Wildman's horse Gift, Mr. Sparrow's horse Venture, Lord Grosvenor's mare Slammerkin, Mr. Adams's horse Cleveland and Mr. Smith's mare Smiling Molly, which was won by Gift, he winning the 2nd and 3rd heats. Venture won the first, therefore was entitled to the stakes. Smiling Molly [with]drew after the first heat, as did Cleveland and Slammerkin after the second, so that the third heat was contested only by Venture and Gift, which was run in nine minutes and won very easy by Gift. Meeting with my friend Mr. Tipper on the hill, he and my brother Slater came home with me about 9.30, very sober . . .

Mon. 27 Aug. . . . After dinner I rode to Chiddingstone (my brother in my absence came over to stand the shop for me). I arrived at Chiddingstone parish about 7.30 and lodged at James Knight's, the person who uses the farm belonging to my brothers and myself.[26]

Tues. 28 Aug. . . . After breakfast Master Knight and myself walked to Chiddingstone Town, where there was a court held for the manor of Chiddingstone Burghurst, a manor in which part of our lands are situated. We dined at Mr. Whapham's at *The Castle*, where the court was adjourned to, on a buttock of beef boiled, a quarter of lamb roasted, carrots, cabbage, a currant pond butter pudding and a currant bread pudding. After dinner we proceeded to business, when myself

[26] The farm, called Buckhurst, had been bequeathed by Turner's father, John, to his mother subject to the payment from the income of various lump sums and annuities to children of his, John Turner's, two marriages. When Turner's mother died she left the property in common to her four children. Turner's quarter was charged with the payment of an annuity to his half-sister Elizabeth. The two wills are in E.S.R.O., South Malling wills register D 9, pp. 233 and 236.

and several more were sworn upon the homage[27] and I was admitted to a fourth part of the land, and paid the steward all quitrents etc. up to St. Michael last. We came back to Mr. Knight's about 7.20, where I supped on a piece of bacon, a piece of beef and plum puddings, carrots and cabbage. I stayed and took another night at Master Knight's. As this supper was my friend James Knight's harvest supper we drank rather too much cider.

Weds. 29 Aug. . . . After breakfast I set out for home, but after drinking too much elderberry wine . . . I rode several miles out of my road, and my surcingle breaking I lost my bags from under me, but however thank God I arrived at home about 6.30, and in the even after I came home my brother went home. My brother and servant dined today on a shoulder of lamb roasted and cucumbers. Gave our tenant's servant 12*d.* and his son a silk handkerchief. Oh, my unhappy temper, that a few glasses of liquor intoxicates me, and yet I am so stupidly foolish as not to absolutely refrain from drinking anything.

Thurs. 30 Aug. . . . Oh, the effect of intemperance—my head, my conscience!

Weds. 5 Sept. In the forenoon accompanied Mr. French and Mr. Porter to Hailsham, this being the day appointed for Mr. Fagg and Mr. Porter to settle the affair between Mr. Beckett and Mr. French relating to Mr. Beckett's post chariot killing Mr. French's horse the 29th June last, of which accident I was a spectator and therefore went as a witness; but however our journey was to no purpose, for the affair could not be settled, both parties being rather obstinate, and I am afraid Master Beckett's clan are hardly good principled and honest. We dined at *The George* in Hailsham on a forequarter of lamb roasted, a piece of beef roasted, a giblet and pigeon pasty, boiled and roasted chicken, ham, plum pudding, carrots, cabbage, French beans etc., in company with a great many people, there being a sitting of the justices for granting beer licences, as also a meeting of the commissioners of the land tax for hearing and determining all appeals to the land and window tax . . .

Thurs. 6 Sept. . . . At home all day and thank God pretty busy. This day came to Jones's a man with a cartload of millinery, mercery, linen drapery, silver etc. to keep a sale for two days. This must undoubtedly be some hurt to trade, for the novelty of the thing (and novelty is surely the predominant passion of the English nation, and of Sussex in particular) will catch the ignorant multitude, and perhaps not them

---

[27] The jury of tenants in a manorial court.

only, but people of sense who are not judges of goods and trade, as indeed very few are, but however as it is it must pass . . .

Sat. 15 Sept. . . . In the afternoon paid the following persons for about ¾ of a day picking the hops growing on Mr. Vine's land (on account of myself and the other assigns of the said Mr. Vine): Lucy Williams 6*d.*, Eliz. Akehurst 6*d.*, Dame Caine 6*d.*, Mrs. Tamkin 6*d.*, Dame Thomson 6*d.*, Dame Day 6*d.*, Dame Vinal 4*d.*, Dame Prall 6*d.*, Ann Prall 6*d.*, Molly Tamkin 2*d.*, Eliz. Burrage 6*d.*, Susan Swift 6*d.* . . .

Weds. 19 Sept. . . . Went up to Master Piper's, his hopping supper being tonight, where we supped on a forequarter of lamb roasted, a loin of lamb roasted, a mutton pie, plum pudding, carrots and cucumbers (in company I dare say with 20 people) . . .

Tues. 25 Sept. . . . This day were married at Laughton Church by the Rev. Mr. Herring, vicar of Chiddingly, my late servant Sally Waller and Mr. John Hastings, a gentleman who I think may justly be called a fortune expectant, for at present they have neither of them any fortune . . .

Fri. 28 Sept. Gave the Rev. Mr. Tho. Porter in cash £65 for . . . [3] bank bills . . . I then received of Mr. Porter in cash £30 in order to pay (in part) of a house and land I am to be admitted to at Whitesmith Court today for Mr. Coates.[28] Mr. Long came to see me today and stayed with my servant during my absence. About 12.30 Mr. Porter and I set out for Whitesmith, where there was a court-leet and baron held for this manor by Mr. William Michell, steward to His Grace the Duke of Newcastle, lord of the said manor. We dined at *The Chequer* on part of a rump of beef boiled, a leg of mutton boiled, a goose roasted, a giblet and raisin and currant suet pudding, turnips and potatoes (my servant at home dining on the remains of yesterday's dinner) in company with Mr. Abra. Baley, Mr. Nich. Gilbert, the steward and his clerk, Mr. John Goldsmith, Joseph and James Fuller, Mr. Carman, Mr. Peters, Mr. Newington, Mr. Carman and Mr. Sam. Gibbs; and I dare say as many dined in another room. Paid Mr. Ben. Shelley, who called on me at the court, in cash ten pounds in full for the same sum he paid Mr. John Crouch the 26th instant for me. After dinner came James Marchant and Eliz. his wife, James Thorpe and

[28] The activities over the next two days show how complicated the transfer of copyhold premises using the machinery of the manorial court could be, especially when, as in this case, three married co-heiresses wished to sell their interest in a heavily mortgaged property to a third party. In this transaction Turner acted as 'attorney' for the buyer.

Martha his wife, and William Williams and Lucy his wife, the women being sisters and co-heiresses of Thomas Mepham, late of East Hoathly, and supposed by many corroborating circumstances to be dead, and died possessed of a house and about 5 roods of land (in two tenements) situate near the church in East Hoathly, to which the said sisters was his heirs, but instead of being admitted (as Mr. Porter had a deep mortgage thereon) they in open court resigned all their right and title thereto into the hands of the lord of the said manor by his steward, upon which he admitted Mr. Porter to the same, who claimed the same upon virtue of his mortgage. Then Mr. Porter resigned the same into the hands of the lord of the manor by his steward, and I was admitted to the same for Mr. Coates, as his attorney. I paid the steward's fees for fines, admittances, examining the women privately, surrenders etc. £12; the beadle 4s.; and to the homage 2s.: in all £12 6s. 0d. I paid James Thorpe and Martha his wife £10, and William Williams and Lucy his wife £10, the sum they agreed to take to release all their right thereto to Mr. Porter, in order to prevent its coming into six tenements, it being now no more than two. We came away about 8.20, but I cannot say thoroughly sober. Spent only 12d. today for my dinner, though I think either Mr. Coates or John Piper should repay me that, as I went purely to transact Mr. Coates's business and to assist John Piper to examine some writings, which were to have been brought to court today but were not. Mr. Long went away about 6.20. A very cold day, but otherwise a very fine day.

Sat. 29 Sept. Received of Mr. Coates by the payment of Mr. Porter in cash £7 4s. 0d. in full as under, *viz.*,

Mr. Coates Dr.

To the house etc. free of all expenses	£70	0s. 0d.
Per contra Cr.		
By money received of Mr. Porter yesterday	£30	0s. 0d.
By a mortgage due to Mr. Porter	£30	0s. 0d.
By almost two years' interest, it being due at 2nd October next, on the mortgage at 4½ percent	£ 2	13s. 6d.
By money left in Mr. Coates's hands to pay the arrears of quitrent and if there is not so much due the overplus to be returned to James Marchant		2s. 6d.
By cash received today	£ 7	0s. 0d.
	£70	0s. 0d.

Paid James Marchant in cash £4 18s. 0d. in full for his and his wife's
share of the money due out of the house, as they had before agreed
that Mr. Coates should have the house for £70, and that each of the
other sisters should have ten pounds apiece clear, and the rent due
they would outset against the interest and the fees etc. if it exceeded
the ten pounds; if not, they would have been the gainer, provided it had
come under that sum; *viz.*,

> Myself Dr.
>
> To money the house was sold for and for which
>   I was to account   £70   0s. 0d.
>
> Per contra Cr.
>
> By money paid James Thorpe and Martha his
>   wife yesterday   £10   0s. 0d.
>
> By money paid to William Williams and Lucy
>   his wife yesterday   £10   0s. 0d.
>
> By Mr. Porter's mortgage   £30   0s. 0d.
>
> By money due on the mortgage to Mr. Porter
>   for interest   £ 2 13s. 6d.
>
> By money paid at the court yesterday for
>   fines, admittances etc.   £12   6s. 0d.
>
> By money left in Mr. Coates's hands to pay
>   the arrears of quitrents   2s. 6d.
>
> By cash paid him today   £ 4 18s. 0d.
>
>   £70   0s. 0d.

My old and worthy friend Mr. Fran. Elless came to see me in the
forenoon . . . Mr. Elless stayed and drank tea with me and in the even
we walked down to Whyly and spent the even with Mr. Blackman.
Came home very sober about 10.30. My friend stayed all night. Very
busy all day. I think my friend Mr. Elless is as agreeable [a] companion
as any among my acquaintance, he being sober and virtuous, and a
man of a great deal of good sense, and endued with good nature. He
has improved his natural parts with a great deal of useful learning.

Sun. 30 Sept. . . . After dinner Tho. Durrant and I set out for
Lewes, I to consult Mr. Henry Burtenshaw, attorney at law, about the
following business (and he for the company), *viz.*, There has been time
immemorial given at Halland at the expense of his present Grace the
Duke of Newcastle and his ancestors to this parish and Laughton a
certain quantity of bread and beer to eight poor persons in this parish
and I think to 12 in the parish of Laughton. The quantity of beer to
each person in both parishes was 3 gallons every Thursday and

Sunday, and a bushel wheat made into bread was distributed every Sunday between the said poor in both parishes, though it was not divided equal, some having a claim as supposed to double the quantity another had, and as such it was distributed. Now, the Duke having lately put in a new steward who is famed for economy and frugality[29] (though I should rather think it deserved the name of niggardliness and done to gain self-applause); but be that as it will, he has given both parishes notice that no more shall be distributed after old St. Michael. Now, as the parishes will sustain a loss upon the whole of £50 annually, they seemed determined to see the end of it, and accordingly we have agreed to pay the charge between both parishes according to their respective quotas, and myself to transact the business. Though at first sight it should appear we have an undoubted right to it, as it has been continued without any alteration that I ever could find in the same manner it now is (and has been looked upon as our right) for time immemorial, and I should think we could trace it back upwards of hundred years, yet, as we have nothing to show for it of any deed or writing, we can have no other claim than a prescriptive right, which, considering the greatness of the person we have to deal with, might prove too great an undertaking for us to be crowned with success. Therefore, I thought proper first to be certain of our right by examining at Doctors' Commons[30] the wills of the ancestors of the Pelham family, and if we are successful in our attempt, then to contest it to the utmost, and I accordingly spoke to Mr. Burtenshaw about searching them . . .

Tues. 9 Oct. After breakfast Mr. Verral and I went up to Mr. Vine's and sold his furniture or at least part of it, there being a public sale, and Mr. Verral was salesman and I wrote and took the money. We went all day without dinner, and came home about 8.20. Mr. Verral supped with me on a cold shoulder of mutton. Sam. Jenner took part of my bed. I paid Mr. Verral 14*s.* for his day's work . . .

Fri. 19 Oct. After breakfast I walked up to Mr. Vine's, in order for the last appraisement of his stock etc., Mr. Robt. Turner of Chiddingly appraising for the assigns and Mr. King of Alfriston for Mr. Stace . . . We stayed there spending words in vain, for we did no business till about 4.20. The reason of our doing no business seems to proceed

---

[29] Abraham Baley had replaced Christopher Coates in September 1764. See below Appendix B, p. 329.
[30] A society of Doctors of Civil Law formed in 1509 which lasted until 1858. Its building, also known as Doctors' Commons, housed five courts one of which dealt with testamentary matters: hence the need to search the records there for the Pelham will.

from a desire in Mr. King and Mr. Stace to over-reach and cozen Mr. Turner our appraiser in every article, for whatever price Mr. Turner asked Mr. King would immediately endeavour to beat him down, though he had made a price perhaps below the real value. Upon the whole I am thoroughly convinced that neither Mr. Stace or Mr. King are men of any honour, and I am afraid as little honesty . . .

Thurs. 25 Oct. . . . In the even went over to Mr. Burges's in order to treat once more with Mr. Stace about Mr. Vine's affairs, and we at last came to the following agreement: that he should take the household furniture at Mr. Geo. Verral's appraisal, and that we would take upon ourselves to dispose of the timber, carriage, two wagons, two carts and one plough. Myself, Mr. Grigson, Mr. Stace and his son stayed and spent the evening very amicably till past one . . .

Thurs. 1 Nov. . . . I this day heard of the melancholy news of the death of my old acquaintance and worthy friend Mr. John Long, who died a-Monday last of the smallpox under inoculation; a very sober and worthy young man, but from a bad constitution had the smallpox excessively full, which proved mortal.

Sat. 10 Nov. . . . In the afternoon walked down to the house of Francis Turner, where the coroner for the Eastern part of this county took an inquisition on the body of Matthew Lewer, a poor man and lath cleaver, who was last night found dead in a wood in this parish where he had been at work the whole day a-lath cleaving in company with Matthew Lewer his nephew and John the son of his nephew, John Hook and John Lavender. The jury was myself as foreman, George Banister, Robt. Hook, Tho. Durrant, John Piper, Thomas Bristow, Edwd. and Richd. Hope, Henry Godley, Stephen Weaver, Francis Gibbs, Will. Henly and Sam. Washer. The verdict we gave was that the person died a natural death by the visitation of God, for upon the examination of Matthew Lewer, John Hook and John Lavender it appeared the deceased had been ill with a cold for some days, and then even complained more particularly and left his work half an hour before any of the rest of them, on account of his being so ill. He was found dead in about an hour by the people at whose house he lodged, who went in search of him, he not coming home so soon as usual, and being a pretty old man. Upon the examination of Fra. and Isaac Turner, the persons who found him, they declared that there did not seem any appearance upon the ground where he lay of more people's footing or that there had been a scuffle with him and any other person. Nor upon the jury's examining the dead body could we find any cut, bruise, stab or any other appearance of hurt, so that I should presume and hope to all probable appearance he died a natural death, agreeable to the jury's verdict . . .

Sun. 11 Nov. My brother, sister and myself at church in the morn . . . During the time they was singing psalms, James Hutson, head-borough, and myself went out and searched the alehouses and the barber's shop. The barber we found exercising his trade, but as it was the first time, we forgave him. The alehouses was clear of tipplers[31] . . .

Tues. 13 Nov. . . . In the even went down to Jones's, there being a public vestry in order to consult what was best to be done relating to John Jones's affair, which was this: Mr. Chester of Lewes having a bill of sale of Jones's goods to the amount of £61 and having been lately obliged to pay £11 for arrears of rent due from Jones (to prevent the landlord's distraining, which would have been attended with more expense, and must have been paid from the effects), now Mr. Chester, having lately advanced the above sum, was desirous there might be a vestry called, that he might attend it to know if the parish would assist Jones with money to pay a part of it. Upon debating the thing, it was the unanimous opinion of all present (who were myself, Mr. Carman, Mr. Blackman, James Fuller, Richd. Hope, John Cayley and Richd. Page) not to yield him any assistance, but for Chester to put his bill of sale in execution when he pleased. I next represented to the vestry that there was due from Tho. Tester to myself and the other executors and devisees in trust of Mr. Will. Piper 2 years' and half rent, which the man had informed us he was unable to pay, and begged they would take the same into consideration whether we should distrain the goods or the parish to pay part or all the rent, when it was determined by 5 voices to 1 (myself as concerned on the other party giving my voice neither way) that we should distrain the goods. We stayed till about 9.30 and then broke up. Spent on the parish account 4s. 3d. A very cold day and not very busy in the shop. I think, of all the company I ever spent the evening with in my life, that of James Fuller is the most disagreeable, he being stupidly ignorant and withal prodigiously abusive.

Weds. 14 Nov. . . . This day was married at our church Mr. Simonds Blackman and Mary, his wife (*alias* Mary Margerison), the reason of which was, she being under age, some months ago they went into Flanders and were married at a place called Ypres, but as this marriage was not in all respects agreeable to the laws of England in

---

[31] For the parish officers' duty to prevent tippling see above p. 38, note 14. The barber was presumably caught under the statute preventing 'persons exercising their worldly calling on the Lord's Day'. *Burn*, p. 131.

regard to their issue enjoying the gentleman's estate without a possibility of a great deal of trouble to prove the fact of their marriage, the young lady's father in consideration of this has given his consent for this second marriage (she not being now 19 years of age) and they was agreeably married this day by a licence which styled her 'Mary Margerison, otherwise Blackman.' In my own private opinion I think instead of making laws to restrain marriage it would be more to the advantage of the nation in general to give encouragement to it, for by that means a great deal of debauchery would in all probability be prevented and a greater increase of people might be the consequence, which I presume would be one real benefit to the nation, and I think it is the first command of the parent and governor of the universe: Increase and multiply. And the observation of St. Paul is that marriage is honourable in all men.

Fri. 16 Nov. . . . In the afternoon my sister went to pay Mrs. Carman a visit, and about 6.30 Tho. Durrant and I went down to accompany her home. We together with my sister and Bett. Fuller stayed and supped at Mr. Carman's in company with their own family . . . We played some time at a game called 'Laugh and lay down';[32] I lost 7*d.* at it . . .

Fri. 23 Nov. . . . At home all day and thank God very busy . . . A very cold day. Oh, how pleasant was the even spent after a busy day in my dear Peggy's time, but now it's all unpleasant, nothing to soothe the anxious mind, no pleasing companion, no sincere friend, nor agreeable acquaintance, or at least amongst the fair sex.

Sat. 8 Dec. . . . At home all day and very busy. A very wet afternoon. In the even Joseph Fuller Jr. sat with me some time, as did also Tho. Durrant. Oh, how low and depressed are my spirits, quite low indeed, almost to a stupid dullness!

Sun. 9 Dec. . . . After dinner Sam. Jenner and I walked to Lewes, I in order to see a girl which I have long since had thoughts of paying my addresses to, and he for company. I was not so happy shall I say to see her, or was I unfortunate in having only my walk for my pains, which perhaps was as well . . .

Weds. 19 Dec. . . . At home all day; a very cold day. In the afternoon my brother Moses and cousin Tho. Ovendean came to see me and stayed and spent the even and supped with me on the remains of

[32] A card game in which each player has six cards and six are dealt on the table; each player in turn tries to make a pair or fifteen with two cards, either with both in his hand or with one on the table. When a player can no longer make such a combination, he lays down his remaining cards and is said to 'laugh and lay down'.

today's dinner and bread and cheese and apple pie. They stayed till about 9.40 and then went away. I think I never saw a more stupid young fellow in my life than my cousin Thomas Ovendean; his discourse is one continued flow of oaths almost without any intermission . . .

Fri. 21 Dec.   Got up early in the morn and posted part of my day book before I opened the shop. This being the day on which the poor in this parish go round the parish to ask for something against Christmas, I relieved [40 paupers] with one penny each and also a draught of beer . . .

Mon. 24 Dec.   . . . At home all day and very busy. Gave the following boys . . . for box money: John Gasson, apprentice to Joseph Durrant, 6*d.*, John Jenner, the hatter's son, 6*d.* Mr. Banister, our officer of excise, having lately made a seizure of some brandy, brought in 2 bottles of it to my house, and myself, Sam. Jenner, Tho. Durrant and Joseph Fuller Jr. clubbed[33] for lemons and sugar, and we had an agreeable bowl of punch in the even, and spent the even till near 12 o'clock . . .

Tues. 25 Dec.   . . . This being Christmas Day myself at church in the morn . . . James Marchant and his wife and Joseph Burges dined with me on a bullock's tongue boiled, a piece of beef roasted, a raisin suet pudding, turnips and potatoes; and they, together with Tho. Durrant and Joseph Fuller Jr. drank tea with me . . .

Weds. 26 Dec.   . . . In the even went to Mr. Burges's, there being a vestry to choose new surveyors, when the old ones made up their accounts and the new ones chose. They were Mr. Tho. Carman and myself, electioners, Jas. Hutson Jr., and Geo. Stace . . . Gave Simon Diggens, apprentice to James Marchant, 6*d.* to his box; gave John Fuller, the chandler's nephew, 6*d.* to his box . . .

Thurs. 27 Dec.   . . . Gave Robt. Hook, son to Robt. Hook, shoemaker, 3*d.* to his box. Gave the post boy 6*d.* for his box . . .

1765

Fri. 4 Jan.   . . . After breakfast, in company with Mr. Wm. Jenner at Chiddingly, set out for Maresfield to meet the general receiver of the land tax for this county in order to pay in the two first quarterly

---

[33] Shared the cost.

payments of the land tax . . . I paid Mr. Tho. Gerry, the deputy Receiver-General, in cash £76 in full for half a year's land tax . . .

Sat. 5 Jan. Immediately after breakfast I walked up to Mrs. Piper's, where I and Arthur Knight against Sam. Durrant and John Piper measured for John Piper 79 oak trees, containing 9 loads, 23 feet . . . Gave John Russell cash for the following bill received of him today:[1]

Frant, 2nd January 1765

Sir,

£24 17s. 10d.

Twenty-five days after date pay to Mr. John Russell or order twenty-four pounds, seventeen shillings and ten pence value received in plank etc., as by advice from

Your humble servant
Ed. Budgen

To Mr. John Corke,
    Timber Merchant,
    at Will's Coffee House,
    Cornhill, London.

For exchanging this he gave me half a crown . . .

Fri. 11 Jan. . . . In the day wrote out Mr. Coates's bill, and in the evening went to Mr. Coates's to carry them in, and accordingly delivered in four bills: one for bedding, one for the boy's clothes, one for nails etc. used for repairing the gatehouse, and the other the running account for the house, the aggregate sum of which four bills was £26 17s. 7¾d.

Sat. 12 Jan. Sent Mr. Robt. Plumer, grocer in the Cliffe near Lewes, the bill value £24 17s. 10d. I received of John Russell the 5th instant, which bill I sent to Mr. Plumer in a piece of paper not sealed, by Charles Vine . . . Received back in the even by Charles Vine a receipt from Mr. Plumer that the bills sent him today came safe to hand . . .

Tues. 15 Jan. . . . In the afternoon wrote out several bills. In the even Joseph Fuller Jr, having an inclination to give me and some more of his most intimate friends (or at least acquaintances) a little treat, and not having the opportunity to do it at his father's, he brought some brandy etc. to my house and he, my brother, and Tho. Durrant and myself spent the evening till near 11 o'clock, but I have not the pleasure to say I was quite sober.

---

[1] For an explanation of the bill system see Appendix C.

Fri. 25 Jan. My brother[2] walked with Mr. Banister his round today. I dined on the remains of yesterday's dinner. Dame and Tho. Durrant drank tea with me. My brother came about 6.30, very much in liquor and having spent every farthing of money in his pocket, and, having fallen down, had the misfortune to dislocate the great bone of the arm, but his companion Mr. Geo. Banister had the presence of mind immediately to pull it into place again. So that of a misfortune I think this may be deemed good fortune. At home all day. The trouble I have with my relations and my own folly makes me quite insane.

Sat. 26 Jan. My brother's arm excessive bad. Mr. Stone paid him a visit . . . I walked over to T. Davy's, from whence I went to church and stood sponsor for his infant daughter . . .

Mon. 28 Jan. In the morn wrote my London letters. I dined at Mr. Porter's in company with himself, wife and two sons, Mrs. Atkins, John and Mrs. French, Mrs. Fuller, and Mr. and Mrs. Walls on a fine dish of carp, a green neat's tongue and turnips, a rump of beef à la daub,[3] a hot chicken pasty, a roast turkey and bread puddings. My family at home dined on the remains of yesterday's dinner. We stayed and spent the afternoon there, and drank tea and coffee, and in the even played at cards, Mrs. Walls and myself playing at whist against Mrs. French and Mrs. Atkins, but we neither won nor lost. We also stayed and supped with Mr. Porter on 2 chickens boiled, 2 ducks roasted, the remains of the tongue and pasty left at dinner and gooseberry tart. We stayed till about 11.15 and came away sober. I gave Mrs. Porter's servant one shilling.

Thurs. 31 Jan. . . . Sent Mr. Richard Stone, surgeon and apothecary at Blackboys, by his servant Christopher Vine, in cash £18 18s. 6d., and 1s. 6d. I kept back for exchange makes £19, which is in full for [a] bill received from him today . . . At home all day and pretty busy . . .

Tues. 5 Feb. . . . Sent Mr. Will Margesson enclosed in a letter by Ben. Shelley delivered to his servant the bill value £19 received from Mr. Stone the 31st ult. . . .

Weds. 6 Feb. Sent Mr. Richard Stone . . . in cash £9 19s. 0d. and one shilling I kept back for exchange makes together the sum of £10 and is in full for [a] bill I received by [his] servant today . . .

Fri. 8 Feb. Exchanged the following draft for Mr. Richard Stone . . . for the doing of which he gave me 6s. 6d.:

[2] Richard Turner.
[3] The meaning of this description is obscure.

£66 0s. 0d.

Lewes   8 Feb. 1765

Gents.,

Fourteen days after date please pay to Mr. Richard Stone or order sixty six pound for value received by

Your humble servant
John Burtenshaw

To Messrs Cardin and Day
 Merchants in
 Bucklebury
 London

Endorsed by the said Richard Stone

Sat. 9 Feb. ... I dined on a sausage batter pudding baked (which is this: a little flour and milk beat up into a batter with an egg and some salt and a few sausages cut in pieces and put in it and then baked) ...

Tues. 12 Feb. Sent Mr. Will. Margesson . . . the bill value £10 I received of Mr. Richard Stone the 6th instant on Mr. William Baldwin. Also the bill value £66 I received of him the 8th instant on Messrs Cardin and Day . . . Exchanged [a] bill [of £8 7s. 6d.] for Mr. Richard Stone . . .

Sat. 16 Feb. ... This day received by the carrier enclosed in a letter from Mr. Will. Margesson the bill value £19 I sent him the 5th instant, as also the bill value £10 sent him the 12th instant, both drawn by Stone on Baldwin and both returned for non-payment. My brother went to Mr. Stone's to talk with him about it, but he was not at home ... This affair of Stone's bill makes me very uneasy lest the whole of his bills should be returned, which amounts to £103 7s. 6d.

Sun. 17 Feb. In the morn my brother went again to Mr. Stone's, whose answer was that Mr. Will. Baldwin (the person he drawed on) had effects in his hands and that he would wait on me tomorrow ...

Mon. 18 Feb. This day received of Mr. Richd. Stone the following bill, instead of the bill value £19 received of him the 31st ult., the bill value £10 received of him the 6th Feb., and also the bill value £8 7s. 6d. received of him the 12th instant, all of which bills together amount to the sum of £37 7s. 6d.

£37 7s. 6d.

Sir: Framfield   Feb. 18, 1765.

Five days after date pay Mr. Tho. Turner or order thirty-seven pounds seven shillings and sixpence value received, and place it to the account of, Sir,

Your most obedient servant
Richard Stone

To
Mr. Will. Baldwin,
Southwark

Tues. 19 Feb. . . . Sent Mr. Will. Margesson enclosed in a letter as above the bill value £37 7s. 6d. I received yesterday of Mr. Richard Stone . . . I dined at Mr. Coates's on a shoulder of mutton roasted and pancakes and fritters. I stayed and chatted after dinner with Mr. Coates some time, my family at home dining on the remains of yesterday's dinner. Bett. Fuller and Tho. Durrant drank tea with me. At home all the even and very cold. Mr. Coates showed me today an original letter wrote from one of the Pelham family to another of the same in the year 1620. The writing was a very neat pretty hand, and the spelling much the same as we use now, and the colour of the ink hardly altered or changed in the least. And I think a prettier letter could not be wrote.

Ash Wednesday, 20 Feb. After breakfast I at the request of Mr. Joseph Burges walked with him to Lewes where we dined at *The White Hart* on a hog's haslet baked . . . The reason of my journey today was this: About four years ago Mr. Porter bought of Mr. Burges a farm. The house he has now just taken down, in doing of which a bricklayer or bricklayer's servant, in digging up the foundation, found several pieces (about 4) of old gold coin, of which one was a piece called a Jacobus (which I bought the 14th instant for 20s. and used to pass for 25s.), [and] some few pieces of silver, which I think is all that I have heard of being found. Now Mr. Porter, as proprietor of the premises (and I doubt spurred on by self-interest, a vice very predominant in the breast of too many of us), claimed the same, but however upon more mature deliberation and persuasion he has been brought over to think it belonged to the lord of the franchise (or manor), as undoubtedly it does by the common law. Now after Mr. Porter is cool in his claim and has freely given up the thoughts of any right, it appears that about 37 years ago the father of the present Mr. Burges (who then lived in the house) was robbed of several such gold pieces of gold and silver, several gold rings and about £5 in crown pieces, none of which could ever be found or heard anything of to this day, notwithstanding several people were at the time of the robbery taken up on suspicion. Therefore it is conjectured the money now found is in all probability a part of that which was then taken, and according to many circumstances it appears to be so. Mr. Burges therefore applied to a justice in order to try if he could obtain any of this treasure trove, but, alas, all in vain. As there could be no oath made to anything that has hitherto been found, no warrant could be granted. But however, Mr. Shelley, one of His Majesty's justices of the peace, did grant a summons to have the man examined, a good-natured action indeed, but what Mr. Justice had no business to grant. For I assure him it was common law business, and His Worship had no business with it. Mr. Burges paid all expenses and thanked me for my company; so I sped well enough. Sam. Jenner, coming in the evening, and being very much in liquor, stayed all night.

Thurs. 21 Feb. Sam. Jenner assisted me in bottling out a cask of raisin wine in the forenoon . . .

Sat. 23 Feb. I received a letter by today's post from Mr. Will. Margesson acquainting me the bill value £37 7s. 6d. sent him the 19th instant on Baldwin was refused acceptance . . . About 11.20 I set out for Lewes in order to see Mr. Stone about the bill on Baldwin; who repaid the money (a good job) . . . I stayed with my brother Moses, who was also in town, till I was really quite in liquor and took up my lodging at *The Cats.* I abominate and loathe myself for not having a more firm resolution, as a very trifle makes me drunk. Therefore I am determined once for all to refrain from drinking.

Thurs. 28 Feb. . . . In the day we had a great deal of snow fell, and had there been any frost for it to have lain on, it would have been an extreme deep snow, it continuing snowing almost the whole day. Received by the post today enclosed in a letter from Mr. Will. Margesson the bill value £37 7s. 6d. sent him the 19th instant, which was returned for non-payment.

Mon. 4 Mar. After breakfast Mr. Porter and myself walked round the parish to collect a brief which he read the 24th ult. for the loss by the hailstorm which happened in Sussex, Aug. 19, 1763. We collected £1 6s. 1½d. but not meeting all at home, we could not complete the collection[4] . . .

Tues. 12 Mar. . . . In the afternoon my brother set off upon a wild ramble. His first place was Chiddingly, then Laughton Pound where Master Hook, whom I sent in pursuit of him, found him and brought him home about 1.45. My brother Moses came over in the even, I having sent to him on the account of Richard's ramble. He stayed with me till past 12 and then went home. A prodigious wet even . . .

Weds. 13 Mar. . . . My brother quite ashamed of his last night's behaviour; kept his bed all day, and may the Supreme Being give and endue him with his preventing grace, strength and resolution of mind enough never to be guilty of drunkenness again. Paid Robert Hook 5s. 3d. in full for the same sum he paid, and for his trouble a-going after my brother yesterday.

[4] For briefs see p. 158, note 37. The severity of the storms in south east England on 19 August 1763 are reported in *The Gentleman's Magazine*, XXXIII (1763), pp. 411–12. At Brighton 'the oldest fishermen say, nothing had been seen like it in their memories'.

Mon. 18 Mar. In the morn walked down to Mr. Porter's where we signed the brief which he read the 24th ult. upon which we had collected upon the whole the sum of £1 12s. 0d., *viz.*,

	£	s.	d.		£	s.	d.
The Rev. Mr. Porter	5	0		Sarah Bridger	0	3	
Mary Overing	0	3		Ann Gear	0	1½	
Eliz. Akehurst	0	2		Mr. John French	0	6	
Mr. Tho. Walls	1	0		Mrs. Eliz. French, widow	0	6	
Mr. Simonds Blackman	2	6		Miss Molly French	0	6	
Mary Cottington	0	1		Eliz. Stredwick	0	3	
Mrs. Mary Piper, widow	0	6		Mary Carpenter	0	3	
Mr. James Fuller	0	6		Charles Vine	0	3	
Tho. Olliver	0	2		Mrs. Eliz. Fuller, widow	0	2	
Edward Hope	0	2		Thomas Reeve	0	6	
John Cayley	0	6		John Nutley	1	0	
Richard Hope	0	6		George Stace	0	3	
Samuel Harmer	0	3		William Gasson	0	3	
Edward Foord	0	6		Mr. Coates	1	0	
Henry Bray	0	3		Ann Nicholass	0	3	
Mary Wheeler	0	3		Joseph Fuller	1	0	
John Durrant	0	6		Thomas Durrant	0	6	
Joseph Burges	0	6		Robt. Hook	0	6	
Thomas Hervey	0	3		Thomas Davy	0	6	
Mrs. Hannah Atkins	1	6		Mrs. Eliz. Hicks	1	0	
Mrs. Eliz. Browne	0	6		James Hutson	0	6	
William Alcorn	0	3		William Henly	0	2	
William Burrage	0	2		John Gasson	0	1	
Henry Godley	0	6		John Hutson	0	3	
John Piper	0	6		Richard, John and Anne Page, widow	1	0	
Thomas Carman	1	0		John Watford	0	6	
Ellen Pierce	0	3		Hannah Marchant	0	2	
Mr. George Banister	0	1		Myself	1	2½	

. . .

Thurs. 21 Mar. . . . In the even went to the vestry at Mr. Burges's, where we made a poor rate at the rate of two shillings in the pound . . . We spent on the parish account 8s. 3d., and 6d. each of our own money, and all parted very near quite sober . . .

Fri. 22 Mar. . . . I dined on some salt fish, egg sauce, parsnips and potatoes. At home all the afternoon and very little to do. In the even read part of Homer's *Odyssey*, translated by Pope, which I like very well, the language being vastly good and the turn of thought and expression beautiful . . .

Sat. 23 Mar. . . . Charles Diggens coming over to take up a suit of clothes for Mr. Porter, he stayed and dined with me on the remains of yesterday's, Thursday's and Wednesday's dinner . . .

Sun. 24 Mar. . . . After tea my brother Richd. and I took a walk (Molly Hicks, my favourite girl, being come to pay Mrs. Atkins a visit in the even, went home to her father's, and I along with her, my brother going with her companion for company.) We came back about 8.10. This is a girl which I have taken a great liking to, she seeming to all appearances to be a girl endued with a great deal of good nature and good sense, and withal so far as has hitherto come to my knowledge is very discreet and prudent.

Thurs. 28 Mar. . . . Dame Akehurst, a-washing for me all day, dined with me on a loin of mutton boiled and turnips. In the afternoon rode over to Chiddingly to pay my charmer, or intended wife or sweetheart or whatever other name may be more proper, a visit at her father's, where I drank tea in company with their family and Miss Ann Thatcher. In the even we played at loo[5] in company with their own family and Nanny and Thos. Thatcher; I won 1*d*. I also supped there on some rashers of bacon. It being an excessive wet and windy night I had the opportunity—sure I should say the pleasure, or perhaps some might say the unspeakable happiness—to sit up with Molly Hicks, or my charmer, all night. I came home about 5.40 in the morn, I must not say fatigued—No! No! that could not be. It could be only a little sleepy for want of rest. Well to be sure she is a most clever girl, but, however, to be serious in the affair I certainly esteem the girl and I think she appears worthy of my esteem. Paid Dame Akehurst 18*d*. for her 2 days' work.

Sun. 31 Mar. . . . I received of Mr. Shoesmith in cash £3 15*s*. 0*d*. in full for the share of the expense due from Laughton Parish attending the expenses I was at for searching the wills of the Pelham family, relating to a gift formerly given to this parish and the parish of Laughton, but has lately been discontinued. As it was a thing of some value the parishes concluded to have the wills of the Pelhams searched at Doctors' Commons,[6] and to pay the expenses jointly between them.

Mon. 1 Apr. . . . At home all day and thank God very busy. In the

[5] See above, p. 171, note 1.   [6] See above, p. 306, note 30.

even wrote my London letters. Mr. Shoesmith and Joseph Fuller, coming in to my house in the even by accident, as it should appear, though by them a contrived thing, spent the even with me till about 10.20. My friend Joseph Fuller I must own is rather too fruitful in his invention to contrive some way to get a little liquor or a pipe or two of tobacco.

Good Friday 5 Apr. . . . In the even met with Molly Hicks (by appointment) and walked home with her, where I stayed with her, the weather being excessive bad, till morn, past 5, and then came home . . .

Sat. 6 Apr. . . . In the even very dull and sleepy; this courting does not well agree with my constitution, and perhaps it may be only taking pains to create more pain.

Mon. 8 Apr. . . . Paid Sarah Prall in cash 7s. 1d. in full for post letters. Also lent her in cash 14s., for which she left in my hands a broad piece of gold, value according to weight 18s. . . .

Weds. 10 Apr. . . . I dined on the remains of yesterday's dinner. In the afternoon went to Mr. Joseph Burges's, where there was a public vestry holden for to settle the yearly accounts of the overseers and choose new overseers etc., when Mr. Carman and myself made up our accounts with the inhabitants, and there was due to the parish the sum of £17 7s. 3¼d. The new officers chosen: myself churchwarden, Simonds Blackman electioner; Tho. Carman overseer, Ed. Foord electioner . . .

Thurs. 11 Apr. . . . My brother out upon the wild order again today. Paid Mrs. Burges 2s. 11d. in full for what he spent.

Sun. 14 Apr. . . . After dinner I set out for Malling to pay Molly Hicks my intended wife a visit, with whom I intended to have gone to church. But there was no afternoon service there. (My brother and servant at church in the afternoon). I spent the afternoon with a great deal of pleasure, it being very fine pleasant weather and my companion very agreeable. I drank tea with her and came home about 9.30. Spent today 8d.: *viz.*, turnpike 2d. and the servant 6d. Now perhaps there may be many reports abroad in the world of my present intentions, some likely condemning, other approving my choice. But as the world cannot judge the secret intentions of my mind and I may therefore be censured for want of knowing the true motives of my proceedings, I will take the trouble to relate what are really and truly my intentions and the only motive from which they spring (which may be some satisfaction to those who may happen to peruse my memoirs). To do the first I shall begin with the latter, as thinking it most necessary; and as to the motives which spur me on to think of marriage, first I think it is a state agreeable to nature, reason and religion and in some manner the indispensable duty of Christians. For I think it the duty of every Christian to serve God and perform his religious services in the most calm, serene and composed manner, which if it can be performed more so in a married state than in a single one, it must then be an indispensable duty. Now as to my own present situation, my house is not at all regular, neither is there any family devotion performed in that serious manner as formerly in my wife's time, nor have I one friend in the world; that is, I have not anyone whom I can thoroughly rely upon or confide in. Neither have I anyone to trust the management of my affairs to that I can be assured in their management will be sustained no loss. I have not one agreeable companion to soften and alleviate the misfortunes incident to human nature. As to my choice I have only this to say: the girl I believe as far as I can discover is a very industrious, sober woman and seemingly endued with prudence and good nature, and seems to have a very serious and sedate turn of mind. She comes of reputable parents and may perhaps one time or other have some fortune. As to her person I know it's plain (so is my own), but she is cleanly in her person and dress (which I will say is something more than at first sight it may appear to be towards happiness). She is I think a well-made woman. As to her education, I own is not liberal, neither do I think it equals my own, but she has good sense and a seeming desire to improve her mind, and, I must in justice say, has always behaved to me with the strictest honour and good manners, her behaviour being far from the affected formality of the prude, nor on the other hand anything of that foolish fondness too often found in the more light part of the sex. Now for my real intentions: [it] is that of marriage and of the strictest honour, having nothing else in view but to live in a more sober and regular manner, and to be better able to perform my duty to God and man in a more suitable and truly religious manner, and with the grace of the Supreme Being to live happy and in a sincere union with the partner of my bosom.

Mon. 15 Apr. . . . In the forenoon I set out for the sitting at Maresfield in order to verify the parish accounts[7] for the preceding year . . . After dinner I verified the accounts upon oath. Paid the justice's clerks 6s., *viz.*, warrant 2s., summons 2s., verifying account 2s. . . .

Thurs. 18 Apr. . . . In the even went to Lewes where I called at Mr. Plumer's and some other places where I had business and then back to

---

[7] The accounts of the overseers of the poor.

Malling where I drank tea with Molly Hicks and many others. I stayed and spent the evening with her till past 5 in the morning. Came home about 6.45 . . .

Fri. 19 Apr. . . . At home all day and very sleepy.

Weds. 24 Apr. . . . In the afternoon Mr. Banister and I walked down to the Nursery, I to see the widow Trill and my friend Banister for company. We stayed and drank tea with Mrs. Browne and came home about 7.20. Sam. Jenner, meeting with us on the road, came home with us, and he and Joseph Fuller sat with us some time in the even. A very pleasant even and quite delightful, nothing wanting to make it so except the company of my dear Molly and an easy mind.

Fri. 26 Apr. . . . After breakfast I set out for Mayfield where I went to attend at a club[8] feast, of which I am a member. I got there about 10.30, and after refreshing myself with a glass of wine, we went in procession to church where we had a sermon preached by the Rev. Mr. Roger Chalice (one of our members), vicar of Mayfield, from *Matthew* 25.40: ' . . . Verily I say unto you, inasmuch as ye have done it unto one of the least of these my brethren, ye have done it unto me,' from which words we had a very good sermon, the theological part being sound divinity and treated on in a very pretty manner, and the application to the discourse relating to the society was excellent, being concise and very much to the purpose. I dined at *The Star* (where our stewards had provided a dinner suitable to the customs and company of the society) in company with about 48 members more of the society (the whole number being 71) on two buttocks of beef, 2 pieces of roast beef, 2 fillets of veal, pond, currant, and raisin suet puddings, greens, butter etc. I stayed and spent the afternoon there vastly agreeable and came home about 11.50 in company with Tho. Davy and Robt. Hook, thank God very sober . . .

Sun. 28 Apr. . . . My friend Tipper dined with me on a piece of boiled beef and greens and a plain rice pudding. After dinner I rode with my friend to bring him on his road to Lewes, but instead of going into the town I went to see my intended wife, with whom I drank tea and spent the evening, and indeed stayed great part of the night with her. I came [home] about 4.40. Spent 2*d.* for the turnpike.

Mon. 29 Apr. . . . At home all day and pretty busy. In the even wrote my London letters. A very pleasant day, and last night spent so agreeably that sleep today has not taken possession of eyes . . .

Weds. 1 May. In the forenoon being at work in my garden I had an

[8] The Mayfield Friendly Society; see above, p. 99, note 25.

accident to bruise my leg very much . . . My leg very bad in the even.
Mr. Read, surgeon, paid me a visit and dressed my leg. At home all day
and a pretty deal of business in the shop . . .

Thurs. 2 May. . . . Mr. Read paid me a visit and dressed my leg,
which is really very bad, and stayed and drank tea with me. Sam.
Jenner called in the evening to see me and stayed all night. At home all
day and my brother very busy in the shop . . .

Thurs. 9 May. . . . At home all day; busy a-making a bed. In the
even Mr. Pellin and I played a few games of cribbage; I won 6d. My leg
continues very bad.

Sat. 11 May. . . . At home all day; my leg very painful. In the even
my intended wife and her sister called to see me and sat with me some
time. This may possibly be imputed to the girl as fondness, but I must
do her the justice to say I esteem it only as friendship and good
manners. For I have never met with more civil and friendly usage from
any one of the fair sex than I have from this girl.

Sat. 18 May. . . . At home all day. In the even Mr. Read paid me a
visit and dressed my leg, which continues very bad. Sam. Jenner sat
with me some time in the evening.

Mon. 20 May. . . . At home all day. My leg mending.

Tues. 21 May. . . . Received of Thomas Davy half a hog weighing
12 stone[9] 2 lb . . .

Fri. 24 May. In the fore part of the day I was out on the roads as
surveyor, the parishioners being warned out to work on the high
roads.[10] There were out John Cayley's, Mrs. French's and Edwd.
Hope's carts and the following labourers: John Watford the elder for
Mrs. Atkins, John Watford the younger for himself, Joseph Burges for
himself, Thomas Burfield for Joseph Fuller, John Gasson for Joseph
Durrant, and Richard Hope for himself . . .

Sat. 25 May. . . . Sam. Jenner, a-drawing off a cask of cider for me,
dined with me on the remains of yesterday's dinner with the addition
of some hog's feet and ear . . . At home all day and pretty busy. My leg
thank God something better . . .

Sun. 26 May. . . . I dined on a sparerib roasted, and just as I had
dined, Mr. Read came to dress my leg. He dined on the remains. In
the afternoon took a ride to Lewes to pay my intended wife a visit, with
whom I drank tea and spent the afternoon and indeed part of the night.
Came home about 2.45 . . .

[9] See above, p. 1, note 2.

[10] For the duties of parishioners with respect to the repair of the roads see *Tate*,
Ch. XI.

Weds. 29 May. In the forenoon out at work in the highways as surveyor. I dined on some pork bones boiled and greens. After dinner rode to Newhaven in order to put up a bed for my friend Tipper, which I had made before. I stayed with him a little in the even and took part of his bed. Spent 2*d.* for the turnpike.

Thurs. 30 May. In the morn I got up and put up Mr. Tipper's bed . . . After dinner I set out for home . . . Called at Malling on my intended wife, with whom I stayed and spent the evening till past 12 o'clock. Came home about 1.10 . . .

Fri. 31 May. On the highways as a surveyor . . .

Sat. 1 June. . . . In the afternoon Mr. Joseph Fuller and myself went out a-window-viewing[11] . . .

Sun. 2 June. . . . After dinner I took a ride to Malling to see my intended wife, with whom I took a serious walk till about 9 o'clock when we parted . . .

Weds. 5 June. Out at work on the highways as surveyor . . .

Thurs. 6 June. . . . At home all day and but very little to do. In the even Sam. Jenner sat with me some time. Excessive hot weather. In the afternoon Mr. Banister and I went to fishing some little time, but had not any luck . . .

Fri. 7 June. In the morn and all the forenoon out at work on the high roads . . . In the even took a ride to pay my intended wife a visit, with whom I took a serious walk and spent the even till about 10 o'clock. After parting with her I went to take my horse, and happening into company, I stayed till about 12.10 . . .

Tues. 11 June. . . . In the afternoon rode up to Hawkhurst Common, where one part of the parish was met to play a game of cricket with the other part. I stayed and played one innings, but being taken very ill came home about 7.30. Had a violent fever in the evening. Paid my shilling as a gamester[12] . . .

Weds. 12 June. . . . At home all day and mighty piteous. Joseph Fuller sat with me some time. Excessive hot all day.

Thurs. 13 June. . . . In the even or more rather in the afternoon I had a very smart return of my ague, or more properly fever, though I am in hopes went off lighter than before. At home all day and very busy.

Fri. 14 June. . . . About 4.30 Tho. Durrant and I set out for Lewes where we spent the remainder of the day at Lewes, part of it with my intended wife . . .

[11] Presumably assessing houses for the window tax.
[12] See above, p. 9, note 9.

Sat. 15 June. . . . At home all day and in the afternoon had my ague very severe.

Sun. 16 June. . . . My brother went in the morn to Cuckfield where he bargained as a yearly servant and came home about 3.30, very sober. He then walked over to Chiddingly on business, but did not stay . . . An excessive hot day.[13]

Weds. 31 July. From the day last mentioned, I have been so embarrassed with a multiplicity of business that I was not able to continue my journal, being on the 19th day of June married at our church (to Mary Hicks, servant to Luke Spence Esq. of South Malling) by the Rev. Mr. Porter, and for about 14 days was very ill with a tertian ague, or rather an intermitting fever. Then the ceremony of receiving visitors and again the returning of them has indeed, together with the business in my trade, taken up so much of my time that I was obliged to omit that which would have given me the greatest pleasure imaginable to have continued. But however thank God I begin once more to be a little settled and am happy in my choice. I have, it's true, not married a learned lady, nor is she a gay one, but I trust she is goodnatured, and one that will use her utmost endeavour to make me happy, which perhaps is as much as it is in the power of a wife to do. As to her fortune, I shall one day have something considerable, and there seems to be rather a flowing stream. Well, here let us drop the subject and begin a new one.

[13] The regular diary ends here. The entry for 31 July was added later.

Appendix A The Turner Family [1]

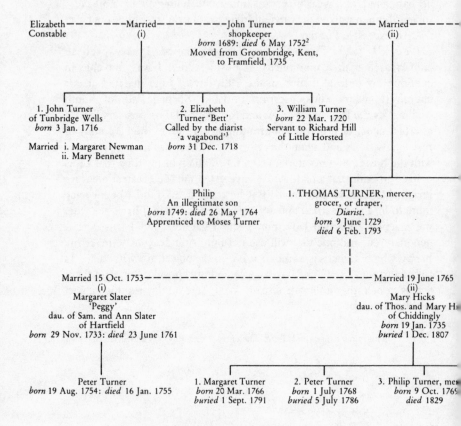

Elizabeth —————Married————— John Turner ————————— Married—————
Constable   (i)   shopkeeper   (ii)
                                born 1689: *died* 6 May 1752[2]
                             Moved from Groombridge, Kent,
                                    to Framfield, 1735

1. John Turner   2. Elizabeth   3. William Turner
of Tunbridge Wells   Turner 'Bett'   *born* 22 Mar. 1720
born 3 Jan. 1716   Called by the diarist   Servant to Richard Hill
                                  'a vagabond'[3]   of Little Horsted
Married i. Margaret Newman   *born* 31 Dec. 1718
        ii. Mary Bennett

                                      Philip   1. THOMAS TURNER, mercer,
                               An illegitimate son   grocer, or draper,
                          born 1749: *died* 26 May 1764   *Diarist.*
                           Apprenticed to Moses Turner   *born* 9 June 1729
                                                                 died 6 Feb. 1793

Married 15 Oct. 1753 —————————————————————————— Married 19 June 1765
         (i)   (ii)
    Margaret Slater   Mary Hicks
       'Peggy'   dau. of Thos. and Mary H
dau. of Sam. and Ann Slater   of Chiddingly
    of Hartfield   *born* 19 Jan. 1735
born 29 Nov. 1733: *died* 23 June 1761   buried 1 Dec. 1807

       Peter Turner   1. Margaret Turner   2. Peter Turner   3. Philip Turner, me
born 19 Aug. 1754: *died* 16 Jan. 1755   *born* 20 Mar. 1766   *born* 1 July 1768   *born* 9 Oct. 176
                             buried 1 Sept. 1791   buried 5 July 1786   *died* 1829

[1] This pedigree is constructed from notes written by Turner himself and printed as an appendix in *F. M. Turner*, pp. 105–12 and *Jennings*, pp. 79–84. Additional information comes from the diary itself, from the East Hoathly parish register, and from L. F. Salzman, 'Philip Turner', *S.N.Q.*, XVI (1963), no. 2, p. 37.

[2] John Turner's will, dated 1 May 1752, is in E.S.R.O., South Malling wills register D 9, p. 233. In it he refers to himself as John Turner *alias* Fann.

[3] She was apprenticed to a mantuamaker in Westerham, Kent, in 1738. See R. Garraway Rice, *Sussex apprentices and masters, 1710 to 1752* (S.R.S., XXVIII, 1922) p. 193. When her stepmother wrote her will on 9 October 1754 she referred to Elizabeth

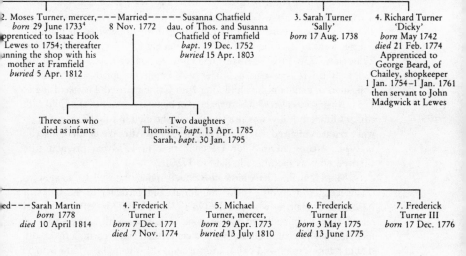

─ ─Elizabeth
Ovenden
of Rotherfield
697: *died* 1 April 1759

2. Moses Turner, mercer,─ ─ ─ Married ─ ─ ─ ─ ─ Susanna Chatfield
born 29 June 1733[4]   8 Nov. 1772   dau. of Thos. and Susanna
pprenticed to Isaac Hook   Chatfield of Framfield
Lewes to 1754; thereafter   *bapt.* 19 Dec. 1752
unning the shop with his   *buried* 15 Apr. 1803
mother at Framfield
buried 5 Apr. 1812

3. Sarah Turner
'Sally'
born 17 Aug. 1738

4. Richard Turner
'Dicky'
born May 1742
died 21 Feb. 1774
Apprenticed to
George Beard, of
Chailey, shopkeeper
1 Jan. 1754–1 Jan. 1761
then servant to John
Madgwick at Lewes

Three sons who
died as infants

Two daughters
Thomisin, *bapt.* 13 Apr. 1785
Sarah, *bapt.* 30 Jan. 1795

ed─ ─ ─Sarah Martin
born 1778
died 10 April 1814

4. Frederick
Turner I
born 7 Dec. 1771
died 7 Nov. 1774

5. Michael
Turner, mercer,
born 29 Apr. 1773
buried 13 July 1810

6. Frederick
Turner II
born 3 May 1775
died 13 June 1775

7. Frederick
Turner III
born 17 Dec. 1776

─────For the children of this family
(through whom the diary descended),
see L. F. Salzman, 'Philip Turner',
S.N.Q., XVI (1963), no. 2, p. 37.

as 'wife of Archibald Blare', but the diarist makes no mention of this. The stepmother's will is in E.S.R.O., South Malling wills register D 9, p. 236.

[4] This is the date given in Turner's notes. The diary entry for 7 July 1757, however, implies that that was his birthday. His apprenticeship in 1749 is recorded in R. Garraway Rice, op. cit., p. 194. He was unmarried during the diary period. He married in 1772, had five children (two of whom survived him), and died, a widower, in 1812. His will, in which he described himself as 'mercer and draper', is in E.S.R.O., South Malling wills register D 11, p.298. Details of his marriage and children are from the Framfield parish registers: E.S.R.O., PAR 343/1/1/4–5.

Appendix B

Principal persons figuring in the diary other than members of the diarist's immediate family[1]

ADAMS, Peter, labourer. A feckless, though evidently charming, individual who was moving downwards in society. He had inherited 20 acres of land from his mother in 1752,[2] yet by 1763 he was in receipt of parish relief.[3] In 1752 when he sold some of his land to Thomas Carman he was described in the Laughton manor court book as 'yeoman'. He was a good deal older than Turner. He and his wife, Hannah, had ten children between 1733 and 1749, and an eleventh in 1756. Only four of them survived to maturity. Turner went to the funeral of one of the daughters (14-year-old Mary) on 29 January 1756. In 1752 Adams had fathered an illegitimate child on Ann Caine. Turner as a parish officer was much occupied with Adams's obligation to maintain the child after Ann Caine married Thomas Ling in 1755. Adams was also widely suspected of being the father of the child with which Elizabeth Elless was pregnant when she died in July 1756, and for a while it was rumoured that he had murdered her. After the death of his wife in 1768, Adams married for a second time in 1773 but his new wife survived only six months. He died in 1780.

ATKINS, Mrs Hannah. The widow of Samuel Atkins, the well-to-do steward ('housekeeper' according to his memorial) and gardener to the Duke of Newcastle at Halland who had died in 1742. In his will he left the interest on £100 to be distributed to the poor of East Hoathly once a year. A memorial plaque commemorating this generosity, and recording his death, that of a nine-month-old child, and subsequently that of his widow is to be found on the exterior south wall of East Hoathly church. His legacy was in the event subsumed into the general parish funds when the church was repaired in the 1760s. Hannah was in a slightly higher social category than Turner, but she was one of the group including Turner who all went to each other's parties in the first two months of each year. She was

---

[1] The information given in these notes is derived from the diary itself with additional genealogical details from the East Hoathly parish registers in E.S.R.O., PAR 378/1/1/1–4. Other sources are separately noted. Some of the characters are more fully drawn in Dean K. Worcester Jr., *The life and times of Thomas Turner of East Hoathly* (1948).

[2] E.S.R.O., Laughton manor court book 1737–65, S.A.S. A 663, p. 293.

[3] E.S.R.O., East Hoathly overseers accounts 1762–79, PAR 378/31/1/1.

godmother to Margaret, the first child of Turner's second marriage.[4] She died, aged 79, in 1769.

BANISTER, George, excise officer. Not a member of an old East Hoathly family, he came to live as Turner's next-door neighbour (occupying the northern end of the building of which Turner's house formed part) in 1763 when he succeeded Lawrence Thornton in the excise. He had a short temper, twice assaulting other villagers, but was a useful man to know—one capable of setting a dislocated arm in an emergency, and one who would share with his neighbours the brandy which he confiscated from the smugglers. His wife, Elizabeth, was evidently a bit of a hoyden, disporting herself in the street in an advanced state of pregnancy. The births of their two daughters are recorded by Turner, who regarded Banister as a close friend, though he was somewhat surprised, as the godfather, that Banister should christen the eldest Molly, which was usually the familiar form of 'Mary'.

BEARD, George, shopkeeper at Chailey. Beard had been Turner's servant in earlier years but by the time the diary begins he was in business on his own account in Chailey. He was a good friend of the Turner family and took Turner's young brother Richard as an apprentice at the age of eleven in 1754. He married his servant in 1755, and Turner never liked her, though he thought the marriage only right since Beard had 'got to bed to her beforehand'.

BENNETT, Moses and John. Turner refers to these men as 'my cousin Bennett'. They were presumably the sons of 'my uncle Ed. Bennett' who was related through Turner's mother. The relationship is obscure.

BROWNE, Mrs. Elizabeth. A substantial landowner in East Hoathly and elsewhere. She was born Mittell, and was related by marriage to the Hutsons. She owned East Hoathly mill and the blacksmith's shop. Turner rented from her the pew in which he sat in church and bought from her the coffins which he used as undertaker. She employed him to do her books for her, and Turner treated her with considerable respect.

BURGES, Joseph, farmer, alehousekeeper and victualler. Burges was the brother of Mrs Mary Virgoe and of John Burges of Rotherfield. From 1764 onwards he kept a public house called *The Maypole* (later *The King's Head*)—for which Turner lent him the purchase price, and was one of those villagers who served his turn in parish offices and attended the vestry meetings. After the failure of Jones, vestry meetings were held at *The King's Head*. He was wealthy enough to buy and sell property and in fact sold the pub to James Fuller in 1768. Turner bought it from Fuller in 1772.[5] He married Sarah Seymer in January 1750 and they had five children, the youngest of whom survived only eight months. Turner thought that he

[4] *F. M. Turner*, p. 110.
[5] E.S.R.O., Laughton manor court book 1766–79, S.A.S. A 664, pp. 93, 197. See also above, p. 293, note 8.

spoiled his children. He died in March 1780, his wife having died eight months previously.

CALVERLEY, Edward. A shadowy figure in the diary who figures principally as a guest at all the main social functions. He was an old man and died, aged 76, in December 1760. He was, according to Blencowe and Lower, a descendant of the Calverleys of The Broad in Hellingly.[6]

CARMAN, Thomas, farmer. Another shadowy figure in the diary. He was not one of Turner's social circle though he served parish office as headborough, surveyor and overseer. Parish money was put out to him to earn interest and he was active in parish affairs. Turner, who was godfather to one of his daughters, regarded him as unnecessarily officious as a surveyor and as one who acted largely out of self-interest in setting the parish rate. In October 1763 when Mrs Carman gave him a present of a duck, Turner worried about what people might think: he noted however that it must be out of pure gratitude since 'I never kissed the woman but once'. On 10 June 1757 when he took over copyhold land previously held by Peter Adams, the Laughton manor court book[7] records him as a bargeman of the Cliffe near Lewes. His name disappears from the church rate books in 1796 and he may have been the Thomas Carman of Ringmer recorded in the South Malling parish register as being buried on 11 June 1796.[8] Part of the Cliffe is in that parish.

CAYLEY, John, husbandman. A tenant of Mrs Browne and to some extent acting as her agent, Cayley was one of the middling sort in the village who served office as overseer, churchwarden, and surveyor during the period of the diary, though he was not one of the circle who regularly ate in each other's houses. Four years older than Turner, he married his servant, Ann Smith, in July 1759, and they had four children in the 1760s. Both he and his wife lived on into the nineteenth century. She died, at the age of 74, in 1800: he survived until 1807 when he died aged 82.

COATES, Christopher, steward at Halland House, gentleman. Coates, by virtue of his position and of his seniority, was an influential man in East Hoathly. He was normally styled 'gentleman' or 'esquire', and came from Wensley in Yorkshire.[9] He was also the father-in-law of the rector, the possessor of considerable patronage (and therefore very important for retailers such as Turner from whom on at least one occasion in August 1755 he withheld the Halland account because Turner would not pay him an inflated price for his wool), and the repository of considerable experience which was frequently drawn upon in difficult matters of parochial administration. He and his wife, Mary, gave parties and were present at many of the parties

[6] *Blencowe and Lower*, p.196, note 2.

[7] E.S.R.O., Laughton manor court book 1737–65, S.A.S. A 663, p. 389.

[8] E.S.R.O., PAR 419/1/1/1. I am very grateful to Christopher Whittick for help with this point.

[9] See, for instance, the monument to the Rev. Thomas Porter and his wife in East Hoathly parish church.

given by others around the Christmas period. He gave up the stewardship at Halland towards the end of the diary period, and died in September 1766: his widow died four years later.

Abraham Baley, the receiver-general of the Duke of Newcastle's rents who succeeded Coates at Halland, indicated in a letter dated 15 October 1764, that perhaps all had not been in order in Coates's administration. He queried the whereabouts of a horse which had been impounded as a stray by Coates and which should have become the property of the Duke. He also queried entries in the accounts for tending the gardens—no regular gardener having been appointed since the death of John Vernon the previous March. The letter says that, despite the expense, Coates 'has put it out of his Grace's power to command so much of [the gardens'] produce as a small apple-pye. For he has strip'd the orchard of every apple in it and gave them away to his servants who hawk'd them about the country for sale. All this was done so quick that I could not possibly have any intelligence of it, nor could I indeed have supposed it possible that any man living could have attempted (to give it no other name) so strange an act. Sure it is very extraordinary.'[10]

CORNWELL, Thomas, husbandman. A member of the Cornwell family of Framfield and East Hoathly, Thomas Cornwell married a girl from Hellingly in 1758. He, his wife, and his brother Peter frequently made presents of freshwater fish, eels, and the occasional rabbit to Turner and his sick wife. He could also beat Turner at cricket. He was a few months younger than Turner, had at least six children in the late 1750s and 1760s, and survived Turner by fourteen years, dying, aged 77, in 1807.

COURTHOPE, George, JP of Uckfield. Courthope was the JP who lived nearest to East Hoathly and it was thus to him that the parish officers turned when they needed a justice's signature on documents, a summons issued, or advice on points of law particularly with regard to the poor. If he could not be contacted then Turner and his fellow officers turned to Luke Spence, JP of South Malling, or John Bridger, JP of Offham. Molly Hicks, whom Turner married as his second wife, was one of Luke Spence's servants for a time.

DAVY, Thomas, shoemaker. During the earlier years of the diary Davy was Turner's best friend. He was exactly the same age as Turner and shared many of his interests. He liked books and cricket. Turner taught him how to use the slide-rule, and they made many visits together on pleasure or on business. Davy's mother, like Turner's, lived in Framfield. Having courted Turner's servant, Mary Martin, Davy suddenly in July 1761 married Mrs Mary Virgoe—property owner and widow of a well-to-do Lewes butcher—and did not ask Turner to the wedding. The first baby, Sarah, was born exactly six months later. Davy succeeded John Streeter as parish clerk

---

[10] E.S.R.O., Abraham Baley's letter book 1763–73, S.A.S., HA 310.

in 1777 and lived until 1801, reaching the age of 72 and outliving Turner by eight years.

DIGGENS, Charles, tailor. Another person who during the earlier years of the diary was a close associate and friend of Turner. He came from Framfield where he had been christened on 5 April 1730.[11] He was thus much the same age as Turner. He was a constant visitor, helping to pack Turner's wool for sale, assisting at funerals, and tailoring for Turner's customers. They went to cricket matches together and on one occasion Turner failed to bet on him when he easily won a race at Piltdown. Early in 1756 when Turner was considering a move to Framfield he offered Diggens a partnership in the shop there: but the scheme came to nothing. Diggens had a wife whose name was Sarah but she scarcely figures in the diary.[12] He died and was buried at Framfield on 15 September 1776.

DURRANT. A very common surname in the Lewes area: many individuals with this name occur in the diary, the principal ones are:

DURRANT, Joseph, blacksmith, and his wife, Sarah, often called by Turner 'Master' and 'Dame' Durrant. Soon after their marriage in 1732, Joseph's uncle, another Joseph, had settled on his nephew the East Hoathly smithy. Joseph served parish office, and the couple were frequently at parties. Sarah was on occasion very generous to the Turners. Joseph died in 1777. They had two sons, the elder of whom, yet another Joseph, is scarcely mentioned in the diary unless he was the one who married the pregnant Elizabeth Reeve in 1763.

DURRANT, Thomas. Younger son of Joseph and Sarah and an especial friend of Turner, though eight years his junior. Turner taught him to read and write in the 1750s when he was in his teens, and frequently employed him on errands or about parish business. He could turn his hand to most things: he could even cut Peggy Turner's hair. He was a sportsman, good at cricket and a fast runner. He married in 1776, and lived to see the century out. He died in July 1800.

DURRANT, Ann, John, Robert. Three of a very large number of children born to John and Elizabeth Durrant. They were lower in the social scale than the children of Joseph and Sarah, ex-smugglers, and frequently in trouble in bastardy cases.

DURRANT, Samuel, mercer of Lewes. From a different branch of the family, Samuel Durrant was one of Turner's suppliers who also acted as his banker in Lewes (see Appendix C). When he died he was reputed to be worth £140,000.[13]

ELLESS, Francis, schoolmaster. He took over the school in East Hoathly from Turner in 1756 and kept it until he left for Alfriston in 1759. He subsequently moved to Uckfield. While in East Hoathly he was a special

[11] E.S.R.O., PAR 343/1/1/3.
[12] E.S.R.O., South Malling wills register D 9, p. 205.
[13] *Sussex Weekly Advertiser*, 23 December 1782.

friend of Turner. They learnt surveying together, and Turner taught him decimals and the use of the slide-rule. Turner always enjoyed visiting him after he had left the village, considering him 'polite and genteel', and 'as agreeable [a] companion as any among my acquaintance.'

FRENCH, Jeremiah, yeoman. The tenant farmer of Whyly and thus the largest farmer in the parish after the Duke of Newcastle. French was a man whose character and activities figure large in the diary. He was twenty years Turner's senior, and Turner went in awe of him. Forceful, quarrelsome, drunken, loud, and at times obscene, he did his best to dominate parish affairs, serving all the important offices, collecting the taxes, and interfering in matters even when not officially involved. An indefatigable party-goer and heavy drinker, he was the scourge of those less fortunate than himself, especially keen to pursue paupers out of the parish if he could ensure their settlement elsewhere. When it came to paying the poor-rate he did his best to shift his responsibilities to others.[14] His wife, Elizabeth, whom he had married in 1740, was well-connected. She was a Virgoe, sister of Isaac, the Lewes butcher, and daughter of Samuel who for a while was Turner's landlord. She was godmother to the first child of Turner's second marriage.[15] The Frenches had four children: Mary ('Molly'), born in 1741 who died aged 30 in 1771; John, born in 1743 and whose death at the age of 18 Turner records in 1761; Elizabeth, who died aged 5 in 1750; and Samuel, born in 1750, who carried on the line. Jeremiah died in 1763 at the age of 55[16] and his widow lived on until 1780.

FULLER. The Fullers were members of the middle order of East Hoathly society: very much on a par with Turner, but below the Frenches or the Porters. There was scarcely a year when one member or other of the family was not churchwarden, overseer, or surveyor and they were at all the social functions. There were two families, descending from a pair of brothers in the generation before Turner.

FULLER, Joseph, senior, butcher. He and his wife, Mary, had eight children in the years between 1730 and 1750, and the six eldest were alive when Turner wrote his diary. They were: Joseph, junior, who was six months younger than Turner and his constant companion in the 1750s and 1760s; John, who became a butcher at Lewes; Thomas; Mary ('Molly'); Richard; and Elizabeth. Turner did not always get on with the elder Joseph and

---

[14] A letter written after French's death by Abraham Baley who was looking into the terms upon which Whyly had been leased to him, shows that probably since 1752 French had been avoiding payment of the poor rate by claiming that it was a charge on his landlord and thus deducting it from his rent. The terms of the lease were, apparently, that the landlord was liable only for the first £20 *per annum*, yet French had been deducting from his rent as much as £38 *per annum*. E.S.R.O., S.A.S. HA 310.

[15] *F. M. Turner*, p. 110.

[16] His will shows that he owned freehold property in Ripe and copyhold land in Chalvington. E.S.R.O. Lewes Archdeaconry wills register A 61, p. 68.

found his wife loud and raucous, but he shared many interests with Joseph junior who stayed a bachelor throughout the period of the diary. He did not marry until 1769. It is not altogether clear whether it was the father or the son who swindled Turner over the purchase of his horse in June 1757. If it was the son, Turner seems not to have borne him a grudge.

FULLER, Thomas, senior, tallowchandler. He married a widow, Elizabeth Driver ('Bett'), and four of their seven children born between 1730 and 1740 were alive when the diary was being written. They were: Thomas, junior, six months younger than Turner; James who married Ann Newenden in 1764; Elizabeth who was married to Samuel Gibbs; and Ann ('Nanny') who married Charles Vine in 1759. Turner was never as friendly with Thomas, junior, as he was with Joseph, junior, and seems actively to have disliked James ('stupidly ignorant and withal prodigiously abusive' he called him at one point); but he was called in on the death of Thomas, senior, in 1762, to read the will and draw up the probate inventory. Thomas junior's wife was called Philadelphia.

GIBBS, Samuel, yeoman, gamekeeper at the Broyle. Another shadowy figure in the diary, though evidently an important man in the parish. Turner, while schoolmaster, was responsible for the education of one of his sons. A hard-bargainer and a heavy drinker, he and his wife Elizabeth (the daughter of Thomas Fuller) were present at many of the social functions, though at times Turner considered him, like Jeremiah French, a person whose 'whole discourse seems turned to obscenity, oaths, gaming and hunting'. He died in 1784 and was buried at Laughton on 8 August.[17]

HILL. The Hills of Little Horsted were related to the Turners through Elizabeth Constable, the first wife of Thomas Turner's father, John. The principal members of the family mentioned in the diary were Richard Hill and his two children.

HILL, Richard. Referred to as 'my uncle Hill', he was a frequent visitor to East Hoathly and to Turner, and Turner occasionally stayed with him. On one visit there he had a really excruciating meal which he described at some length: it included 'sauce which looked like what is vomited up by sucking children'. Turner's half-brother William was employed as a servant to him at Little Horsted. Besides his son Charles, Richard also had a daughter, Molly.

HILL, Charles. Sometimes called Dr. Charles Hill, he was a surgeon who served in the navy on board the *Monarque* and thus was able to send home eye-witness reports of the court-martial and the death of Admiral Byng. On visits to his relatives he had many a story to tell of his travels to far-flung corners of the world which Turner found entertaining, but at other times, as when Turner visited him in April 1764, he was not keen to repay the hospitality and Turner considered him downright mean. In the circum-

---

[17] E.S.R.O., PAR 409/1/1/3.

stances Turner was wise to make sure that the £60 he lent him (at 5%) was secured by a bond.

HOOK, Robert, shoemaker. Not one of Turner's close friends, but a fellow tradesman. He was one of those who sponsored Turner for the Mayfield Friendly Society; he bought books and brandy from Turner and made shoes for him and his wife. He was surveyor of the highways in 1756 and headborough in 1758. It was with him that Turner placed his nephew Philip apprentice at the age of ten in 1759, and it was Hook's daughter Mary ('Molly'), 'a poor, wild girl' whom Turner tried as a servant (aged 12) for a month in 1758. Hook died early in 1775.

HOPE, Edward, farmer. Edward Hope was one of the older men in the parish when the diary was being written. His wife, Mary, had died in 1750. He retained a roving eye however and at the age of over 70 he got his servant, Catharine Jenner, pregnant in 1763. He died in 1766.

HOPE, Richard, brickmaker. The son of Edward and Mary Hope. He married Sarah Weller in April 1755. Both he and his wife outlived Turner and saw the century out. He died aged 75 in 1803 and she, aged 83, in 1802.

HUTSON, James, senior, husbandman. When the diary begins he was married to Mary. Their children had been born in the 1740s and had included twins who had both died within three weeks of their birth in 1748. Mary Hutson died in 1756, and their 19-year-old son, Joseph, followed her to the grave in 1764. Turner was in charge of the funeral. In 1758 Hutson was thought to be the father of an illegitimate child of his servant Mary Evenden, and a year later he married her. They had further children in the 1760s and 1770s. He died in 1788.

HUTSON, James, junior. Married to Sarah Stevens in 1763 and had six children in the 1760s and 1770s.

JENNER, Samuel. A bachelor and, following the death of Peggy, a true friend and constant companion of Turner. He helped Turner about the house, garden and shop, carried messages for him, stayed with him as extra security when there was money in the house, and even accompanied him on courting expeditions. At times (e.g. 9 and 23 March 1764) Turner was worried by what people might say of their constant companionship but valued him highly as a 'very sincere friend, a worthy man, a learned and agreeable companion'. He died and was buried at East Hoathly on 18 August 1767.

JONES, John, alehousekeeper and victualler. Jones's pub, *The Crown*,[18] was the one to which Turner and his friends resorted most often, both privately and in an official capacity since vestry meetings were held there. Jones's house was one of the centres of the village's social life: travelling salesmen set up their stalls there, curiosities were exhibited, cock-fights took place, and teams of cricketers resorted to it for refreshment. But for Jones it provided

---

[18] The name was recorded by Turner in the East Hoathly overseers account book, E.S.R.O., PAR 378/31/1/1.

a precarious existence: he was in debt to a Southwark distiller and, in 1762, he was forced into mortgaging his property in order to pay the debt.[19] Things did not improve, however, and in November 1764, the vestry, meeting under his roof, refused to make any contribution towards the payment of his debts. His furniture was sold and he was forced out of business. Jones's wife, Grace, died in January 1756. Like Turner, however, he remarried: his second wife in October 1761 was Mary Heath.

LONG, John, schoolmaster and excise officer. Long was one of Turner's successors as schoolmaster in East Hoathly. Turner's immediate successor was Francis Elless who was followed by Alexander Whitfield. John Long took over in January 1760 and became a firm friend and intellectual companion to Turner: he helped him, too, with his duties as collector of the land tax from time to time, and Turner stood surety for him as a hop assistant. He eventually went into the excise early in 1764 but did not live long thereafter: he died of smallpox in November 1764 having apparently contracted it from an inoculation.

MADGWICK, John, linen draper of Lewes. One of Turner's principal local suppliers but also an influential friend. Turner always made a point of visiting him when in Lewes. He was godfather to Peter, Turner's son by his first marriage who died at the age of five months. He also took Turner's 18-year-old brother, Richard, as a servant in his shop when he emerged from his apprenticeship in 1760.

MARCHANT, James, tailor. The Marchants were a large family. When Richard Marchant was buried in May 1757 many members came to the funeral. The family was close to Turner's: one of the older ladies (probably James's aunt) always dined with Turner on Christmas Day, though they were not usually of the group who entertained each other during the party season. James Marchant, junior, the tailor, had come to East Hoathly as a young man as apprentice to his uncle Richard, but had not served out his time. As well as making up clothes for Turner's customers he was also one of the younger friends with whom Turner went to cricket matches and to the alehouse. He married Elizabeth Mepham, a seamstress who worked for Turner and was a friend of his first wife (it was she who laid her out). After their marriage in 1763 the East Hoathly parish authorities went to law to ensure that they were legally settled at Ticehurst even though they lived in Hoathly (see Plate 6). James and Elizabeth had two daughters, and died in 1777 and 1779 respectively.

MARTIN, Mary, servant. Mary Martin was a favourite servant of Turner. She lived with the Turners from Lady Day 1754 until Lady Day 1758 and then left for reasons unstated, but it was Peggy Turner who gave her notice. She

---

[19] The distiller was Thomas Neatby. In subsequent dealings about this property in the manorial court Turner acted as Neatby's attorney. E.S.R.O., Laughton manor court book 1737–65, S.A.S. A 663, p. 485, and Laughton manor court book 1766–79, S.A.S. A 664, p. 132.

frequently returned to visit, and after Peggy's death in 1761 she often came over to stay with Turner, sometimes for a week at a time, to help him out or to 'stand the shop' while he was about parish or other business. She was evidently devoted to Turner though he thought that Thomas Davy had let her down badly, when, having kept her company for many years, he married the widow Virgoe.

The MEPHAM family. Members of this family are constantly referred to in the diary. They were lower in the social order than Turner and were often in and out of his house, helping him cut out Sussex 'round frocks' or doing similar jobs. They were all children of Thomas and Elizabeth Mepham both of whom had died before the diary begins. Their children were: Elizabeth, born in 1729, who married James Marchant in 1763; Martha, born in 1731, who married James Thorpe, a collier from Goudhurst, Kent; Thomas, born in 1733, who became a sailor on HMS *Torbay* (Turner drafted his will for him in March 1756) and who by September 1764 was presumed dead; Joseph, born in 1736 and who died in September 1756; Lucy, born in 1738, who married William Williams, a tailor, and was removed from the parish in 1764; and two other children, Hannah and John, who died as infants in the 1740s. The parents had owned a house and land in East Hoathly but it had been heavily mortgaged to the Rev. Thomas Porter in 1760 and Turner was involved in a complicated transaction in the manor court involving the property in September 1764.[20]

PIPER, William, farmer. Piper is one of those (like Jeremiah French, and the Revd. Thomas Porter) whose personalities come alive in the pages of the diary. He was a substantial farmer in the parish, a regular attender at the vestry, and one who served office as overseer, surveyor and churchwarden. He was also for some years collector of the land and window tax. He was Turner's social equal and both men attended and gave parties with the Porters, Frenches, Coateses, and Fullers. His principal characteristic however was his meanness. Time and again Turner complained of this: he expected jobs to be done without payment, and his parties were penny-pinching affairs even when they were celebrating a christening—yet he was always keen to get drunk at the expense of others. To get a bottle of beer from him, Turner considered, was like getting blood from his heart. He was an old man at the time of the diary. His first wife, Mary, died childless in March 1755, and in December of that year he married his servant, Mary Denmall, a 'good-natured, poor, harmless, ignorant, inoffensive creature' according to Turner. They had three children, and Turner was godfather to the elder son William. Piper died in April 1761 and Turner acted as a trustee (along with the Revd. Thomas Porter and Piper's brother John) for the young family, though he had declined to draft the will.[21]

[20] See above, pp. 303–5, and E.S.R.O., Laughton manor court book 1737–65, S.A.S. A 663, pp. 512–4.
[21] See above, p. 222.

PORTER, The Revd. Thomas, MA, rector. Porter was rector of East Hoathly throughout the period of the diary and for many years afterwards. Previous editors of the diary, from Blencowe and Lower onwards, have confused him with his brother and immediate predecessor as rector—The Revd. Richard Porter; and even John and J. A. Venn in *Alumni Cantabrigienses*[22] have the wrong one as the 'convivial' rector of Turner's diary.

Richard Porter, according to a note in the East Hoathly parish register,[23] was inducted into the living at East Hoathly on 8 December 1741. He resigned on becoming Vicar of Mayfield on 9 January 1752. He succeeded his uncle, another Richard Porter, as rector of Chailey on 30 October 1753, and held both livings until his death at the age of 46 on 3 February 1762.[24]

Thomas Porter, who, like Richard, had been at Jesus College, Cambridge, was inducted at East Hoathly on 10 February 1752 and remained rector until his death aged 74 on 30 September 1794. There is a memorial to him on the south wall of the chancel of East Hoathly church. He also held the living of Ripe in plurality. He was succeeded at East Hoathly by the Revd. E. R. Langdale, and at Ripe by his son-in-law the Revd. E. R. Raynes. Thomas Porter was one of the two dominating personalities in the parish. The other was Jeremiah French. A leader, as might be expected, in parochial affairs, he was also an inveterate (and not always sober) party-goer, and a major land-owner always seeking to acquire more. In matters of land-owning he drove a hard bargain, and was not universally popular. In 1762 someone set fire to his woods and opened one of his fishponds, killing all the fish. The *Sussex Weekly Advertiser* reported this on 13 June, and it is odd that Turner did not mention it. He seems to have been a reasonably conscientious, though worldly, parson, conducting services and preaching regularly if sometimes repetitiously. He was ten years older than Turner and had married, on 4 April 1749, Mary Coates, the only daughter of Christopher Coates, the steward at Halland House.[25] Their first child was baptized on 3 January 1750 and in the following ten yeaars they had eight more children. Five of the children died in infancy.[26] Of those who survived, Harriet (born in 1752) subsequently married the Revd. Edward Robert

[22] Part I (to 1751), III.

[23] E.S.R.O., PAR 378/1/1/1.

[24] E. H. W. Dunkin, 'Contributions towards the ecclesiastical history of the deanery of South Malling, in the county of Sussex', *S.A.C.*, XXVI (1875), p. 71. The will of the uncle is in E.S.R.O., Lewes Archdeaconry wills register A 59, p. 7. He left the advowson of Chailey to 'my sister-in-law Elizabeth Porter widow of my brother Thomas Porter' on condition that within three months of the testator's death his nephew Richard should be presented to the living. Other property was left to the testator's other nephew, the Revd. Thomas Colepepper Porter, who, if Richard was not living, was to be presented to Chailey. This was the Revd. Thomas Porter of the diary.

[25] Coates, according to the Porter monument, was from Wensley in Yorkshire. This was, presumably, why *Blencowe and Lower* (p. 189) referred to Mary as 'the daughter and coheiress of a Yorkshire gentleman of fortune'. Mary died, aged 67, on 1 August 1792.

[26] They are also commemorated on the Porter monument.

Raynes, a fellow of Jesus College, Cambridge, who became Archdeacon of Lewes; and three of their sons became priests. The Raynes's were, by the 1820s, very substantial landowners in East Hoathly. Mary Porter was godmother to Turner's son, Frederick, in 1771.[27]

SLATER, Samuel and Ann, of Hartfield. Turner's parents-in-law. Turner got on well with Samuel. He was a frequent visitor to East Hoathly, usually enlivening his visits with a present, and was especially solicitous following Peggy Turner's death. Turner advised him over the purchase of wool, and likewise he advised Turner when it came to buying a horse. By contrast Turner had little patience with Mrs. Slater: she evidently doted on her daughter and suspected Turner of ill-treating her (even, apparently, of causing her death). Turner likened her to Xantippe, and thought her unsuited to civilized society, though she 'might do well to sell oysters at Billingsgate'. The Slaters had three children who all predeceased them: Ann who died aged 18 in July 1758, Margaret (Turner's wife) who died in June 1761, and Samuel who died in September 1771. Mrs Slater died and was buried at Hartfield on 28 July 1773, and Samuel was buried there on 26 September 1779.[28] He left all his property to his brother John and to John's four sons, John, Fortune, Isaac and William.[29]

SNELLING, John, of Alfriston, doctor. The local surgeon who tended to be called in and consulted about matters which were beyond the power of the apothecary Richard Stone. Turner consulted him about his brother's deteriorating eyesight and about his own, and underwent a painful small operation to try and relieve the problem. Snelling was called in, along with Dr. Davy, a man-midwife, to conduct the post-mortem examination on the unfortunate Elizabeth Elless in July 1756; and it was he, too, whom rumour suspected of having hastened Peggy Turner's death by attempting to perform an operation to 'castrate' her. He was a godfather to one of Turner's sons by his second marriage in 1773.[30]

STONE, Richard, apothecary and surgeon of Blackboys. A very frequent visitor to the Turner house, especially during the lifetime of Peggy Turner. He regularly bled both Peggy and Thomas (his apprentice bled the servants): he extracted teeth, and performed minor surgery. One issue which he cut in Turner's back caused him considerable pain for a long time and in the end Turner had to consult the physician Dr. Snelling about it. He occasionally drank to excess and on one occasion nearly came to blows with Turner in a pub when they were both in their cups.

STREETER, John, parish clerk and postman. A much older man than Turner, he was born in 1694 and married late. He married Constant Neeves in 1751. She died aged 70 in 1764. He lived to be over 80, dying in 1777. He

---

[27] *F. M. Turner*, p. 110.
[28] E.S.R.O., PAR 360/1/2.
[29] E.S.R.O., Lewes Archdeaconry wills register A 64, p. 142.
[30] *F. M. Turner*, p. 111.

succeeded Robert Baker, another octogenarian, as parish clerk in 1762. As the postman on the Mayfield route, he was a useful man to Turner, constantly carrying goods and messages. He evidently felt that his worth to Turner was sometimes undervalued and in December 1758 actually refused his Christmas box as being too small. This stung Turner, who simply bore it in mind and gave it to him again the following Christmas. He was none the less friendly with Turner, betting with him on occasion and acting as his sponsor to the Mayfield Friendly Society.

THORNTON, Lawrence, excise officer and publican. Thornton first figures in the diary in December 1758 when he came to East Hoathly as excise officer. He soon became a firm friend of Turner: they played cards, drank, and went to cricket matches together. The excise officer was an important man for someone like Turner, dealing in tea, coffee, spirits, hops and other taxed commodities; and Turner wrote letters for him and acted as surety for him when he married Mary Gibbs, a girl who was under age in May 1762. Thornton left the area in 1763 to become an excise officer in Derbyshire, but had returned within a year to become landlord at *The Black Boy*, where his father-in-law, Joseph Gibbs, had been a previous licensee.

VINE, John, senior and junior. John Vine, senior, was at the beginning of the diary a farmer and the senior tenant of the Pelham properties in East Hoathly. Turner rented property from him to use as a schoolroom. This was in the recently erected building which subsequently became *The King's Head*.[31] He was, however, not a good businessman and in 1764 went bankrupt. Turner, together with Joseph Fuller and Abraham Baley of Lewes, were made trustees for the payment of his debts and a great deal of the administration of the affair fell to Turner.[32]

John Vine, Junior, the man who actually built *The King's Head*, was much more involved in village life during the period of the diary: he served parish office, joined in the social life and was responsible for the new decorations of the church and, apparently, for building a new house for the Revd. Thomas Porter.

The Vines were a prolific breed and many of that name move in and out of the pages of the diary.

VIRGOE, Mrs Mary. Mary Virgoe was the sister of Joseph Burges and, at the beginning of the diary, the widow of Isaac Virgoe, a well-to-do butcher at Lewes. She was thus the sister-in-law of Jeremiah French. She had married in 1747, and Isaac Virgoe had died in 1753, leaving her with two children whom, Turner considered, she spoiled. Turner, who rented property from Samuel Virgoe, her father-in-law, was much occupied in quasi-legal business for Mrs Virgoe: he dealt with the sale of a house for her in Lewes, with a mortgage, with her will, with her insurance, and with getting in and settling her debts. He also wrote letters for her and on one

[31] See above, p. 293, note 8.

[32] See above, p. 290, note 5; Abraham Baley considered Vine 'a poor but I believe honest man'.

occasion a petition. It was a considerable surprise to Turner when in 1761 she married one of his best friends, Thomas Davy, the shoemaker. She died in 1775 and was buried at East Hoathly on 3 February.

WATFORD, John, senior and junior. John Watford senior was an elderly man when the diary was written: a widower who had been born in the seventeenth century, he was lower in the social order than Turner. He did gardening work for Turner who regarded him as a 'mighty honest, good, sort of man', harmless, innocent and artless. He died in 1769. His son, John, was a farmer in a small way, older than Turner (he was born in 1718) and going up the social ladder but not yet at Turner's level. He was electioner to the overseer for four years during the period of the diary and headborough for one year. He sold a pig to Turner from time to time and lent him a horse. Turner liked him and compared his generosity, when Turner did some writing for him, with the penny-pinching attitude of his social superiors. He and his wife Elizabeth had three children between 1754 and 1760. He died in 1796, aged 77.

WELLER, Mrs Rebecca, widow. The widow of a mercer, Francis Weller, who died in 1749. It was to his shop in East Hoathly that Turner had come at Michaelmas 1750, and his widow, Rebecca, was thus Turner's landlord. She died in July 1763 by which time her son, Francis, who had been aged ten at his father's death, was of age. Just after the diary ends, Turner bought the premises outright from Francis junior.[33]

---

[33] E.S.R.O., Laughton manor court book 1766–79, S.A.S. A 664, p. 9.

Appendix C

Thomas Turner's financial dealings

TURNER was one of the few people in East Hoathly who thoroughly understood financial matters and how to draw up accounts. He spent his evenings making sure that the daily record of transactions in his shop was transferred item-by-item into the ledger which recorded the state of each customer's account with him. This was the activity known as 'posting my day book'. He was used by others (Mrs. Browne, for example) to keep their books in order. In the years before he himself became collector of the window and land taxes in 1760, he was called upon to write out the assessments to be taken for the approval of the commissioners in June/July of each year and to prepare the returns to be delivered to the receiver with the money collected then and in January. If a shopkeeper or farmer failed in business it was as often as not Turner who would be called in to make an inventory and to act as accountant at any subsequent sale. He it was who was trusted to distribute charity moneys and to transmit money collected upon charitable briefs; and year after year his skills in casting up accounts were required by those who served the parish office of overseer of the poor involving as it did the collection of the local poor rate and its disbursement to the large number of paupers. He was himself either churchwarden or overseer in the years 1756–7, 1757–8, 1762–3, 1763–4, 1764–5 and 1765–6 (indeed in 1764–5 he was churchwarden, overseer and surveyor of the highways all at the same time), and the only surviving eighteenth-century overseer's account book for the parish, which covers the period from 1762 to 1779, is entirely in his hand.[1]

Aside from his skills in accounting Turner also fulfilled another important role in a community such as East Hoathly. He traded in an age when an adequate supply of legal tender coin could not always be taken for granted, and when there was no network of banks (at any rate beyond the purely local level) by means of which money could be transferred from place to place by cheque. His shop was the point at which a large number of customers, and a very wide spread of suppliers (from as far away as Manchester), came together, and as such it played a key part in the transfer of cash and credit. There were, generally speaking, four ways in which goods were bought and sold, debts settled and paid: by barter, by cash including the as yet uncommon banknote, by promissory note of hand, and by bill of exchange. The diary shows all four of these ways in operation.

Barter was common in the community at large, and even in the commercial

[1] E.S.R.O., PAR 378/31/1/1.

world was frequently used. Turner often paid his servants not in cash but in goods: this was the case for instance with the devoted Mary Martin whose yearly salary was paid in goods on 11 July 1755. He exchanged goods for butter with Mrs. French on 21 January 1760, and then used the butter, as an account with Thomas Scrase of Lewes on the same day shows, to barter for wigs for himself and his brother. But perhaps the commonest form of direct barter was of rags for paper: having collected up rags in the neighbourhood, Turner packed them and sent them off to the paper mills at Hawkhurst and Loose in Kent or elsewhere, taking paper and a cash adjustment in exchange. At a lower level there were those who paid their parish rates in kind: Mrs. Elizabeth Browne, for example, settled hers on one occasion not in cash but by supplying a coffin for a pauper's funeral; and the parish occasionally acted similarly—for instance when Thomas German's fee of 7s. 6d. for keeping the church clock wound up was settled by giving him a hat.

Turner often complained of the lack of money in the area, but this was usually an exasperated comment on his inability to get his customers to pay their debts, rather than a reflection on the lack of coin. The diary does not indicate any permanent shortage of hard cash in the locality, and though there were occasions when in an emergency Turner had to borrow ready money from the Revd. Thomas Porter, Mrs. Atkins, Dr. Snelling or others, he was normally able to pay out and to receive cash for all small (and for some very large) transactions, and so were his customers and friends. At times he had great sums of money in the house and was worried by it: on New Year's Day 1763, for instance, he got Samuel Jenner to stay with him overnight to help him guard it. This was an occasion when he was about to send the accumulated tax contributions from the parish to the receiver at Maresfield. These sums were large: £76 in 1763, when Jenner carried the money there; over £110 when he went himself in May 1764; over £121 when Thomas Durrant took it in June 1761. It must have made the carrier feel very vulnerable on the road. Banknotes were few and far between, had only a local circulation for the most part because of the difficulty of cashing, say, a Lewes banknote in Liverpool, and also, since they were payable to bearer, there was always the danger of theft. Nevertheless they occasionally came Turner's way: the Revd. Thomas Porter, for instance, gave him one along with some cash as his contribution towards repairing the church in April 1763. The risks involved in dealing in coin however were greater than that of straight theft. As the diary shows there was a great variety of coins in circulation. Besides the guineas, half-guineas, crowns, half-crowns, shillings, sixpences and pence, there were coins of foreign origin: double pistoles also known as doubloons, and pistoles (generally agreed to be worth 36s. and 18s. respectively), half-pistoles (9s.), moidores (27s.), pieces of eight (4s. 6d.) and others. On 8 April 1765 Turner accepted 'a broad piece of gold' from Sarah Prall which he valued by weight at 18s. though he lent her only 14s. on its security. In that transaction Turner showed the importance of a coin's weight, for coins could be clipped and therefore lose their value and be refused. He normally equipped his messenger with some spare coins should any be refused by the receiver of the land tax (three guineas in May 1760, eight guineas and one half guinea in June 1761); and on one occasion, on 30 July 1762, one of his guineas was rejected by William Margesson in Southwark, 'it being not weight'. At other times Turner evidently accumulated too much cash in the way of business: then he would parcel it up and despatch it either to one of his local suppliers, such as Samuel Durrant in Lewes (to whom, for instance, he sent £26 via the postman, John Streeter, in May 1758), William Margesson in Southwark, or John Madgwick in Lewes. It was to Madgwick that on 27 July 1760 he gave £70 in cash in exchange for a far less vulnerable means of exchange—a bill drawn on a third party and payable at a fixed date.

Though cash along with barter was the normal means of exchange between Turner and those who used his shop or paid their taxes and rates, the commonest means of payment between traders was the bill of exchange which had great advantages over both cash and local banknotes. It was payable to a particular named person and therefore no great harm could ensue if it were lost; it was backed by commodities and not by cash alone and was therefore felt to be more secure than a local bank note; it was a safe, and, in many cases, an increasing, asset; and it could always be exchanged by a banker or another trader for cash (at a discount) should cash be needed before the bill became due.

The workings of the bill system in Turner's time have been clearly explained by T. S. Ashton and by T. S. Willan in their studies of two contemporary north-country tradesmen: Peter Stubs of Warrington and Abraham Dent of Kirkby Stephen.[2] A simple example of its operation can be seen on 11 May 1756. Turner met in Lewes Stephen Fletcher, travelling salesman for Samuel Ridings of Manchester, supplier of haberdashery to Turner. Turner's acount with Ridings was in debit £14 19s. 0d., and to settle it he drew a bill for that amount on the Southwark merchants, Margesson and Collison, payable at some date in the future, probably thirty days later. Margesson and Collison, who knew Turner well and respected his credit, would redeem the bill for cash on the due date. Ridings could in the meantime use the bill either to settle one of his own debts, simply by endorsing it to his creditor, or to raise cash by selling it at something less than its face value (this process is called 'discounting') to a local banker or anyone who would give him cash for it. In reverse it can be seen on 15 July 1756: a Richard Waite had evidently settled an account with George Kemp by drawing a bill on Turner (with whom he was presumably in credit). Turner sent Kemp the necessary cash and expected to receive the bill through the post for cancellation.

Another, slightly different, example occurs on 17 and 18 January 1755. Thomas Diplock needed to send payment to his landlord but did not wish to

[2] T. S. Ashton, *An eighteenth-century industrialist: Peter Stubs of Warrington, 1756–1806* (Manchester, 1939, reprinted 1961); T. S. Willan, *An eighteenth-century shopkeeper: Abraham Dent of Kirkby Stephen* (Manchester, 1970).

send cash. He therefore gave Turner the cash; Turner paid the cash, along with his other payments, to John Collison who happened to be in the area at the time, then drew a bill on Margesson and Collison payable to Diplock. Diplock presumably endorsed it and sent it to his landlord. Again, on 23 September 1760 Turner and his brother Moses, having bought up wool from the local farmers, sold it to Thomas Friend in Lewes; Friend paid them with a bill which had been drawn by someone in Warminster, Wiltshire, on a merchant in Basinghall Street, London, payable to Friend. Friend endorsed it to Turner who presumably gave Moses cash for his part and sent the bill to Margesson and Collison for them to redeem and to credit his account. At times Turner is seen discounting bills, as on 5 January 1765 when he charged John Russell 2s. 6d. for cashing a bill on a London timber merchant worth £24 17s. 10d. (which he transcribed verbatim and then used to pay his own account with Robert Plumer, the Lewes grocer); and at other times, for example in June 1762, he used a bill instead of cash to pay the parish's land tax quota.

The system is seen at its fullest on Turner's trip to London in March 1759. Small sums he paid in cash but the larger accounts were settled by bills drawn on Margesson and Collison, while he paid into his account there a £20 banknote and a large bill of £130 drawn on Sir Joseph Hankey and partners by the Lewes merchant Samuel Durrant, for which on the previous Saturday Turner had paid partly in cash and partly by giving an interest-bearing promissory note. By acting as clearing-houses for bills certain merchants were performing all the functions of bankers. To all intents and purposes Margesson and Collison were Turner's London bankers, and Samuel Durrant of Lewes was his Sussex banker. In this way the foundations of modern banking were laid.

This whole system of course depended on mutual respect and trust amongst traders. Occasionally, for one reason or another, this trust broke down and things could be serious for the tradesman left holding the bills. This nearly happened to Turner in February 1765 when, having redeemed at a discount bills to the value of over one hundred pounds drawn by the apothecary Richard Stone on various tradesmen in London, Turner found that one of the tradesmen refused to pay—presumably because he was not satisfied as to Stone's creditworthiness. This happened twice, but, after fearing that he might lose his money, Turner pursued Stone and recovered the debt from him in cash.

Most of Turner's business was done on credit; almost all his customers ran up debts with him, to be paid when harvest came, or when the sums could be bartered away, and most of these debts were unsecured. Occasionally Turner would accept a note of hand—a promise to pay (sometimes with interest) at a future date or on demand but not guaranteed by a third party; and once, in the case of his cousin Charles Hill, he insisted that a bond at 5% interest should be substituted for two of his notes of hand. But for the most part it was up to Turner to get in the money owed to him by his customers and it was a task which he did not much enjoy. Often he would have to wait to settle at one of the many local fairs where dealers met one another; and many fruitless days were spent travelling from customer to customer, failing to get in what was due to him, and he chided himself for having been too indulgent when a book debt of £40 was likely to be lost in February 1764. A year earlier, on 25 February 1763 he complained that 'I am so confined with . . . some large debtors in my parish that I hardly know which way to act or extricate myself out of so great a dilemma'. In the absence of any of Turner's ledgers it is not always easy to see how long he allowed debts to remain, but Thomas Darby, for instance, who ran into financial difficulties in 1758 had evidently been allowed to run up a debt amounting to £17 or £18 over a four-year period. The thought of pursuing an old acquaintance into the courts filled Turner with guilt and made him wish 'that I lived in solitude and had not the occasion to act in trade'; so he put the matter into the hands of a Lewes attorney and eventually took a £22 bond from Darby backed by a bill of sale on his chattels. Nevertheless he was still trying to get in £16 of it four years later, and hesitating over the idea of putting the bill of sale into execution.

His suppliers on the other hand regularly visited him to settle accounts, usually staying the night with him if they came from any distance. John Collison of Margesson and Collison was often in East Hoathly; Richard Sterry, the London oilman, stayed overnight on 30 July 1762, and riders from other suppliers called when they were in the area. Those from the more immediate locality were, like John Madgwick of Lewes, or Thomas Tipper, the Newhaven brewer, numbered amongst Turner's friends and settling with them was part of the normal social intercourse.

*

The suppliers whom Turner visited on his trip to London in March 1759 can, for the most part, be identified from contemporary directories[3] as follows:

Mr. John Albiston	John Albiston, tobacconist, Friday Street.
Barlow and Wigginton	Barlow, Wigginton & Francis, linen-drapers, Gracechurch Street.
Mr. James Blake	James Blake, linen-draper, Gracechurch Street.
Mr. Calverley and Son	Calverley and Sons, druggists, Borough, Southwark.
Mr. Corderoy	John Corderoy, horse-millener, Upper Thames Street.
Mr. John Crouch	John Crouch, grocer, Borough, Southwark.
Mrs. Crowley and Co.	Theodosia Crowley & Co., ironmongers, 151 Upper Thames Street.
Cruttenden and Burgess	Cruttenden & Burgess, hosiers, Borough, Southwark.

[3] For London directories of this period see C. W. F. Goss, *The London directories, 1677–1855* (1932).

Daker and Stringer	?
Mr. Farnworth	William Farnworth, warehouseman, Cullum Street.
Michael Gatfield	of Gedaliah Gatfield, hat warehouse, Blowbladder Street.
Gore and Perchard	Gore and Perchard, metal button and hardware warehouse, Cannon Street.
Mr. Joseph Hillier	Joseph Hillier & Co., woollen-drapers, Borough, Southwark.
John Levy	John Levy, chinaman, Borough, Southwark.
Mr. Margesson	William Margesson of Margesson and Collison, haberdashers, Borough, Southwark.
Mr. Thomas Neatby	Thomas Neatby, distiller, Borough, Southwark.
Mr. Norfolk	Richard Norfolk, pewterer, St. Margaret's Hill, Southwark.
Mr. Nuns	? Moses Nunes, merchant, Bury Street.
Mr. George Otway	George Otway, cheesemonger, St. Margaret's Hill, Southwark.
Mrs. Mary Reynolds	?
Ruston and Kendall	Rushton & Kendall, wholesale mercers, Gracechurch Street.
Mr. Richard Sharp	Richard Sharp, Fish Street Hill, Gracechurch Street.
Smith and Bickham	Smith & Bickham, haberdashers, Fish Street Hill, Gracechurch Street.
Mr. Thomas Standing	Thomas Standen of Hargrave and Standen, Gracechurch Street.
Mr. Sterry	Richard Sterry, oilman
Messrs. Thomson and West	Thompson and Son, hop-factors, Borough, Southwark.
Mr. John Wathin	John Wathen, hatter, the Three Conies, in Cannon Street.

The original diary contains a great deal of information which has had, unfortunately, to be omitted here, about Turner's other suppliers and the payments made to them. Those mentioned in these extracts are:

Beard, George, of Chailey, grocer.
Bentley, Thomas, broom-maker.
Blackwell, Edward, of Hawkhurst, Kent, paper-maker.
Bristow, Faulkner, of Lewes, confectioner and grocer.
Brooker, William, of Mayfield, wigmaker.
Burfield, Thomas, of Steyning, supplier of beehives.
Burtenshaw, of West Peckham, Kent, paper-maker.

Chittenden, Thomas, of Hawkhurst, mercer.
Cleaver, Robert, Richard and Joseph, woollen-drapers in Cornhill, London.
Clymson, of Wannock, glover.
Crowhurst, James, Bayley's Lane, earthenware manufacturer.
Durrant, Samuel, of Lewes, mercer.
Fieldcox, James, of Battle, clog and patten-maker.
Fielder, John, tile maker.
Freeman, Thomas, of Mayfield, clog and patten-maker.
Friend, Thomas, of Lewes, wool-factor.
Funnell, William, brickmaker.
Gilbert,   , of Waldron, carpenter.
Godfrey, Dr.   , of Lewes, cordial supplier.
Gosling, John, junior, of Maidstone.
Hammond, John, of Battle, gunpowder supplier.
Harman, Thomas, of Lewes, clay pipe maker.
Hart, Percival, of Uckfield, ?tailor.
Horsecraft, widow, of Chiddingly, supplier of spirits.
Howard,   , of Hartfield, gingerbread baker.
Jenner, John, of Hailsham, hatter.
Lashmer, Edward, of Chiddingly, grocer.
Madgwick, John, of Lewes, grocer and mercer.
Merrick,   , of Southwark, grocer.
Minifie, John, partner of James Jordan, draper.
Paris, Francis, supplier of spirits.
Plumer, Robert, of Lewes, grocer.
Pope, Thomas, junior, of Maidstone, mercer.
Reeder, Cocks and Co., wool-factors.
Relfe, Edward, of Lewes, saddler.
Ridings, Samuel and John, of Manchester, haberdashers.
Roase, William, of Lewes, grocer.
Roase, Mrs., widow, of Lewes, grocer.
Scrase, Thomas, of Lewes, victualler.
Smith, Edward, supplier of cards.
Swainstone and Hedges, hop-factors.
Tomlin, George, of Red-lion-street, Borough, Southwark, hop-factor.
Verral, George, of Lewes, supplier of soap.
Wilbar, John, brushmaker.

Appendix D

Thomas Turner's reading[1]

TURNER was used to having books in his house; he bought books from time to time; he sought solace in books; and he regularly read aloud to his wife and his friends in the evenings. His tastes ranged from theological controversy and exhortatory sermons to works of history, from jest books to works of great piety, from the philosophy of John Locke (which he found hard going) to works of history and travel. He read Shakespeare's plays and the *Complete Peerage*, the works of James Hervey and the *Night Thoughts* of Edward Young, and had a selection of practical works—on medicine, on surveying, on farriery, on parish administration and on other matters. He seems to have been able to read the latest issue of both the *London Gazette* and the *Sussex Weekly Advertiser* (which he normally read on a Monday), and may well have subscribed to certain periodicals such as *The Universal Magazine of Knowledge and Pleasure* and *The London Magazine*, since his references are always to the most recent issue.

The books mentioned in these extracts as amongst his possessions are as follows (the phrase 'and subsequent editions' means prior to Turner's time):

Almanacs, unspecified.

Addison, Joseph, *The Freeholder* (1715–16); *The Spectator* (1711–12, 1714).

——*The evidences of the Christian religion . . . to which are added several discourses against atheism and infidelity, and in defence of the Christian revelation.* (First published separately 1730; and subsequent editions).

——*Cato, a tragedy . . .* (1713, and subsequent editions). [See also Steele].

Ainsworth, Robert, *Thesaurus linguae latinae compendarius: or a compendious dictionary of the latin tongue.* 2 vols. (First published 1736).

Allestree, Richard, *The new whole duty of man, containing the faith as well as the practice of a Christian.* (Many editions prior to Turner).

Bally, George, *The wisdom of the Supreme Being.* (A Poem: 1756)

Beckford, Richard, see *The Monitor*.

Beveridge, William, *Private thoughts upon religion digested into twelve articles, with practical resolutions form'd thereupon* (1709, and subsequent editions).

Bracken, Henry, *The traveller's pocket-farrier: or a treatise upon the distempers and common incidents happening to horses upon a journey . . .* (1743. At least six editions prior to Turner).

[1] The identification of many of the works mentioned in the diary was first undertaken at Yale by Harold P. Melcher Jr., over thirty years ago. I am most grateful to him for allowing me to use his notes and to draw on the work he did then. Mr. Melcher's work is now with the Thomas Turner Collection at the Stirling Library at Yale, series II, box 9, folder 31.

Budgen, Richard, *The passage of the hurricane from the seaside at Bexhill in Sussex to Newingden-Level, the twentieth day of May 1729* (1730).

Burkitt, William, *The poor man's help and young man's guide* (1694. In its 30th edition in the year that Turner read it).

Burn, Richard, *The justice of the peace and parish officer* (1755).

Burnet, Gilbert, *The history of the reformation of the Church of England* (1679–1714, and subsequent editions).

Butler, Samuel, *Hudibras*. (Many editions from the 1660s onwards).

[Byng]. *An appeal to the people: containing the genuine and entire letter of Admiral Byng to the Secr. of the Ad——y . . .* (1756).

Centlivre, Mrs. Susanna, *The busie body*. (A play: five editions prior to Turner).

Cibber, Colley, *The careless husband*. (A play: many editions prior to Turner).

Collins, Arthur, *The peerage of England*. (Three editions prior to Turner).

Congreve, William, *Love for love*. (A play: many editions prior to Turner).

——*The way of the world*. (A play: many editions prior to Turner).

[Delap] 'The examination of Francis Delap Esq., late Provost Marshal General of the Island of Jamaica . . . Nov. 10th, 1756'. (Not identified).

Derham, William, *Physico-Theology: or, a demonstration of the being and attributes of God, from his works of creation. Being the substance of XVI sermons preached . . . at the Honble Mr. Boyle's lectures, in the years 1711 and 1712*. (1713, and subsequent editions).

Dobel, D., *Primitive Christianity propounded; or an essay to revive the ancient mode or manner of preaching the gospel* (1755).

Drelincourt, Charles, *The Christian's defence against the fears of death. With seasonable directions how to prepare ourselves to die well*. (Many editions prior to Turner).

Gay, John, *Fables* (1727, and subsequent editions).

Gibson, Edmund, *The evil and danger of lukewarmness in religion: being the first part of the Bishop of London's late pastoral letter . . .* [of 1739]. (Many editions prior to Turner).

——*Trust in God the best remedy against fears of all kinds: a sermon . . .* (Many editions prior to Turner).

Gordon, Patrick, *Geography anatomized: or a compleat geographical grammer* (1693, and subsequent editions).

Haworth, Richard, Sermon 'preached in this church on 1 August 1716'. (Not identified).

Hervey, James, *Meditations among the tombs: in a letter to a lady* (1746, and subsequent editions).

——*The time of danger, and the means of safety; to which is added, the way of holiness. Being the substance of three sermons preached on the late public fast-days* (1757).

——*Theron and Aspasio: or, a series of dialogues and letters upon the most important and interesting subjects* (1755).

Horneck, Anthony, *The great law of consideration; or, a discourse, wherein the* nature, usefulness and absolute necessity of consideration, in order to a truly serious and religious life is laid open (1677, and subsequent editions).

Josephus, Flavius, 'The antiquities of the Jews'. Probably from *The genuine works of Flavius Josephus*, translated with notes by W. Whiston (1755).

Leadbetter, Charles, *The royal gauger; or gauging made perfectly easy* (fourth edition, 1755).

'Lewes newspaper', 'Lewes journal', 'the newspaper': *Sussex Weekly Advertiser, or Lewes Journal.*

Locke, John, *An essay concerning humane understanding* (1690, and subsequent editions).

Martin, Benjamin, *The general magazine of arts and sciences* (in 14 volumes between 1755 and 1765).

Mead, Richard, *A mechanical account of poisons; in several essays* (1702, and subsequent editions).

Medical essays and observations, revised and published by a society in Edinburgh (5 vols. 1733–44, and subsequent editions).

Milton, John, *Paradise Lost* and *Paradise Regained* (Many editions prior to Turner).

Nicholl, John, *The execrable practice of buying and selling livings commonly called Simony: in a sermon . . . preached at a visitation held by the . . . Archdeacon of Lewes . . .* (1764).

Pope, Alexander, *Odyssey* (1705, and subsequent editions).

Pitton de Tournefort, Joseph, *A voyage into the Levant . . . to which is prefix'd the author's life, in a letter to M. Begon: as also his eulogium pronounc'd by M. Fontenelle . . .* (3 vols, 1741).

Richardson, Samuel, *Clarissa. Or, the history of a young lady* (1747–48, and subsequent editions).

Rowe, Nicholas, *Tamerlane* (A play: 1702, and subsequent editions).

Salmon, Thomas, *A critical essay concerning marriage . . . by a gentleman* (1724).

A serious address to the public, concerning the most probable means of avoiding the dangers of innoculation . . . (1758)

Shakespeare, William, *The merry wives of Windsor; As you like it; The taming of the shrew; Works.*

Sherlock, Thomas, *Sermons on various subjects, moral and theological, now first published* (1747)

Sherlock, William, *A practical discourse concerning death* (1689; and subsequent editions).

Smart, Christopher, *On the eternity of the Supreme Being: a poetical essay* (1750, second edition 1752).

——*On the immensity of the Supreme Being: a poetical essay* (1751, second edition 1753)

——*On the omniscience of the Supreme Being: a poetical essay* (1752).

——*On the power of the Supreme Being: a poetical essay* (1754).

Steele, Richard, *The conscious lovers, a comedy in five acts.* (A play: 1723, and subsequent editions).

————(with Addison), *The Guardian* (1713), *The Tatler* (1709–11). [See also Addison].

Sterne, Laurence, *Tristram Shandy* (only six of the nine volumes had appeared when Turner came across it in 1763).

'The history of England' (not identified).

The London gazette.

The London magazine; or, gentleman's monthly intelligencer (1732–85).

The Universal magazine of knowledge and pleasure (1747–1814).

The Monitor; or the British freeholder (1755–63).

'The West country clothier' (not identified).

Tillotson, John, *Sermons* (1694, and subsequent editions).
    6th 1756 . . . (2nd ed., 1756).

Torriano, Nathanael, *A sermon preached at Hooe and Ningfield in . . . Sussex . . . previous to the day of public fasting . . .* (1756).

Vanbrugh, Sir John, *A journey to London* (A play: many editions prior to Turner).

Wake, William, *The principles of the Christian religion explained in a brief commentary upon the church catechism* (1699, and subsequent editions).

Walder, James, *The ax laid to the root; or, a preservative against the erroneous doctrines of the Methodists; candidly offered to the consideration of all Christians. In three discourses* (1763).

Wharton, Henry, *A defence of pluralities, or, holding two benefices with cure of souls* (1692; second edition 1703).

Wilkes, John, *The North Briton* (1762).

Wiseman, Richard, *Several chirurgical treatises* (1676, and subsequent editions).

Young, Edward, *A vindication of providence; or, a true estimate of human life* (1728, and subsequent editions).
    ————*The complaint: or night thoughts* (1747, and subsequent editions).
    ————*The revenge* (A tragedy: 1721, and subsequent editions).

Elsewhere in those parts of the diary not reproduced here, Turner mentioned the following works as in his possession or as read by him (unspecific references such as 'a political pamphlet', 'some old magazines' and books of the Bible have not been noted):

Addison, John, *Remarks on several parts of Italy &c. in the years 1701, 1702, 1703* (1705, and subsequent editions).

An apology for the life of Bamfylde-Moore Carew . . . commonly known . . . by the title of King of the Beggars . . . (1749, and subsequent editions).

Baker, Thomas, *Tunbridge-walks: or, the yeomen of Kent* (A play: 1703, and subsequent editions).

Beveridge, William, [various sermons from] *The works of the right reverend Father in God Dr. William Beveridge . . .* (1720).

Boyle, Roger, Earl of Orrery, *The Black Prince* (A play: perhaps in Boyle's *Dramatic works*, 1739).

Bracken, Henry, *Farriery improved: or, a compleat treatise upon the art of farriery* (1737, and subsequent editons).

Brookes, Richard, *The general practice of physic* . . . (1754).

Bunyan, John, *The pilgrim's progress* (1684, and subsequent editions).

Burkitt, William, *Expository notes, with practical observations, on The New Testament* (1700, and subsequent editions)

Butler, Samuel, *Posthumous works in prose and verse* (1715, and subsequent editions).

Cheselden, William, *The anatomy of the humane body* . . .(1713, and subsequent editions).

Colman, T. and Thornton, Bonnell, *The Connoisseur* (1754–56).

The complete letter-writer; or, a new and polite English secretary (1755).

Courtail, John, *The expedience and necessity of wisdom and innocence towards a due discharge of the ministerial office, preached at the visitation of the Archdeacon of Lewes* (1760).

Defoe, Daniel, *A tour thro' the whole island of Great Britain* . . . (1724, and subsequent editions).

Delany, Patrick, an unspecified sermon by.

Dryden, John, *Religio laici or a layman's faith* (1682, and subsequent editions).

Dunton, John, *Athenian sport: or, two thousand paradoxes merrily argued, to amuse and divert the age* (1707).

Euclid, *Elements of geometry* (Many editions).

Farquhar, George, *The recruiting officer, a comedy* (A play: 1706, and subsequent editions).

Fletcher, John, *The humorous lieutenant* (A play; 1620, and subsequent editions).

Frewen, Thomas, an unidentified answer to Giles Watts.

Gibson, Edmund, an unidentified sermon.

Gordon, Thomas and Trenchard, John, *The independent whig* (1720–1721, and subsequent editions).

Gray, John, *Poems on several occasions* (1720; and subsequent editions).

Huxham, John, *An essay on fevers and their various kinds as depending on different constitutions of the blood* (1750).

Letters from a Moor at London to his friend at Tunis (1736).

Locke, John, *Some thoughts concerning education* (1693, and subsequent editions).

Love, John, *The whole art of surveying* . . . (1716).

Mauriceau, François, *The accomplish't midwife, treating of the diseases of women with child, and in child-bed* . . . (1673).

Mead, Richard, *A treatise concerning the influence of the sun and moon upon human bodies and the diseases thereby produced* . . . (1748).

Medulla medicinae universae; or, a new . . . *dispensary* . . . (1747).

Nelson, Robert, *A companion for the festivals and fasts of the Church of England; with collects and prayers for each solemnity* (1704, and subsequent editions).

Nelson, William, *Lex maneriorum: or, the law and customs of England relating to manors and lords of manors, their stewards, deputies, tenants and others* . . . (1724, and subsequent editions).

Perronet, Vincent, an unspecified book by.

Philips, Ambrose, *The distrest mother. A tragedy* (A play: 1712, and subsequent editions).

Pope, Alexander, *The dunciad. An heroic poem* (1728, and subsequent editions).

———*The Iliad* (1716, and subsequent editions).

Pringle, John, *Observations on the diseases of the army in camp and garrison* (1752, and subsequent editions).

Russell, Richard, *Oeconomia naturae in morbis acutis et chronicis glandularum,* and its English version *The Oeconomy of human nature* . . . (1755).

Salmon, Thomas, *The modern gazetteer: or, a short view of the several nations of the world* . . . (1746, and subsequent editions)

———*A new geographical and historical grammar* (1749, and subsequent editions).

Scott, David, *Every man his own broker* (1761).

Shakespeare, William, *Hamlet, King Lear, Macbeth, Othello.*

Sharp, John, *The works of the most rev. Dr. John Sharp* (1754).

Sherlock, Thomas, *A letter to the clergy and people of London and Westminster* . . . *on occasion of the late earthquakes* (1750).

Sherlock, William, *A practical discourse concerning a future judgment* (1692, and subsequent editions).

Smart, Christopher, *On the goodness of the Supreme Being* (1756).

Smollett, Tobias, *The adventures of Peregrine Pickle* . . . (1751).

The Sports of the muses, or a minute's mirth for any hour of the day . . . (1752).

Steele, Richard, *The Christian hero* (1701, and subsequent editions).

Sterne, Laurence, *The sermons of Mr. Yorick* (1760 onwards).

Swift, Jonathan, *A tale of a tub* . . . (1704, and subsequent editions).

Thomson, James, *The seasons* (1730, and subsequent editions).

Warren, Robert, *The devout Christian's companion* . . . (1733).

The whole duty of man, and *The whole duty of woman.*

PERIODICALS

The gentleman's magazine.
Owen's weekly chronicle.
Read's weekly journal; or the British gazetteer.
The Royal magazine; or, gentleman's monthly companion.

UNIDENTIFIED WORKS

Sentence of the court . . . at Lisbon . . . on the Duke of Aveiro and Marquis Tavora with several more, who conspired and made an attempt against the life of his most faithful majesty on the night of the 3rd of September last (Feb. 1759).

Castalio and Julia.

Discourses on difficult texts by unknown hands.

Every man his own lawyer.

Joel the prophet.

New view of London and Westminster.

The prudent jester.

Appendix E

The manuscript and its history
by Dean K. Worcester Jr.

THE manuscript of the diary of Thomas Turner is among the Thomas Turner Papers in the library of Yale University, New Haven, Connecticut. It consists of 111 memorandum books, or notebooks, approximately 4 inches wide and 6¼ inches tall, centre-sewn, with pages varying in number between 64 and 96. Some volumes have a thin grey or blue paper cover. They are worn at the edges, suggesting that they may have been carried in a pocket. They are written in Turner's big, bold hand with few erasures or crossings-out: they are not therefore jottings made as and when events happened or ideas occurred; they appear to have been written carefully and sometimes, evidently, several days' entries were copied in at the same time. To judge from Turner's comments on his family (especially his first wife during her lifetime), his friends, and his enemies, the diary was a record for his eyes alone, and yet more than once he envisages others reading it. On 11 December 1761, for example, feeling that a pauper had been shabbily treated by his neighbours, he set down his own views 'for the future satisfaction of anyone who may happen to see my memoirs', and a similar motivation lay behind his comments on 15 April 1765 on his prospective second marriage.

There were originally at least 116, and possibly 117, volumes. The first three were labelled A, B, and C, and there may have been one labelled D. The remainder were numbered from 1 to 113. Volume A starts with an entry for 21 February 1754 and ends with that for 13 April. Volume B is now missing, though it was available to the first editors and the entry of 26 June 1754 can be supplied from their edition. Volume C covers the period from 26 July to 25 August 1754, and it is possible that a volume D followed this since there is a gap until volume 1 begins on 1 January 1755. The series numbered from 1 to 113 lacks four volumes: 40 (23 July to 24 August 1758), 50 (2 August to 2 September 1759), 89 (29 July to 1 September 1763), and 91 (14 October to 29 October 1763). Volume 113 ends on 16 June 1765 but contains one final entry made six weeks later on 31 July. All together they comprise Series I of the Thomas Turner papers.

Series II of the Thomas Turner papers consists of the papers produced in the late 1940s when I worked on the diary with a view to publication; and Series III consists of a small group of papers which were acquired by Yale University with the diaries. Amongst these papers are two letters, both written by Henry Martin Turner, the son of Philip Turner and the grandson of the diarist. The first is unsigned and undated. It begins: 'Herein contains a

manuscript &c. of the late Thomas Turner sacred to the memory of
[   ]¹ consigned to Henry Holman Senior his acceptance as tribute for
his invaluable service rendered the [   ]² Henry Martin Turner and his
family generally. [It] is suggested by H.H. seconded by H.M.T. the Dearest³
name should be more known thro' an extended publication . . . '. The second
letter is on a large sheet of blue paper. One side has been ruled into rectangles
and an effort made to make a chart of the volume dates and numbers. On the
other side is written:

> The diary of the late Mr. Thomas Turner East Hothly Sussex I
> bequeath to my nephew Thomas Holman surgeon son of Henry
> Holman Jr. surgeon of the above nam'd. It is my particular wish the said
> Diary be kept in the family; and doubtless the [perusal?] of which will be
> interesting and amusing especially so to an 'Antiquarian' in comparing
> those days of June AD 1754 up to the present time 1858. I suggest
> should it be thought prudent to lend either the numbers to a friend, not
> without the knowledge or sanction of the second person and member of
> the family.

> Witness my hand
> this [   ] day of July 1858   Henry Martin Turner

While the manuscript was still in the possession of Henry Martin Turner it
was lent to R. W. Blencowe and M. A. Lower, and they used excerpts from it
in an article entitled 'Extracts from the diary of a Sussex tradesman, a hundred
years ago', published in the volume of the *Sussex Archaeological Collections* for
1859. Their purpose was simply to point up the difference between life in
fashionable mid-nineteenth-century Sussex and that one hundred years
previously when the county had been, in their words, 'one of the most secluded
and uncivilized districts in England',⁴ and the quotations which they took from
the diary were deployed, not necessarily in chronological order, solely to that
end. It is their selection which until the publication of this present edition has
remained the only publicly available diary of Thomas Turner.

Charles Dickens read Blencowe and Lower's article and wrote about it in
All the Year Round (13 April 1861) under the heading 'Thomas Turner's
parlour'. Charles Fleet gave an appraisal of it in *Glimpses of our Sussex Ancestors*
(1882) and Arthur Ponsonby devoted five pages to it in his *English Diaries from
the Sixteenth to the Eighteenth Centuries* (1923).⁵ In 1925 John Lane The Bodley
Head reissued an inaccurate version of the quotations alone, with a few of
Blencowe and Lower's footnotes, a foreword by Florence Maris Turner (Mrs
Charles Lamb), great-great-granddaughter of the diarist, and an introduction
by J. B. Priestley. In 1979 Oxford University Press published a corrected

---

¹ Word or words lost.   ² Word or words lost.
³ i.e. diarist.   ⁴ *Blencowe and Lower*, p.180.
⁵ Other extensive uses of the diary in print are mentioned in Dean K. Worcester Jr.,
The life and times of Thomas Turner of East Hoathly (1948), appendix A.

version of the quotations with an introduction by G. H. Jennings and with the title *The Diary of a Georgian Shopkeeper*.

The manuscript meanwhile had apparently been deposited by the Holman family in Barbican House, Lewes, the home of the Sussex Archaeological Society, and it was there that in the 1930s it was seen by Professor Mildred Campbell of Vassar College, who was preparing what was to become her book *The English Yeoman under Elizabeth and the Early Stuarts* (1942). Realizing that it was vastly more extensive than the published extracts had led her to believe, she brought it to the attention of Professor Wallace Notestein of Yale, a celebrated authority on diaries, most notably of the seventeenth century. Doubtless through his influence Yale University Library acquired the diaries together with the documents in Series III in April 1944 by purchase through the rare book dealer, William H. Robinson Ltd. of Pall Mall, London. The only further documents written by Thomas Turner which have come to light in the meantime are those in the East Hoathly parish records, now in the East Sussex Record Office in Lewes,[6] and a group of miscellaneous bills and parish accounts found in the roof of Turner's house at the end of the nineteenth century, given to the Sussex Archaeological Society in about 1940, and now also in the East Sussex Record Office.[7]

In a seminar at Yale in 1946 Professor Notestein mentioned the diary and the enormous wealth of detail it might be assumed to contain (for he had not been able to read it all), and suggested it as a research project for those inclined to do the necessary work. Two students responded to the suggestion, Harold P. Melcher Jr. and myself. Over the next nine months we transcribed it, omitting about one-third in the process. Harold Melcher concerned himself mainly with Turner's voluminous reading, in the hope that he might have made significant or noteworthy comments on the current literature—a hope which was in fact doomed. I submitted an essay called *The life and times of Thomas Turner of East Hoathly*, which was published by the university presses of Yale and Oxford in 1948 as Volume VI of the Yale series: *Undergraduate Prize Essays*. I then prepared our manuscript for the press, but was unable to find a publisher. Rather than throw away the work of several years, I presented Harold Melcher's notes and index, and my own papers, to the Yale Library, where most of them remain as Series II of the Thomas Turner papers. Both Harold Melcher and I are very happy that 35 years later our work should provide much of the raw material from which this new edition is constructed.

Series III of the Thomas Turner papers consists, besides the letters quoted above, of seven letters from Turner to his children Margaret and Philip written between 1786 and 1790, a group of miscellaneous medical prescriptions, a few odd East Hoathly parochial documents, copies of the East Hoathly census returns for 1811 and 1821, and a copy of Thomas Turner's will, which was made on 14 January and proved on 28 September 1793.

[6] E.S.R.O., PAR 378.
[7] E.S.R.O., AMS 5841. See L. B. Smith and F. A. Hadley, 'Turner trove. A link with a famous Sussex diarist', *S.C.M.* 9 (1935), p. 546.

Index

Neither Thomas Turner nor East Hoathly have been separately indexed. Thomas Turner is referred to as T. throughout. Persons whose names are set in **bold type** have an entry in Appendix B. All indexed English place-names are in Sussex unless otherwise stated.

Hawkhurst, Kent, 10 and n., 164, 261, 345

Blake, James, linen-draper, Gracechurch St., London, 160, 178, 201, 344

Blakeney, William, general, 55 and n.

Blare, Archibald, 324n.

Blencowe, R. W., previous editor of the diary, vii, 355

Boarshead, in Rotherfield, 50, 60

Bodmin, John, of Chalk, Kent, 205

Bonnick, Mr., 12

Bonwick, Mr., butcher, 19

Bonwick, Joseph, overseer of the poor, Waldron, 115, 117, 167, 235

books (*books mentioned in the text are indexed by author, periodicals by title*): list of, read by T., Appendix D, 347–53; importance of, to T., xxv, 143, 282; T. buys, 8, 128, 143; Tomsett's, 39

boots, 29, 30

Boscawen, George, admiral, 156 and n., 161, 194n.

Bowles, Mr., racehorse owner, 300

Bowra, Peter, 261

Box, John, attorney, of Hailsham, 122–3, 208

Boxall, Henry, 107

Bracken, Henry, *Pocket farrier*, 210

'Branan Rules', 299

brandy, 5, 24, 37, 39, 53, 68, 72, 73, 122, 201, 260, 282, 286, 296, 298, 310, 311, 327, 330

Bray, Mr., racehorse owner, 210

Bray, Henry, 316

Brazer, Ann, 39, 57, 121

Brazer, John, 284

Brazer, Lucy, 39, 121

Brazer, Mary, 19

Brazer, Richard, servant to Edward Russell, 19, 74, 120–1

Brede, 140

Breeden, John, of Pevensey, 111, 112, 114, 182, 224, 225, 226

Brest, France, 154

brewing, 6, 145, 183

bricks, 10

Bridger, John, Esq., J. P., of Offham, 52, 59, 60, 119, 144, 329

Bridger, Sir John, of Hamsey, 278

Bridger, Sarah, 316

Bridges, Henry, inventor of the Microcosm, 161n.

Bridgman, Richard, 45, 71, 135, 156, 195, 204

briefs, church, *see* charity

Brighton, xxiv, xxxvii, 43, 157, 213, 264, 315n.

Bristed, Revd. John, vicar of Slaugham, 100

Bristow, Ann, wife of John B., 74, 276

Bristow, Edward, son of John B., 276

Bristow, Fanny, wife of William B., 270, 271, 276–7 and n.

Bristow, Faulkner, confectioner of Lewes, 220, 345

Bristow, John, of Waldron, 276

Bristow, John, son of John B., 276

Bristow, Samuel, son of William B., 270, 271, 276–7 and n.

Bristow, Thomas, 307

Bristow, William, son of William B., 270, 271, 276–7 and n.

Bristow, William, son of John B., 270, 271, 276–7 and n.

Broad Oak, 45, 98

Broadstone, 106

Brooker, David, 192

Brooker, William, of Mayfield, wigmaker, 168, 345

Browne, Mr., of Withyam, 46

Browne, Mrs. Elizabeth, xxx, 5, 30, 117, 123, 223, 236, 248 and n., 249, 269n.; 285, 293 and n., 294, 316, 320, 327, 328, 340, 341

Browne, John: 38, 139–40, 195, 199, 219; death and funeral of, 220; infant son of, buried, 129; infant son of, christened, 212

Browne, John, of Ditchling, 261

Broyle, 10, 96, 100, 118, 168, 332

Brunswick, Prince Ferdinand of, *see* Ferdinand

Buckall, John, of Lewes, 29

Buckhurst, 46

Buckwell, John, 78

Budgen, Edward, bill signed by, 311

Budgen, Richard, *Passage of the hurricane . . .*, 128, 136

Bull, James, of Whitesmith, 111, 128, 145, 189

bullace, T. grows, 65

Buller, Mr., 252

bumboo, 36

Burfield, Thomas, of Steyning, 2, 71–2, 202, 217, 321, 345

Burges, Jael, sister of John B., 110

Burges, John, of Brook House, Rotherfield, 17, 29, 110, 327

Porter, Harriet, (subsequently Raynes), daughter of Revd. Thomas P., xxxii, 257, 336–7

Porter, Mary, wife of the Revd. Thomas P.: birth of children, 45, 128; gives and receives hospitality, 141–2, 169, 171–2, 176, 194, 197, 198, 200, 209, 242, 253, 312; leads revellers, xxxii and n., 137–9; treats T. badly, 93, 98; treats T. well, 132; mentioned, 101; notes on, 336–7

Porter, Revd. Richard, uncle of Revds. Richard and Thomas P., 336

Porter, Revd. Richard, brother of Revd. Thomas P.: death of, 244; mentioned, 336

Porter, Richard, son of Revd. Thomas P., 129, 312

Porter, Revd. Thomas, M.A., rector of East Hoathly:

acquisition, sale or mortgage of property by, xxix–xxx, 62, 221–2, 298, 303–4, 314, 335; new house for, 299–300; gives and receives hospitality, xxxvi, 5, 137–9, 141, 142, 156, 169, 170–2, 176, 194, 195, 197, 198–200, 242, 253, 258, 312; his tithe feasts, 71, 121, 166, 192–3; sermons preached by, 25, 109, 135, 137, 140, 184, 186, 189, 194, 214, 252, 271; said to have preached the same sermon seven times, 271; receives gloves at funerals, 98, 184, 280; at visitation, 297; beats parish bounds, 145–6; administers communion to T. and his sick wife, 227; performs marriage ceremony for T. and his second wife, 323; objection to banns dealt with by, 116; payment for church repair, 269; at vestry, 147, 148, 267; involved in poor law and charity administration, 48, 51–4, 88, 90–1, 117–19, 148, 238, 244, 266, 281, 284 and n., 288, 315, 316; advice sought in parish and other affairs, 17, 43, 56–7; arbitrator in a dispute, 302; petitions composed by, 123, 140; his hop pickers, 64, 162–3; buys herrings, 69–70; T. borrows coal from, 38; buys from T., 9, 37, 93, 98, 180, 242; appeals against window tax and rates, 244, 291; meanness and ingratitude of, xxiii, 60, 139, 155; T. condemns his behaviour, 137, 155; children of, 55, 180, 187, 202, 204–5, 216, 234, 248; trustee of W. Piper, 223, 224,

226, 237; mentioned, 1, 5, 21, 22, 36, 39, 55, 56, 127, 130, 206, 209, 212, 213, 215, 245, 270, 282n., 283, 317, 328, 338, 341; notes on, xxxi–xxxii, 336–7

Porter, Thomas, father of Revd. Thomas P., 336n.

Porter, Thomas, son of Revd. Thomas P.: baptized, 187; death and funeral, 202

Porter, Thomas II, son of Revd. Thomas P., death and funeral, 234

Porter, Thomasin, daughter of Revd. Thomas P., 257

Portland, Duke of, xxiv, 105

Portmore, Charles Colyear, second Earl of, racehorse owner, 12 and n., 209

Portsmouth, Hants., 41, 58n., 84, 85, 92

post mortem, on Elizabeth Elless, xxxv, 51–4

postboy, 310

postman, 13, 17, 30, 37, 42, 150, 168, 199, 337

Potter, T.'s cousin, 107

Potter, John, 58, 75, 135

Prall, Dame, 279, 303

Prall, Ann, daughter of Richard P., 258, 303

Prall, Richard, 74, 75, 258, 285

Prall, Sarah, 318, 341

Prall, Thomas, 60, 118–19, 120

Pratt, Charles, Lord Chief Justice, 278 and n.

preaching, see sermons; field preacher

Presnal, Dorothy, 59

press-warrants, see impressment

Price, Thomas, steward to Mrs. Browne, 123

Priestley, J. B., comments on the diary, vii, 355

prisoners, in Horsham gaol, 130

proclamations: for general fast, 23, 132, 174, 203, 217, 246; for thanksgiving, 194, 270; against vice and profaneness, 236; of peace, 270

Prussia, Frederick II, king of, victory by, 99, 124, 156

public houses: Alfriston, The Star, 103; Battle, The Bull's Head, 114; Broyle Gate, John Martin's, 48, 59, 118, 133; Chiddingly, widow Horsecraft's, 184–5; two unnamed, 93; Chiddingstone, Kent, The Castle, 301; Cuckfield, The King's Head, 48; East Hoathly, The King's Head (otherwise The Maypole, or Burges's), xxii n., xxix, 293 and n., 299,